# 本书编委会

**主　编**　金伟良　张晓林

**副主编**　杨立英　朱苏永

**编委会**（按姓氏汉语拼音排序）

丁洁兰　金伟良　林承亮　马建华
史双青　汤建民　徐晓艺　杨国梁
杨立英　岳　婷　张国昌　张晓林
诸葛洋　朱苏永

**学术顾问**　武夷山　Ronald Rousseau　蒋国华

**主办**　中国科学学与科技政策研究会科学计量学与信息计量学专业委员会
**承办**　浙江大学宁波理工学院
中国科学院文献情报中心
**赞助**　爱思唯尔出版集团
汤森路透科技集团

# 鲁索与中国科学计量学的发展

Ronald Rousseau and the Development of Scientometrics in China

## 第八届科学计量学与大学评价国际研讨会论文集

Proceedings of the 8th International Conference on Scientometrics and University Evaluation

金伟良 张晓林◎主编

科学出版社
北京

**图书在版编目(CIP)数据**

鲁索与中国科学计量学的发展：第八届科学计量学与大学评价国际研讨会论文集/金伟良，张晓林主编. —北京：科学出版社，2014.12

ISBN 978-7-03-042761-8

Ⅰ.①鲁… Ⅱ.①金…②张… Ⅲ.①科学计量学－国际学术会议－文集 Ⅳ.①G301-53

中国版本图书馆 CIP 数据核字（2014）第 291880 号

责任编辑：朱丽娜 张翠霞/责任校对：张凤琴
责任印制：赵 博/整体设计：楠竹文化
编辑部电话：010-64033934
E-mail：fuyan@mail. sciencep. com

科学出版社 出版
北京东黄城根北街 16 号
邮政编码：100717
http://www. sciencep. com
骏杰印刷有限公司 印刷
科学出版社发行 各地新华书店经销
*
2015 年 1 月第 一 版 开本：720×1000 1/16
2015 年 1 月第一次印刷 印张：17 3/4
字数：340 000

**定价：78.00 元**

# 序　言

武夷山　杨立英

自 20 世纪 90 年代以来，我国科学计量学研究与科研量化评价发展迅速，利用定量分析方法开展大学评估工作的国际影响也在不断扩大，这要归功于中国科学计量学研究工作者的艰辛努力和极富成效的国际交流。在与我国科学计量学界密切交往的诸多国际同仁之中，有一位非常杰出的友好人士，他就是国际著名科学计量学家鲁索教授（Prof. Dr. Ronald Rousseau，现任国际科学计量学与信息计量学学会会长，比利时鲁汶大学教授，科学计量学领域的最高荣誉——普赖斯奖获得者）。

鲁索教授自 1998 年首次访华以来的 16 年间，为促进中外科学计量学界的学术交流、培养中国科学计量学研究力量作出了杰出贡献。他到我国各地的学术访问次数近 30 次；他亦分别被中国科学院文献情报中心、河南师范大学、大连理工大学、浙江大学、上海大学、南京农业大学、武汉大学等聘为客座教授或荣誉教授；截至 2013 年 10 月他与国内众多科学计量学者合著发表的论文已达 62 篇。为表达中国同行对鲁索教授的敬意，感谢他不遗余力为推动中国科学计量学研究与应用所作出的卓越贡献，2013 年年初，经国内诸多同行郑重提议，中国科学学与科技政策研究会科学计量学与信息计量学专业委员会决定于 2013 年 11 月 7～8 日召集“第八届科学计量学与大学评价国际研讨会暨鲁索教授与中国学者合作研讨会”。

研讨会在浙江省宁波市宁波理工学院举行。会议吸引了全国各地及比利时、荷兰、瑞典、英国、西班牙等国家的 95 名中外专家与会，其中有三位外国专家是国际科学计量学界的学术最高奖——普赖斯奖获得者，鲁索教授本人亲自到会并发言。研讨会紧密围绕国内外科学计量学研究的最新进展、大学评估与科研量化评价、中国科学计量学研究的发展历程（特别是中国学者与鲁索教授的合作经历）等主题进行了学术交流与研讨。

鲁索教授报告了他关于测度网络节点的桥梁作用的最新成果——Q 指标；国际科学计量学与信息计量学学会秘书长、普赖斯奖得主、比利时鲁汶大学 Wolfgang Glänzel 教授作了《关于应用高级的文献计量学指标和引文表现等级对大学及研究机构进行评价研究》的报告，报告中提出了一个基于“引文表现等级”的新的评价方法，以便为不同层级的评价研究提供一个“无预设参数”；普赖斯奖得主、荷兰阿姆斯特丹大学 Loet Leydesdorff 教授做了《基于三螺旋模型的创新系统协同效应的测度指标》的报告，介绍了他与很多学者（包括中国学者）合作开展的大量实证研究。

本次会议最突出的特色之一是纪念鲁索教授初次访华暨与中国学者合作 15 周年。与鲁索教授有密切合作关系的众多中国学者在这一主题的发言中抒发了感谢、感激、感慨之情。与已故赵红州教授共同开创了中国科学计量学研究先河的蒋国华研究员作了《中国科学计量学界的伟大朋友》的演讲；金碧辉研究员在《我眼中的鲁索教授》报告中回顾了两人长达 15 年的学术合作历程；梁立明教授以 Web of Science 库为数据来源，分析了鲁索教授的学术研究“节律”；叶鹰教授用文献计量学研究方法分析了鲁索教授和国内学者合作发表论文的被引特征；清华同方中国知网万锦堃教授在《鲁索教授与清华》报告中，通过大量的图片展现了鲁索教授对中国知网文献计量学研究的支持和帮助；汤森路透科技集团中国部科学计量学研究顾问岳卫平博士也通过大量珍贵的照片生动地回顾了她和鲁索教授 7 次会面时的情景。鲁索教授在其每份电子邮件的最后都附有一句话“In scientific affairs one can never be too generous”（在科学事务中不存在过分慷慨的问题）。他本人就实践着这一理念，不断给科学计量学界同行提供慷慨的咨询与建议。我们确信，正如中国医学界不会忘记白求恩一样，中国科学计量学界将永远铭记鲁索教授对我们的慷慨帮助。

研讨会为中国学者提供了与国际高水平同行进行学术交流的平台，充分表达了中国科学计量学界对鲁索教授的诚挚感谢。会议的成功举办离不开许多单位与个人的鼎力支持。在此，我们隆重感谢浙江大学宁波理工学院、中国科学院文献情报中心承办会议；感谢爱思唯尔出版集团、汤森路透公司对会议的赞助；感谢浙江省第十届政协副主席盛昌黎女士到会并致辞；感谢宁波大学理工学院金伟良院长为会议提供的诸项便利条件和周到服务；感谢中国科学院文献情报中心张晓林馆长百忙中赴会并发言；感谢中国科学计量学研究前辈蒋国华教授在会议筹办、组织过程中的特殊贡献；感谢国家行政学院史朝教授为会议承办所作的大量协调工作。

在同行评议的基础上，编委会遴选出较优秀的会议论文 22 篇结集正式出版。这些论文在一定程度上反映了当前科学计量学领域主要的研究问题，我们希望论文集的出版能够为中国科学计量学者的研究工作提供借鉴和参考。

国际科学计量学与情报计量学学会是 1994 年正式成立的，迄今已经 20 年了。20 岁，是青春勃发的年龄。我们有幸成为科学计量学研究事业的一员，光荣地与国际科学计量学与情报计量学学会一起成长。我们洒下了辛勤的汗水，也体会到了摘取研究果实的愉悦。我们热望更多的年轻人加入这支队伍，让“我们的队伍越走越长”。

# CONTENTS 目录

## 三、科学计量指标研究



## 四、科学计量与大学评估



## 五、科学计量学在中国

# 一、科学计量理论与方法研究

# 1-1 Gauging the Bridging Function of Nodes in a Network: The Gefura Measure

Ronald Rousseau[①], Raf Guns[②], Liu Yuxian[③]

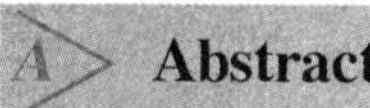

## Abstract

In this article the authors review the study of so-called $Q$-measures in networks. $Q$-measures are related to betweenness centrality but are defined in case the network is subdivided into subgroups. They are indicators gauging the bridging, brokerage or gatekeeper function of a node. Two definitions, a basic and a structural $Q$-measure, are proposed in the case that there are more than two subnetworks and the difference between these two approaches is illustrated. As the term $Q$-measure is not optimal (there exists other $Q$s) we propose the term gefura measure (meaning bridge measure) and the corresponding symbol $\Gamma$ as a better term for this network measure.

**Keywords**: centrality measures; bridging function; subnetworks; gefura measure

## 1 Networks

It is stating obviously when pointing out that graphs or networks (these two words will be considered as synonyms) are everywhere. Indeed, road maps represent the network of cities and highways, and similarly we have other transportation networks such as the worldwide air transport network (Barrat et al., 2004; Guimerà

---

① Institute for Education and Information Sciences, IBW, University of Antwerp, Venusstraat 35, Antwerp, B-2000 (Belgium) & KU Leuven, Leuven B-3000 (Belgium). ronald. rousseau@ua. ac. be.

② Institute for Education and Information Sciences, IBW, University of Antwerp, Venusstraat 35, Antwerp, B-2000 (Belgium). raf. guns@ua. ac. be.

③ Library of Tongji University, Tongji University, Siping Street 1239, Shanghai, 200092 (China). yxliu@lib. tongji. edu. cn.

et al. , 2005) or the shipping and harbors network. Nowadays the Internet is probably the best known network. This network consists actually of a worldwide assemblage of computer networks consisting of local, regional and global academic, business, government, private and public subnetworks (Wikipedia, 2013). Since decades sociologists study friendship and other social networks (Wasserman and Faust, 1994; Otte and Rousseau, 2002), while nowadays the study of gene and disease networks is a hot topic. Indeed, following Goh et al. (2007)—cited already more than 700 times in the Web of Science—medical researchers came to the conclusion that a disease is rarely the consequence of problems with one single gene. Only a study of the intercellular network can lead to advances in the study and ultimately identifying drug targets and biomarkers for complex diseases (Barabási et al. , 2011).

Within the information sciences colleagues study article citation graphs, journal citation graphs, collaboration networks, co-word networks, author co-citation networks and many other co-occurrence networks. Maps of science are constructed based on the complete Web of Knowledge or Scopus. Moreover, a whole new subfield related to the Internet, namely webometrics, has emerged within the field of informetrics.

Generally four types of networks may be distinguished: undirected unweighted networks, directed unweighted networks, undirected weighted networks and directed weighted networks. Directed networks may further be subdivided into acyclic and general directed networks. Examples within the information sciences are given in Table 1.

**Table 1 Examples of different types of networks in informetrics**

| Types | Undirected | Directed |
|---|---|---|
| Unweighted | Author collaboration network, not taking frequencies into account | Article citation network (acyclic, except special cases); journal citation network, not taking frequencies into account |
| Weighted | Author collaboration network, including the frequency of collaboration | Article citation network, including weights related to frequency or importance of use (acyclic); general journal citation network, including citation frequencies |

Networks can be characterized by several different measures. These can be subdivided into two main groups: measures related to the network as a whole, and measures related to nodes or links (separately). The simplest global measure is probably the number of nodes. Density, another global measure, is an indicator for the general level of connectedness of the graph. If every node is directly connected to every other node, we have a complete graph. The density of a graph is defined as the number of links (or arcs) divided by the number of links in a complete graph with the same number of nodes. This indicator as well as those that follow will be defined only for simple, i. e. , undirected unweighted networks, unless stated otherwise.

The next group of network measures, namely centrality measures, consists of node indicators. The most important ones are: degree centrality, closeness centrality, betweenness centrality and eigenvector centrality. Degree centrality of a node is defined as the number of ties that a node has. Closeness centrality of a node is equal to one divided by the total distance (using geodesics in the graph) of this node from all other nodes. Betweenness centrality may be defined loosely as the number of shortest paths that pass through a given node. More precisely this notion is defined as follows (Anthonisse, 1971; Freeman, 1977).

For an undirected network with $n$ nodes the (normalized) betweenness centrality of node $a$, denoted as $C_B(a)$ is defined as:

$$C_B(a)=\frac{2}{(n-1)(n-2)}\sum_{g,h}\frac{p_{g,h}(a)}{p_{g,h}} \tag{1}$$

In this formula, $p_{g,h}$ denotes the number of shortest paths connecting nodes $g$ and $h$; $p_{g,h}$ ($a$) denotes the number of these shortest paths passing through $a$, where $a$ is not one of the endpoints.

The fourth centrality measure can best be introduced using the matrix representation of a graph. For simplicity we consider only simple graphs. Assume that a graph has $n$ nodes (also called vertices). Then the adjacency matrix $\boldsymbol{A}$ is a square ($n$, $n$) -matrix, such that element $a_{ij}=1$ if vertex $i$ is connected to vertex $j$ and $a_{ij}=0$ otherwise (including the diagonal elements $a_{jj}$). As the graph is undirected $a_{ij}$ is always equal to $a_{ji}$ so that the matrix $\boldsymbol{A}$ is symmetric. As the matrix $\boldsymbol{A}$ is symmetric it can be shown that all its eigenvalues are real (Lay, 2003). Recall that an eigenvalue of a matrix is a number $\lambda$ (in general this may be a complex number) for which there exists a vector $\boldsymbol{X}=(x_1, x_2, \cdots, x_n)$ such that $\boldsymbol{A}.\boldsymbol{X}=\lambda\boldsymbol{X}$ or stated otherwise:

$$\sum_{j=1}^{n}a_{ij}x_j=\lambda x_i, i=1,\cdots,n \tag{2}$$

In the case of an adjacency matrix it can be shown that all components of the eigenvector can be chosen to be positive. Then eigenvector centrality of node $i$ is defined as:

$$x_i=\frac{1}{\lambda}\sum_{j=1}^{n}a_{ij}x_j \tag{3}$$

where $\lambda$ is the largest eigenvalue of matrix $\boldsymbol{A}$. We see that the eigenvector centrality of a vertex is proportional to the sum of the eigenvector centralities of the vertices to which it is connected. Although this description sounds like a circular definition, equation (3) and the underlying mathematical theory (the Perron-Frobenius theorem) show that it is not and that it leads to a well-defined notion (Meyer, 2000). Density and degree centrality can also be expressed using the adjacency matrix. Density is equal to $\frac{\sum_{i,j=1}^{n} a_{ij}}{n(n-1)}$ and the degree centrality of node $j$ is $\sum_{k=1}^{n} a_{jk}$.

Degree centrality of a node is related to activity; closeness centrality of a node relates to efficiency; a high value for the betweenness centrality of a node reflects a high control on the flow of information by this node; finally a high eigenvector centrality refers to a node that is highly influential. All four centrality measures can be said to be related to leadership.

## 2 Q-measures for Two Groups: Definition and an Example

In 2004, Flom, Friedman, Strauss and Neaigus, four researchers from New York (USA) introduced a new sociometric network measure, denoted as $Q$, for individual actors as well as for whole networks (Flom et al., 2004). As this measure captures the idea of bridges between two groups, the higher an actor's $Q$-value, the more this actor behaves as a bridge/broker between the two groups. Hence using $Q$-measures leads to an answer of the question: Which nodes are most important in bridging activities between groups in a network? Figure 1 illustrates the type of network Flom et al. (2004) had in mind.

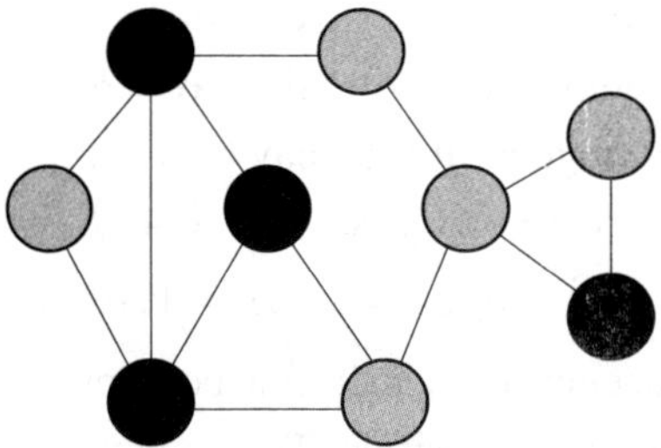

Figure 1 Simple network subdivided into two groups

We note that existing measures such as betweenness centrality did not make a distinction between nodes belonging to different groups. This is the main motivation for the introduction of the $Q$-measure. An example: consider a network consisting of all universities of two countries (the two groups) linked if they (= at least one researcher in each university) collaborate in a given topic using $Q$-measures, answers the question: Which universities are the main facilitators of international collaboration? $Q$-measures are based on geodesics (shortest distances) between nodes.

*The defining formulae*

Assume that there are $T$ actors (nodes) in the network. Group $\boldsymbol{A}$ contains $m$ nodes, while the other group, denoted as $\boldsymbol{B}$, contains $n$ nodes, hence $T=m+n$. If actor $x$ belongs to group $\boldsymbol{A}$, and assuming for simplicity that actor $x$ is $a_m$, then the $Q$-measure for this actor, $Q(x)$, is defined as shown below. In this formula $p_{a,b}$ and $p_{a,b}(x)$ are defined in the definition of betweenness centrality in formula (1).

$$Q(x)=\frac{1}{(m-1)n}\left(\sum_{i=1}^{m-1}\sum_{j=1}^{n}\frac{p_{a_ib_j}(x)}{p_{a_ib_j}}\right) \tag{4}$$

Flom et al. (2004) also introduced a $Q$-measure for the whole network, denoted as $Q_{\text{net}}$, as the normalized average difference between the most central node (in the $Q$-sense), denoted as $Q^*$, and all other nodes. This is:

$$Q_{\text{net}}=\frac{\sum_{i=1}^{m}\left(Q^*-Q(a_i)\right)+\sum_{j=1}^{n}\left(Q^*-Q(b_j)\right)}{T-1} \tag{5}$$

An example: consider Figure 2 consisting of two groups $\boldsymbol{A}=\{a_1, a_2, a_3\}$ and $\boldsymbol{B}=\{b_1, b_2\}$.

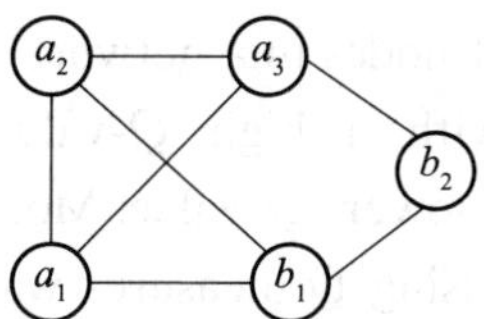


Figure 2 Example network for the calculation of a $Q$-measure

Table 2 shows all nodes that lie on a shortest path (sometimes there are two shortest paths between nodes situated in different groups). When there is no such node this is denoted by "—".

**Table 2 Example paths for the nodes**

| Nodes | $a_1$ | $a_2$ | $a_3$ |
|---|---|---|---|
| $b_1$ | — | — | $a_1/b_2$ |
| $b_2$ | $b_1/a_3$ | $b_1/a_3$ | — |

$$Q(a_1)=\frac{1}{4}\left(0+\frac{1}{2}+0+0\right)=\frac{1}{8};\ Q(a_3)=\frac{1}{4}\left(0+0+\frac{1}{2}+\frac{1}{2}\right)=\frac{1}{4}$$

$$Q(b_1)=\frac{1}{3}\left(\frac{1}{2}+\frac{1}{2}+0\right)=\frac{1}{3};\ Q(b_2)=\frac{1}{3}\left(0+0+\frac{1}{2}\right)=\frac{1}{6}$$

Finally $Q(a_2)=0$. In group $\boldsymbol{A}$, $a_3$ plays a more important role than $a_1$, while $a_2$ plays no role at all as a bridge. In group $\boldsymbol{B}$ all nodes play a bridging role, but $b_1$'s role is more important.

## 3 Literature Review Related to Q-measures for Two Groups

$Q$-measures were introduced in social network theory by Flom et al. in 2004. The idea was picked up by Rousseau and presented at the 68th ASIST Conference (Rousseau，2005). Calculations for theoretical examples (line graphs，star graphs) as well as two co-author networks based on real data were presented. A more elaborated example presented first at the 9th International Conference on Science & Technology Indicators (2006，Leuven，Belgium) was published (Chen and Rousseau，2008). Their result indicated that Cambridge University，Manchester University，Technische Universität Berlin，the Max Planck Institute，Stuttgart University and Forschungszentrum Karlsruhe play the most important roles as bridges between England and Germany (at least within the *Journal of Fluid Mechanics*). It was concluded that having a high degree centrality and being a key node are important factors explaining the ranking of nodes in a network according to $Q$-value. Moreover，it was found that institutes with a high $Q$-value have，on average，a higher production than those with a lower $Q$-value. Moreover，after a presentation by Rousseau in Dalian the idea of using $Q$-measures was worked out—using a dedicated computer program—by Zhang，Yin and Pang (2009) for the COLLNET network，divided in male and female researchers，as an example.

## 4 Q-measures for More Than Two Groups

The next step taken in the study of $Q$-measures was the extension of its definition to more than two groups. This，however，turned out not to be trivial. Depending on the relative importance one gives to groups or to nodes we came up with two definitions (Guns and Rousseau，2009).

### 4.1 Definition 1：the Global Structural Q-measure

Consider a network subdivided into $S$ non-overlapping groups. Each group is denoted as $G_i (i = 1,\cdots,S)$ and contains $m_i$ members. Then the global $Q$-measure of node $a$ is denoted as $Q_G^S(a)$ (the notation is explained further on) and is defined as：

$$Q_G^S(a) = \frac{2}{S(S-1)} \sum_{k,l} \left( \frac{1}{\mathrm{TP}_{k,l}} \sum_{\substack{g \in G_k \\ h \in G_l}} \frac{p_{g,h}(a)}{p_{g,h}} \right) \tag{6}$$

where $p_{g,h}$ and $p_{g,h}(a)$ are used as defined earlier. The symbol $\mathrm{TP}_{k,l}$ refers to the number of possible combinations of elements belonging to groups $G_k$ and $G_l$ . Hence, $\mathrm{TP}_{k,l} = \#(G_k \setminus \{a\}) \cdot \#(G_l \setminus \{a\})$ , where "#" refers to the number of elements in the set between brackets. The term "global" is used because we will further decompose this global measure into a local and an external one. Observe that in this definition two kinds of normalization are applied.

Yet, a global $Q$-measure when several groups are present in the network can also be defined differently.

## 4.2 Definition 2: the Global Basic Q-measure

In this alternative definition, denoted as $Q_G^B(a)$ , only one normalization is applied.

$$Q_G^B(a) = \frac{1}{M} \sum_{\substack{g,h \in V \\ \mathrm{group}(g) \neq \mathrm{group}(h)}} \frac{p_{g,h}(a)}{p_{g,h}} \tag{7}$$

where the symbol $M$ is defined as:

$$M = \sum_{k,l} \#(G_k \setminus \{a\}) \cdot \#(G_l \setminus \{a\}) \tag{8}$$

The sum in formula (8) is calculated over all pairs of different groups and the notation group ($g$) refers to the group to which node $g$ belongs.

When there are only two groups these two definitions (the basic one and the structural one) coincide with equation (4).

## 4.3 The Difference Between These Two Definitions

We illustrate the different behavior of basic and structural $Q$-measures by the following example (Figure 3). In this example we use the following convention: nodes $a_1$, $a_2$, …form a group, nodes $b_1$, $b_2$, …form another group, and so on.

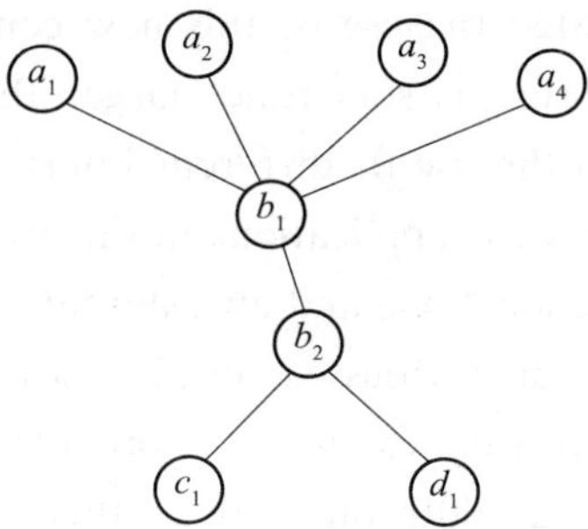


Figure 3 Example: four groups, illustrating the difference between structural and basic Global $Q$-measures

In the example shown in Figure 3, we have four groups: group $A$ with nodes

$a_1$, $a_2$, $a_3$ and $a_4$, group $B$ with nodes $b_1$ and $b_2$, and groups $C$ and $D$, each consisting of a single node. Clearly all $Q$-measures for the nodes in groups $A$, $C$ and $D$ are zero. Group $B$ is the interesting one: $Q_G^S(b_1) = 0.5$; $Q_G^S(b_2) = 0.833$; $Q_G^B(b_1) = 0.8$; $Q_G^B(b_2) = 0.733$. According to the structural $Q$-measure node $b_2$ plays a more important role than node $b_1$, as $b_2$ connects to two different groups and $b_1$ only connects to one group. However, according to the basic definition $b_1$ is more important than $b_2$, as $b_1$ is situated on the shortest paths than $b_2$ (but recall that this measure does more than just counting the shortest paths). This illustrates how the structural $Q$-measure attaches more importance to the group level, while the basic $Q$-measure attaches more importance to the node level.

In Rousseau et al. (2013) it is shown how the global structural $Q$-measure can be decomposed as a convex sum of a local structural $Q$-measure and an external one. Here a local $Q$-measure takes only the shortest paths into account that begin in the group to which node $a$ (the node for which the measure is calculated) belongs, while the external $Q$-measure always begins and ends in a group to which $a$ does not belong. It is easy to obtain a similar decomposition for the global basic $Q$-measure.

## 5 Literature Review Related to Q-measures for More Than Two Groups

The global structural $Q$-measure was introduced by Guns and Rousseau (2009) during the 12th ISSI Conference in Rio de Janeiro (Brazil). In that article also the local $Q$-measure was introduced. Two artificial examples and one based on real data (Leydesdorff, 2007) illustrated the use of this new concept and showed that for a given node the local $Q$-measure is sometimes larger than the global one and vice versa. This reversal is due to the use of different kinds of normalization. The article also contains some suggestions for applications in different fields. The decomposition of a global $Q$-measure into a local one and an external one (and the definition of an external Q-measure) came later in Rousseau et al. (2013).

An extensive and interesting example is provided in Guns et al. (2011), originally written for Rousseau's 60th birthday. In this article the authors studied the collaborating network in the field of bibliometrics-informetrics-webometrics-scientometrics (BIWS) over the period 1990-2009. Subdividing this period into four time slices of five years each makes it possible to include dynamic aspects. Using country groups, the top three authors in terms of their global structural $Q$-values are:

Glänzel, Kretschmer and Rousseau. The authors conclude that BIWS as a field is expanding and intensifying international collaboration. A somewhat similar example (Guns and Liu, 2010) related to China was published earlier (but written later).

Rousseau et al. (2013) focused on mathematical properties of structural $Q$-measures, including a convex decomposition, relations with betweenness centrality and a characterization of nodes with $Q$-measure equal to one (the highest possible value). In Liu et al. (2013) a binary tree was used as a model for a hierarchical structure. All nodes situated at the same height are assumed to form a group. The global and local structural $Q$-values were determined for each node in a binary tree of arbitrary height. This paper was published to honor Ranganathan and the Sarada Ranganathan Endowment for Library Science in Bangalore (India).

The difference between the structural and the basic $Q$-measure is introduced in this article and will be studied further in future articles.

## 6 Further Research and a Remark on Naming

As different types of network can be distinguished it is obvious which investigations await further developments: $Q$-measures must be studied in directed networks (acyclic and general ones) and in weighted networks. A first approach to $Q$-measures in weighted, directed networks has been published in Rousseau and Zhang (2008). Its results have been used in a patenting network by Guan and Chen (2012). Finally, all this should be studied not only in the context of disjoint groups but also in the more realistic context of overlapping subnetworks.

Colleagues working in social network theory and marketing observed that the term $Q$-measure is non-descriptive and, moreover, is also used in other contexts. Indeed, in marketing Tobin's $Q$ is equal to the ratio between the market value and replacement value of the same physical asset (Brainard and Tobin, 1968). $Q$-analysis, a mathematical technique to study and analyze structures, was introduced by Atkin (1972) and described for information scientists by Davies (1985). In 2004 Sakai (2004) introduced a $Q$-measure as an information retrieval metric in a graded relevance context (as opposed to binary relevance); see also (Sakai, 2007). The best known $Q$ in network theory is probably Newman and Girvan's $Q$ denoting modularity in a network (Newman and Girvan, 2004). So, indeed, the use of the symbol $Q$ and the term $Q$-measure to study "bridgeness" is not optimal at all.

As the $Q$-measure is used to gauge the bridging role of nodes the old Greek term gefura (γεφυρα) measure, meaning bridge measure, might be a more descriptive term with universal appeal. Recall that in the Roman times Greek was the lingua franca in the eastern part of the empire. In this submission the term gefura measure may be used as a term unifying English，Dutch and Chinese—the lingua franca of the present day (Garfield，1989) and the authors' mother languages. From now on the authors intend to use the symbol $\Gamma$ (capital gamma：the first letter of gefura) instead of $Q$.

## Acknowledgement

Ronald Rousseau thanks the organizers of the 9th International Conference on Scientometrics and University Evaluation，Ningbo (China) for their invitation. The authors acknowledge support from NFSC grant number 71173154.

## References

Anthonisse J M. 1971. The rush in a directed graph. Technical Report BN 9/71，Stichting Mathematisch Centrum，Amsterdam.

Atkin R. 1972. From cohomology in physics to *q*-connectivity in social science. *International Journal of Man-Machines Studies*，4：139-167.

Barabási A L，Gulbahce N，Loscalzo J. 2011. Network medicine：a network-based approach to human disease. *Nature Reviews Genetics*，12 (1)：56-68.

Barrat A，Barthélemy M，Pastor-Satorras R，et al. 2004. The architecture of complex weighted networks. *Proceedings of the National Academy of the United States of America*，101 (11)：3747-3752.

Brainard W C，Tobin J. 1968. Pitfalls in financial model building. *American Economic Review*，43 (2)：99-122.

Chen L X，Rousseau R. 2008. *Q*-measures for binary divided networks：bridges between German and English institutes in publications of the *Journal of Fluid Mechanics*. *Scientometrics*，74 (1)：57-69.

Davies R. 1985. *Q*-analysis：a methodology for librarianship and information science. *Journal of Documentation*，41 (4)：221-246.

Flom P L，Friedman S R，Strauss S，et al. 2004. A new measure of linkage between two sub-networks. *Connections*，26 (1)：62-70.

Freeman L C. 1977. A set of measures of centrality based upon betweenness. *Sociometry*，40 (1)：35-41.

Garfield E. 1989. The English language：the lingua franca of international science. *The Scientist*，3 (10)：12.

Goh K I，Cusick M E，Valle D，et al. 2007. The human disease network. *Proceedings of the National Academy of Sciences of the USA*，104 (21)：8685-8690.

Guan J C，Chen Z F. 2012. Patent collaboration and international knowledge flow. *Information Processing & Management*，48（1）：170-181.

Guimerà R，Mossa S，Turtschi A，et al. 2005. The worldwide air transportation network：anomalous centrality，community structure，and cities' global roles. *Proceedings of the National Academy of the United States of America*，102（22）：7794-7799.

Guns R，Liu Y. 2010. Scientometric research in China in the context of international collaboration. *Geomatics and Information Science of Wuhan University*，35：112-115.

Guns R，Liu Y X，Mahbuba D. 2011. *Q*-measures and betweenness centrality in a collaboration network：a case study of the field of informetrics. *Scientometrics*，87（1）：133-147.

Guns R，Rousseau R. 2009. Gauging the bridging function of nodes in a network：*Q*-measures for networks with a finite number of subgroups. In：Larsen B，Leta J（eds.），*Proceedings of ISSI 2009—the 12th International Conference on Scientometrics and Informetrics*（pp. 131-142）. Berlin：Springer-Verlag.

Lay D C. 2003. *Linear Algebra and Its Applications*. Boston：Addison-Wesley.

Leydesdorff L. 2007. Visualization of the citation impact environment of scientific journals：an online mapping exercise. *Journal of the American Society for Information Science and Technology*，58（1）：25-38.

Liu Y X，Guns R，Rousseau R. 2013. A binary tree as a basic model for studying hierarchies using *Q*-measures. *SRELS Journal of Information Management*，50（5）：521-528.

Meyer C D. 2000. *Matrix Analysis and Applied Linear Algebra*. Philadelphia：SIAM.

Newman M E J，Girvan M. 2004. Finding and evaluating community structure in networks. *Physical Review E*，69（2）：026113.

Otte E，Rousseau R. 2002. Social network analysis：a powerful strategy，also for the information sciences. *Journal of Information Science*，28（6）：441-453.

Rousseau R. 2005. *Q*-measures for binary divided networks：an investigation within the field of informetrics. *Proceedings of the 68th ASIST Conference*，42：675-696.

Rousseau R，Liu Y X，Guns R. 2013. Mathematical properties of *Q*-measures. *Journal of Informetrics*，7（3）：737-745.

Rousseau R，Zhang L. 2008. Betweenness centrality and *Q*-measures in directed valued networks. *Scientometrics*，75（3）：575-590.

Sakai T. 2004. Ranking the NTCIR systems based on multigrade relevance. In：Myaeng S H，Zhou M，Wong K F，et al.（eds.），*Proceedings of the Asia Information Retrieval Symposium*（*AIRS*）（pp. 170-177）. Berlin，Heidelberg：Springer.

Sakai T. 2007. On the reliability of information retrieval metrics based on graded relevance. *Information Processing & Management*，43（2）：531-548.

Wasserman S，Faust K. 1994. *Social Network Analysis*. Cambridge：Cambridge University Press.

Wikipedia. http：//en. wikipedia. org/wiki/Internet［2013-12-27］.

Zhang W L，Yin L C，Pang J. 2009. The application of *Q*-measure to gender study in cooperation network. *Science & Technology Progress and Policy*，26（15）：100-103.

# 1-2 Measuring the Knowledge-based Economy of China in Terms of Synergy among Technological, Organizational, and Geographic Attributes of Firms

Loet Leydesdorff①, Zhou Ping②

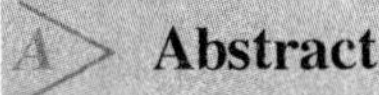

Using the possible synergy among geographic, size, and technological distributions of firms in the Orbis database, we find the greatest reduction of uncertainty at the level of the 31 provinces (not containing Hong Kong, Macau and Taiwan in this paper) of China, and an additional 18.0% at the national level. Some of the coastal provinces stand out as expected, but the metropolitan areas of Beijing and Shanghai are (with Tianjin and Chongqing) most pronounced at the next-lower administrative level of (339) prefectures, since these four "municipalities" are administratively defined at both levels. Focusing on high- and medium-tech manufacturing, a shift toward Beijing, Shanghai, and Tianjin (near Beijing) is indicated, but the synergy is on average not enhanced. High- and medium-tech manufacturing is less embedded in China than in Western Europe. Knowledge-intensive services "uncouple" the knowledge base from the regional economies mostly in Chongqing and Beijing. Unfortunately, the Orbis data is incomplete since it was collected for commercial and not for administrative or governmental purposes. However, we provide a methodology that can be used by others who may have access to higher-quality statistical data for the measurement.

**Keywords**: China; knowledge base; Triple Helix; synergy; entropy

---

① Amsterdam School of Communication Research (ASCoR), University of Amsterdam, Kloveniersburgwal 48, 1012 CX Amsterdam, The Netherlands; loet@leydesdorff.net; http://www.leydesdorff.net.

② Department of Information Resource Management, Zhejiang University, No. 866 Yuhangtang Road, Hangzhou, 310058, China; pingzhou@zju.edu.cn; corresponding author.

# 1 Introduction

Mutual information in three (or more) dimensions can be derived from Shannon's (1948) formulas of information theory (e. g., McGill, 1954; Abramson, 1963), but it can no longer be considered as Shannon-type information because it is a signed information measure that can also be negative (Krippendorff, 2009a). Any two sources of variance in the three- (or more-) way interaction may spuriously correlate given a third variable because of mutual overlaps in the expected information contents of the distributions. The overlaps can be considered as repetitions and therefore redundant or, alternatively, these non-linearities may be considered as loops that are incompatible with the linear framework of Shannon's theory (Krippendorff, 2009b).

Leydesdorff and Ivanova (in press) have recently shown that the information in the overlaps should be counted twice—as in the case of pure sets—so that the total information content is enlarged, and consequently the complementary relation between information and redundancy is shifted in the favor of the redundancy. Unlike Krippendorff's (2009a) interaction information ($I_{ABC \to AB:AC:BC}$), mutual information in more than two dimensions can then be considered consistently as a measure of redundancy or reduction of the uncertainty that prevails at the systems level (Ivanova and Leydesdorff, in preparation).

Leydesdorff (2003) used this measure first as an operationalization of the possible reduction of uncertainty in the Triple Helix of university-industry-government relations (Park et al., 2005; Leydesdorff and Sun, 2009; Park and Leydesdorff, 2010; Ye et al., in press). When all relations are in place, an overlay may reduce uncertainty for the individual players at the systems level. This configurational effect cannot be attributed to specific links or nodes, but is a result of the interaction. These studies used co-authored publications as units of analysis, and considered possible relations as attributes of these units.

In another series of studies of European innovation systems we used three (or more) attributes of firms as a potential source of synergy in the economy. Storper (1997) hypothesized that the relational interaction among technology, geography, and organization can generate synergy in what he called a "Holy Trinity". We operationalize geography as the distribution in terms of geographical addresses, technology in terms of the OECD classification of firms according to the "Nomenclature générale des Activités économiques dans les Communautés Européennes"

(NACE)[①], and organizational size in terms of numbers of employees. Can a latent construct among the three (or more) attributes be indicated that potentially reduces uncertainty in the configuration?

In the Netherlands (Leydesdorff et al., 2006) and Sweden (Leydesdorff and Strand, in press), we found a regional pattern with surplus value at the national level, measurable as a between-regional reduction of uncertainty. In Germany (Leydesdorff and Fritsch, 2006) such national synergy was not found, but the surplus is realized at the level of the federal states, whereas the Hungarian system seems to consist of a western part integrated with neighboring EU-countries, a metropolitan center around Budapest, and an eastern part in which the old (state-led) system still prevails (Lengyel and Leydesdorff, 2011).

Most interestingly, Strand and Leydesdorff (2013) have found that the synergy in the Norwegian economy is concentrated along the shore in relation to marine and maritime (offshore) industries, whereas the university centers of the country (e.g., in Trondheim and Oslo) have not been integrated into the economy, but remain at a distance. In Sweden, 45.3% of the synergy was highly concentrated in the three metropolitan regions of Stockholm, Gothenburg, and Malmö/Lund.

Our design is based on using firms as units of analysis, and the following three variables and their interactions as attributes:

(1) geographical addresses as an indicator of regional or other geographic provenance;

(2) size in terms of numbers of employees as a proxy of economic organization (e.g., small- and medium-sized companies);

(3) NACE codes of the OECD as a technological classification.

In each study, however, we had to make compromises because of possible imperfections in the sample data. In Hungary, for example, we did not have NACE codes for all firms, and Statistics Sweden uses its own classification system. In these various studies, the data about numbers of employees was not always as fine-grained as it was in the original study of the Netherlands (Leydesdorff et al., 2006).

In this further study we make an attempt to analyze the knowledge-based economy of China. Recently we obtained access to the database Orbis of the Bureau van Dijk (BvD; available at https://orbis.bvdinfo.com) containing the necessary data given our design. Although organized at the firm level and in considerable detail, this data is collected worldwide for commercial purposes and not for administrative or

① The NACE code can be translated into the International Standard Industrial Classification (ISIC) that is used, for example, in the USA.

governmental purposes such as by a national bureau of statistics.

The data collection is based on more than one hundred information providers, or, as the BvD claims: "We're experts in company information and business intelligence. We integrate information from numerous sources to create Orbis and complement it with our own research. We combine this unique dataset with our software to create a dynamic global research tool. "① The encompassing database covers company information for more than 100 million firms (including banks) worldwide. The three variables that are core to our research questions (addresses, NACE codes, and numbers of employees) are also included.

Orbis provides information collected during the past ten years, but with a continuously moving window retrospectively from the current date. Regional databases are derived from Orbis (such as Amadeus for Europe), but contain approximately the same data only for the most recent year (and are therefore cheaper). The update frequencies of either database, however, are not specified precisely and may vary among countries and firms. In a study of the coverage of Orbis, Ribeiro et al. (2010) concluded that this coverage is poor, and can vary among countries and sectors, but precise estimates were not provided. In any case, self-employed entrepreneurs are usually not covered, whereas these firms (e. g. , startups) may be most interesting from the perspective of innovation policies. A further issue may be that headquarters and research centers are not always located in the same place, but we have the impression that Orbis tries to correct for such problems.

In a recent study of the Italian innovation system (Cucco and Leydesdorff, in preparation) an almost perfect correlation was found (Pearson $r = 0.98$; Spearman's $\rho > 0.99$) between the distributions of our synergy values based on 462 316 valid observations using Orbis data versus 4 480 473 firms registered by Statistics Italy in 2007. This gives us some confidence in the representativeness of the Orbis data and its usefulness as a sample for our purpose of estimating synergy as a system property. In this study, we explore the usefulness of this data as a sample for investigating the important question of how to measure the knowledge base of China in terms of Triple-Helix relations at the level of firm data. In any case, this study provides a methodology that can be improved if the complete data set for the population is made available.

We asked the National Bureau of Statistics of China for the complete set, but, for legal reasons, data is made available by this office only on the aggregated level at http: //219. 235. 129. 58/welcome. do# . Since no other data is readily available for

① "Orbis: Company Information around the Globe"; available at http: //www. bvdinfo. com/ About-BvD/Brochure-Library/Brochures/ORBIS-brochure.

China, we decided to explore Orbis as a source and thus harvested all firm data for China on December 14, 2012. With the caveat that this data is an incomplete sample because we collected it for business purposes that are different from administrative purposes (such as governmental statistics), the domain enables us nonetheless to address in considerable detail the research question of where synergy is generated in the Chinese economy in terms of relations between geographical addresses, company sizes, and NACE codes.

# 2 Methods and Materials

## 2.1 Data

Retrieval from Orbis provided us with 768 949 records with a Chinese address of which 768 948 could be downloaded. Figure 1 shows the yearly distribution of this data (the data of Hong Kong, Macau and Taiwan is not included in this paper) (As noted, Orbis accumulates data for the last ten years).

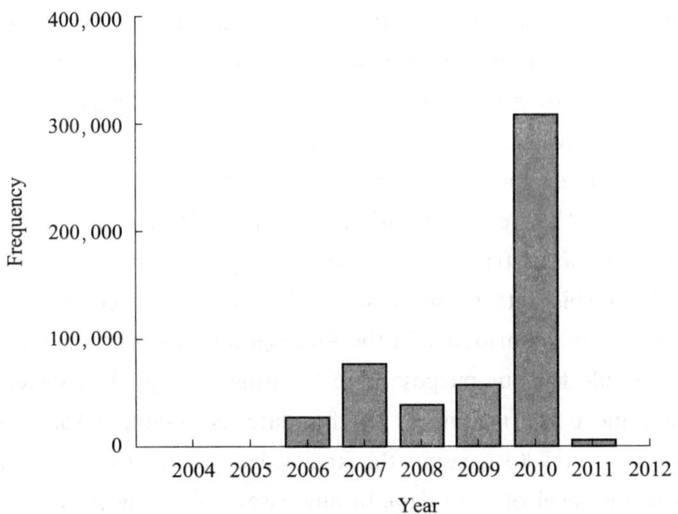


Figure 1 Yearly distribution of firm data for china from Orbis (December 14, 2012; $N = 768\ 948$)

Figure 1 teaches us that data collection in China did not begin until 2006, and that the data for 2011 were not yet included at the time of this download (December, 2012)①.

① When we returned to the database on May 20, 2013, the retrieval was 1 612 309.

We focus on the data for 2008-2010, comprising 402 604 records (52. 4%), of which 379 026 contained valid information in the three fields of interest: city name, NACE code (Revision 2; 4 digits), and firm size.

The attribution of city names to 31 provinces was fully standardized and made complete in terms of name variants etc. by one of us. Name variants of cities were additionally standardized on the basis of computer routines that, for example, relabeled "Beijing Capital City" as equivalent to "Beijing", etc. As could be expected, the distribution of the firms included varies widely by region: from 67 for the autonomous region of Xizang to 62 805 for the heavily industrialized region of Jiangsu on the east coast (Figure 2). The distribution accords with the conventional wisdom that China is both industrialized and in large parts also rural.

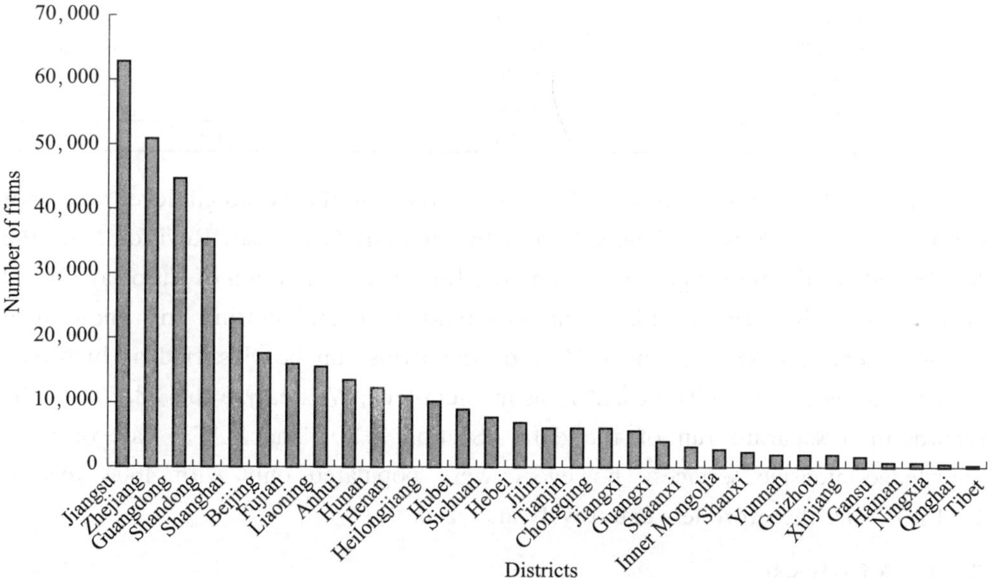


Figure 2 Geographical distribution of the firms (Years 2008-2010; $N = 379\ 026$)

Using SPSS v21 for the geographic mapping, we imported a shape-file for the administrative organization of China from the Internet at http: //www. diva-gis. org/ datadown. The database of this file distinguishes three authoritative layers within China, of which the 31 provinces are the first, and 2410 locations (city names) the third. The in-between level 2 contains 339 units (prefectures). Using dedicated routines, we were able to establish 330 897 records of firms (87. 3%) additionally with this information at level 2.

For the size distribution in terms of numbers of employees, we used the same categories as those used in the first study of Dutch data (Leydesdorff et al. , 2006). Table 1 shows this distribution. As could be expected, medium-sized firms dominate the pattern; the relative absence of very small firms (0.9%) can be considered as an artifact of the data collection by the Bureau van Dijk (Ribeiro et al., 2010). As noted, these missing values probably constitute the main shortcoming of this data because one would expect this class of firms to be relatively large.

**Table 1 Size classes in terms of numbers of employees**

| Size class | Number | Percentage |
|---|---|---|
| 0, 1, or n. a. | 3 761 | 0.9 |
| 2-4 | 1 003 | 0.2 |
| 5-9 | 4 027 | 1 |
| 10-19 | 19 938 | 5 |
| 20-49 | 95 949 | 23.8 |
| 50-99 | 110 662 | 27.5 |
| 100-199 | 83 469 | 20.7 |
| 200-499 | 58 271 | 14.5 |
| 500-749 | 10 676 | 2.7 |
| 750-999 | 4 870 | 1.2 |
| >1 000 | 9 978 | 2.5 |
| Total | 402 604 | 100 |

In 10 074 records (2.5%), the NACE codes at the two-digit level were not valid; these records were excluded from the analysis. Orbis uses Revision 2 of the NACE code at the four-digit level. Using the further classification of Eurostat/OECD (2009, 2011) in terms of high- and medium-tech manufacturing and knowledge-intensive services, 28 659 (or 7.1%) of the firms can be classified as high-tech manufacturing and 105 604 (26.2%) as medium-tech. We analyze this 33.3% of the records in a separate run of the data (Section 3.2). The NACE categories of knowledge-intensive services, however, were populated only with 4604 records (1.1%), and therefore less extensively analyzed.

## 2.2 Methods

As noted above, mutual information in more than three dimensions—the Triple-Helix indicator to be used here—is a signed information measure (Yeung, 2008), and therefore not a Shannon-information (Krippendorff, 2009a, 2009b). However, this measure is derived in the context of information theory and follows from the Shannon formulas (e.g., Ashby, 1964; Abramson, 1963; McGill, 1954).

According to Shannon (1948) the uncertainty in the relative frequency distribution of a random variable $x$ ( $\sum_x p_x$ ) can be defined as $H_X = -\sum_x p_x \log_2 {}^{p_x}$. Shannon denotes this as probabilistic entropy, which is expressed in bits of information if the number 2 is used as the base for the logarithm. (When multiplied by the Boltzman constant $k_B$, one obtains thermodynamic entropy and the corresponding dimensionality in Joule/Kelvin. Unlike thermodynamic entropy, probabilistic entropy is dimensionless and therefore yet to be provided with meaning when a system of reference is specified.)

Likewise, uncertainty in a two-dimensional probability distribution can be defined as $H_{XY} = -\sum_x \sum_y p_{xy} \log_2^{p_{xy}}$. In the case of interaction between the two dimensions, the uncertainty is reduced with the mutual information or transmission: $T_{XY} = (H_X + H_Y) - H_{XY}$. If the distributions are completely independent, then $H_{XY} = H_X + H_Y$, and consequently $T_{XY} = 0$.

In the case of three potentially interacting dimensions ($x$, $y$, and $z$), the mutual information can be derived (e.g., Abramson, 1963) as:

$$T_{XYZ} = H_X + H_Y + H_Z - H_{XY} - H_{XZ} - H_{YZ} + H_{XYZ} \quad (1)$$

The interpretation is as follows: association information can be categorized broadly into correlation information and interaction information. A spurious correlation in a third attribute, for example, can reduce the uncertainty between the other two. The correlation information among the attributes in a data set can be interpreted as the total amount of information shared among the attributes. The interaction information can be interpreted as multivariate dependencies among the attributes.

Compared with correlation, mutual information can be considered as a parsimonious measure for the association. The multivariate extension to mutual information was first introduced by McGill (1954) as a generalization of Shannon's mutual information. This signed information measure (Yeung, 2008) is similar to the analysis of variance, but uncertainty analysis remains more abstract and does not require assumptions about the metric properties of the variables (Garner and McGill, 1956). Han (1980) developed the concept further; positive and negative interactions were also discussed by Jakulin (2005), Leydesdorff and Ivanova (in press), Sun and Negishi (2010), Tsujishita (1995), and Yeung (2008).

One of the advantages of entropy statistics is that the values can be fully decomposed. As with the decomposition of probabilistic entropy (Theil, 1972), mutual information in three dimensions can be decomposed into groups as follows:

$$T = T_0 + \sum_G \frac{n_G}{N} T_G \quad (2)$$

Since we will decompose in the geographical dimension, $T_0$ connotes between-region uncertainty; $T_G$ is the uncertainty prevailing at a geographical scale $G$; $n_G$ is the number of firms at this geographical scale; and $N$ is the total number of firms in the whole set.

The between-region uncertainty ($T_0$) can be considered as a measure of the dividedness. A negative value of $T_0$ indicates an additional synergy at the higher level of national agglomeration among the lower-level geographical units. In the Netherlands, Norway, and Sweden, for example, a surplus was found at the national level; in Germany, this surplus was found at the level of the federal states (Länder). Note that one cannot compare the quantitative values of $T_0$ across countries—because these values are sample-specific—but one is allowed to compare the dividedness in terms of the positive or negative signs of $T_0$ and as a percentage of the total synergy for each country. All values of the contribution of subsets to the knowledge-based economy are based on normalization on the total set (that is, $n_G/N$ in Equation 2).

# 3 Results

## 3.1 Decomposition at the Provincial Level of China

We run the analysis for all firms in the whole set and then for each of the 31 provinces separately. This leads to values for $T$ and $T_G$, respectively, that can be used in Equation 2; the values of $N$ and $n_G$ are known from the download. Normalized values of the contributions of provinces to the national synergy ($\Delta T = n_{G/N} \times T_G$) and the between-region synergy ($T_0$) can then be derived.

According to the data of 31 provinces of China in 2008-2010, and setting $N =$ 379 026, we can get a data map of synergies in the knowledge-based economy of China at the provincial level. And then we can know the 31 provincial units' levels according to their respective contributions to the synergy in the knowledge-based economy. The total synergy for the nation is −196.48 mbits of information, of which 18.0% (35.46 mbits) is realized at the above-provincial level. This is more than we found in the case of Norway (11.7%), but less than for the Netherlands (27.1%) or Sweden (20.4%). As what was said, we found no additional synergy in Germany and Hungary at the national level (for different reasons). China thus functions very much as a unified nation state (Table 2).

**Table 2 Province classification of China**

| | All sectors | | High- and medium-tech | | | KIS | |
|---|---|---|---|---|---|---|---|
| Province (a) | Number of firms (b) | $\Delta T$ (mbit) (c) | Number of firms (d) | $\Delta T$ (mbit) (e) | Percentage of contribution [(e)/(c)] (f) | Number of firms (g) | $\Delta T$ (mbit) (h) |
| Jiangsu | 62 805 | −12.48 | 1 783 | −3.19 | 25.6 | 49 | 0.03 |
| Shandong | 35 152 | −12.23 | 10 862 | −2.66 | 21.7 | 346 | 0.01 |

Continued

| Province (a) | All sectors |  | High- and medium-tech |  |  | KIS |  |
|---|---|---|---|---|---|---|---|
|  | Number of firms (b) | $\Delta T$ (mbit) (c) | Number of firms (d) | $\Delta T$ (mbit) (e) | Percentage of contribution [(e) / (c)] (f) | Number of firms (g) | $\Delta T$ (mbit) (h) |
| Guangdong | 44 692 | −10. 99 | 15 986 | −1. 75 | 15. 9 | 90 | −0. 01 |
| Zhejiang | 50 699 | −10. 92 | 17 265 | −2. 33 | 21. 4 | 24 | 0. 05 |
| **Beijing** | 17 490 | −8. 68 | 6 665 | −2. 41 | **27. 8** | 339 | **0. 08** |
| Hunan | 12 019 | −8. 38 | 3 811 | −1. 80 | 21. 5 | 218 | −0. 12 |
| **Shanghai** | 23 049 | −7. 93 | 10 164 | −2. 37 | **29. 9** | 17 | 0. 01 |
| Hubei | 8 969 | −7. 39 | 2 512 | −1. 66 | 22. 5 | 181 | −0. 02 |
| Sichuan | 7 807 | −7. 34 | 1 930 | −1. 47 | 20. 0 | 161 | −0. 04 |
| Liaoning | 15 565 | −7. 25 | 556 | −1. 44 | 19. 8 | 748 | −0. 01 |
| Anhui | 13 275 | −6. 98 | 4 260 | −1. 61 | 23. 0 | 130 | −0. 01 |
| Henan | 10 899 | −6. 62 | 3 088 | −1. 46 | 22. 0 | 112 | 0. 01 |
| Heilongjiang | 9 993 | −5. 99 | 3 174 | −1. 22 | 20. 4 | 137 | 0. 04 |
| Hebei | 7 062 | −5. 82 | 1 737 | −1. 22 | 20. 9 | 760 | 0. 00 |
| **Chongqing** | 6 015 | −5. 70 | 2 287 | −1. 38 | 24. 3 | 92 | **0. 11** |
| Fujian | 16 001 | −5. 59 | 3 344 | −1. 17 | 21. 0 | 187 | −0. 03 |
| Jilin | 6 190 | −5. 09 | 4 768 | −1. 12 | 21. 9 | 145 | 0. 02 |
| Jiangxi | 5 790 | −4. 01 | 1 977 | −0. 98 | 24. 3 | 140 | −0. 02 |
| Guangxi | 3 888 | −3. 33 | 1 112 | −0. 65 | 19. 7 | 89 | −0. 07 |
| Shanxi | 2 363 | −2. 72 | 698 | −0. 72 | 26. 7 | 75 | 0. 01 |
| **Tianjin** | 6 132 | −2. 69 | 2 774 | −0. 83 | **30. 8** | 13 | 0. 00 |
| Inner Mongolia | 2 605 | −2. 37 | 24 823 | −0. 28 | 11. 6 | 147 | −0. 01 |
| Guizhou | 1 695 | −2. 11 | 433 | −0. 23 | 10. 8 | 31 | 0. 03 |
| Xinjiang | 1 625 | −2. 05 | 270 | −0. 11 | 5. 6 | 87 | 0. 07 |
| Shaanxi | 2 868 | −1. 92 | 971 | −0. 45 | 23. 4 | 96 | 0. 02 |
| Gansu | 1 510 | −1. 82 | 374 | −0. 21 | 11. 7 | 33 | 0. 05 |
| Yunnan | 1 707 | −1. 72 | 419 | −0. 15 | 8. 4 | 49 | 0. 00 |
| Hainan | 431 | −0. 54 | 111 | −0. 05 | 9. 9 | 12 | 0. 01 |
| Ningxia | 409 | −0. 29 | 127 | −0. 02 | 8. 4 | 1 | 0. 00 |
| Qinghai | 254 | −0. 08 | 81 | 0. 00 | — | 2 | 0. 00 |
| Tibet | 67 | 0. 01 | 12 | 0. 00 | — | 6 | 0. 01 |
| Σ= |  | −161. 02 |  | −34. 95 | 17. 7 |  | 0. 23 |
| China $T_0$ | 379 026 | −196. 48 | 128 374 | −41. 75 | 21. 2 | 4 517 | −1. 07 |
|  |  | −35. 46 |  | −6. 80 | 19. 2 |  | −1. 30 |

Notes: 31 provinces of China sorted by their *contribution* to synergy (column c); all sectors (columns b-c; $N$ = 379 026); high- and medium-tech manufacturing (columns d-f; $N$ = 128 374); knowledge-intensive services (columns g-h; $N$ = 4517). Total values of China and between-region values $T_0$ are added in the bottom rows. The four municipalities are boldfaced.

In Table 2, the 31 provinces are sorted in terms of their contributions to the overall synergy (column c). Since the number of firms in a region ($n_G$) is a factor in Equation 2, the correlations with the number of firms are high, and therefore the relatively smaller provinces compared with Beijing and Shanghai (in terms of total numbers of firms) figure less prominently than one might perhaps expect. We return to this issue in Section 3.3 when we analyze the data at the next-lower level of aggregation. However, we first turn to the decomposition of the set in terms of high- and medium-tech manufacturing (columns d and e in Table 1), and knowledge intensive services (columns f and g) in the next section by using the NACE codes (Table 3).

**Table 3 NACE classifications (revision 2) of high- and medium-tech manufacturing, and knowledge-intensive services**

| Code | High- and medium-tech manufacturing | Code | Knowledge-intensive Sectors (KIS) |
|---|---|---|---|
| | High-tech Manufacturing | 50 | Water transport |
| 21 | Manufacture of basic pharmaceutical products and pharmaceutical preparations | 51 | Air transport |
| | | 58 | Publishing activities |
| 26 | Manufacture of computer, electronic and optical products | 59 | Motion picture, video and television programme production, sound recording and music publishing activities |
| 30.3 | Manufacture of air and spacecraft and related machinery | 60 | Programming and broadcasting activities |
| | | 61 | Telecommunications |
| | | 62 | Computer programming, consultancy and related activities |
| | Medium-high-tech Manufacturing | 63 | Information service activities |
| 20 | Manufacture of chemicals and chemical products | 64-66 | Financial and insurance activities |
| | | 69 | Legal and accounting activities |
| 25.4 | Manufacture of weapons and ammunition | 70 | Activities of head offices; management consultancy activities |
| | | 71 | Architectural and engineering activities; technical testing and analysis |
| 27 | Manufacture of electrical equipment | | |
| 28 | Manufacture of machinery and equipment n.e.c. | 72 | Scientific research and development |
| | | 73 | Advertising and market research |
| 29 | Manufacture of motor vehicles, trailers and semi-trailers | 74 | Other professional, scientific and technical activities |
| | | 75 | Veterinary activities |
| 30 | Manufacture of other transport equipment | 78 | Employment activities |
| | | 80 | Security and investigation activities |
| | • excluding 30.1 Building of ships and boats, and | 84 | Public administration and defence, compulsory social security |
| | | 85 | Education |
| | • excluding 30.3 Manufacture of air and spacecraft and related machinery | 86-88 | Human health and social work activities |
| | | 90-93 | Arts, entertainment and recreation |
| 32.5 | Manufacture of medical and dental instruments and supplies | | Of these sectors, 59 to 63, and 72 are considered high-tech services. |

Sources: Eurostat/OECD (2009, 2011); Laafia (2002) and Leydesdorff et al. (2006).

## 3.2 Sectorial Decomposition

### 3.2.1 High- and medium-tech manufacturing

Let us first turn to the subset of 134 263 firms (33.3% of the data) which are classified with the NACE codes (Revision 2) as high- and medium-tech manufacturing (Table 2). The columns d and e of Table 1 provide the corresponding figures, and in column f, the values of $\Delta T$ for this subset (in column e) are compared with those in column c for all sectors. Columns c (for the total set) and e (for high- and medium-tech manufacturing) are highly correlated: Pearson $r = 0.962$ ($p<0.01$); Spearman's $\rho = 0.984$ ($p<0.01$); $N = 31$. However, the relative contribution of high- and medium-tech to the nation provides only 21.2% of the synergy, while 33.3% of the firms were classified as such.

The contributions are most pronounced in the regions of Beijing (27.9%), Shanghai (29.9%), and Tianjin (30.8%). The latter is a province between Beijing and the coast. In the next section, we shall see that these three provinces are also considered as municipalities with Chongqing as a fourth one. However, the knowledge base of this latter province is not so strongly enhanced given this focus on high- and medium-tech manufacturing. This synergies in high-tech and medium-tech manufacturing at the provincial level of China is based on the data of 31 provinces of China in 2008-2010, and $N = 128\ 374$.

### 3.2.2 Knowledge-intensive services

We found an uncoupling effect of the knowledge-intensive services from the regional economy in the Western-European countries studied previously because a knowledge-intensive service can be provided from any location near a railway station or airport; the geographical location ("rooting") is thus less relevant in the case of knowledge-intensive services. This effect is attenuated when R&D facilities ("high-tech knowledge-intensive services") are needed because these activities may require laboratories that are grounded. Since our data is thin in this domain (column g), we did not pursue a further decomposition within the category of knowledge-intensive services.

Column h shows that Chongqing is the strongest in terms of this uncoupling effect with Beijing at the second place. The synergy indicator in these instances is positive which means that one adds to the uncertainty with a localized focus on these provinces. However, this is not the case for all districts. Shanghai, for example, does not seem to play a role from this perspective (although the data is extremely poor; $N = 17$). Guangdong which is represented with 346 of these services, however, does not perform any better on this indicator ($\Delta T = 0.01$ mbit).

In summary, the decomposition in terms of sectors most relevant to the knowledge-based economy did not show the strongly enhanced role of high-and medium-tech manufacturing that could be expected on the basis of previous studies (for European nations), but the uncoupling effect of knowledge-intensive services was confirmed for administrative centers such as Chongqing and Beijing. A further decomposition in terms of high-tech might be interesting, but we were hesitant to pursue this further given the limitations in this data.

## 3. 3 The Second Administrative Level

According to the data from the 339 second-level administrative units in China, and comparing their contribution to the synergy among technology, geography and organization, we can get the pronounced contribution to the synergy at the second administrative level of four units (Beijing, Shanghai, Tianjin, and Chongqing) that have the special status of "municipalities" directly managed by the central government of China. Table 4 provides the numerical breakdown and shows that the data is also mainly collected from these municipalities and the prefecture of Dezhou (neighboring to Tianjin, but in the province of Shandong) that follows at a next-lower level. However, Cangzhou—south to Tianjin—is larger in terms of the number of firms, but indicated as much less synergetic in terms of the three dimensions studied here.

**Table 4 Administrative units at the second level sorted in terms of their contributions ($\Delta T$) to the synergy among technology, geography, and organization**

| Item | $\Delta T$ in mbits | Number of firms |
|---|---|---|
| **Shanghai** | **−3. 91** | **12 742** |
| **Chongqing** | **−3. 65** | **13 488** |
| **Beijing** | **−3. 32** | **4 394** |
| **Tianjin** | **−2. 60** | **6 316** |
| Dezhou | −1. 46 | 6 630 |
| Nanping | −0. 96 | 1 823 |
| Yantai | −0. 90 | 3 127 |
| Cangzhou | −0. 82 | 7 628 |
| Zhangzhou | −0. 75 | 4 830 |
| Fuzhou | −0. 72 | 2 894 |
| Tieling | −0. 67 | 2 349 |
| Weifang | −0. 58 | 4 427 |
| Yuncheng | −0. 51 | 735 |
| Zhengzhou | −0. 51 | 1 900 |
| Hengyang | −0. 47 | 998 |
| Deyang | −0. 44 | 913 |
| Luohe | −0. 44 | 2 994 |
| Yichun | −0. 43 | 765 |
| Luoyang | −0. 41 | 1 675 |
| Yanbian | −0. 38 | 227 |
| (…) | | |
| Σ | −40. 84 | 310 974 |
| China (adm _ 2) | −183. 42 | |
| $H_0$ | −142. 58 | |

Because we can evaluate only 87. 3% of the data at this level, the reduction of uncertainty at the national level is −183.42 as against −196.48 mbits in Table 1 (93.4%). Table 4 shows that only 22.3% of this synergy (40.84 mbits) is found at the second level of the administration. The provinces (at level 1) are thus the relevant units for studying the synergy in the Chinese economy, but the role of the national level is considerable. The country is far more centralized than, for example, Norway; but the distribution is less skewed towards the metropolitan centers than in Sweden.

## 4 Discussion

As noted, the major point for discussion is the data collection by the Bureau van Dijk that fills the Orbis database on the basis of information provided by more than one hundred information suppliers and by its own research. The methods of data collection are discretionary since the information is controlled by the companies in question. This data, therefore, provides an incomplete sample. However, the selection of journals for example by Thomson-Reuters and Elsevier for their databases (Web of Science and Scopus, respectively) is also not public. Thus, such a state of affairs is more common when using commercial databases for scientometric research.

The main remaining question concerns the extent to which one can expect biases in the data collection to influence the results. We noted that small-sized enterprises are under-represented in this data, and that self-employed entrepreneurs seem not to be included at all. Consequently, startups are presumably not included. This may bias the results against university-based entrepreneurship that may not be distributed equally across the country. However, the industrialized provinces of China all have a considerable number of universities with potentially industrial activities of graduates. This effect, however, may disfavor the largest metropolitan areas such as Beijing and Shanghai (Hong Kong is not included in this data).

The synergy indicator is not an output but a structural indicator, although—as noted—the number of firms is also registered in Equation 2 as the units of analysis. Most governmental (OECD) statistics are (linear) output indicators (e.g., Schaaper, 2009). For example, we found a report entitled "High-tech Statistics China 2012" on the website of the National Bureau of Statistics① which provides a table (1-7) with "gross industrial output value of high-tech industries by region" (in RMB ¥100 million). Not surprisingly, this indicator correlates significantly with our

① Available at http://www.sts.org.cn/sjkl/gjscy/data2012/data12.pdf.

indicator using the first-administrative level of 31 provinces; Spearman's $\rho = 0.867$ ($p < 0.01$). All distributions over the Chinese provinces are skewed and tend therefore to be correlated.

If we focus within the data only on high- and medium-tech, Spearman's $\rho$ further increases only marginally to 0.879. However, the regions indicated as most productive in terms of output are different from the ones signaled by our methods, namely Jiangsu and Guangdong. These latter regions are among those with the largest numbers of firms in our set: 62 805 and 44 692 firms, respectively. Zhejiang (50 699 firms in our data), however, is categorized with Beijing, Shanghai, and Shandong in the second group by this report, and with Chongqing among the third category.

As against these rankings in terms of size, our indicator is a measure of synergy or resonance among the distributions in three (or more) dimensions. The decomposition in terms of sectors and levels taught us that the four municipalities carry a specific function in the knowledge-based economy that was not anticipated. High- and medium-tech is more concentrated in and around the metropolitan areas of Beijing and Shanghai, but setting this filter does not lead to a higher synergy across the provinces (as we had expected on the basis of previous studies). The roles of Chongqing and Beijing as centers of knowledge-intensive administration are notable. In other words, the design allows us to fine-tune in terms of the three relevant dimensions (technology, geography, and size) as different projections of the three-dimensional data. The quality of this data, however, remains beyond our control.

## 5 Conclusions

Most synergy is generated in the knowledge base of the Chinese economy at the provincial level (administrative level 1), but the national level adds a substantial contribution (>18%). The level of the prefectures seems much less relevant except in the case of Shanghai, Beijing, Chongqing, and Tianjin as "municipalities" that are defined also at the (first) level of the provinces. In general, the synergy is generated above the city level; that is, at the level of regions.

A focus on high- and medium-tech manufacturing shows that the differences in terms of this selection can be between 20 and 30 percent for provinces. Tianjin joins Beijing and Shanghai as the "winners" when taking this perspective, whereas large industrial regions (such as Guangdong) may be less profiled in terms of high- and medium-tech firms. Decomposition in terms of knowledge-intensive services is based on relatively small sets, but BvD claims that services are also included in Orbis and thus this data suggests that the Chinese economy is more manufacturing than service-oriented. The knowledge-intensive services are geographically located in the administrative centers and perhaps associated to the government (Perevodchikov et al. ,2013).

## Acknowledgement

We thank Inga Ivanova, Fred Y. Ye, and two anonymous referees for comments on a previous version of this manuscript. The study was supported by the National Natural Science Foundation of China (NSFC) with grant number 71073153.

## References

Abramson N. 1963. *Information Theory and Coding*. New York: McGraw-Hill.

Ashby W R. 1964. Constraint analysis of many-dimensional relations. *General Systems Yearbook*, 9: 99-105.

Eurostat/OECD. 2009. Aggregations of manufacturing based on NACE Rev. 2 (January 2009). http://epp.eurostat.ec.europa.eu/cache/ITY_SDDS/Annexes/htec_esms_an3.pdf.

Eurostat/OECD. 2011. High technology and knowledge-intensive sectors, December 2011. http://epp.eurostat.ec.europa.eu/cache/ITY_SDDS/Annexes/hrst_st_esms_an9.pdf.

Garner W R, McGill W J. 1956. The relation between information and variance analyses. *Psychometrika*, 21 (3): 219-228.

Han T S. 1980. Multiple mutual information and multiple interactions in frequency data. *Information and Control*, 46 (1): 26-45.

Ivanova I A, Leydesdorff L. 2014. A simulation model of the Triple Helix of university-industry-government relations and the decomposition of the redundancy. http://link.springer.com/article/10.10076ZFS11192-014-1241-7 [2014-08-13].

Jakulin A. 2005. Machine learning based on attribute interactions. Unpublished PhD Thesis, University of Ljubljana.

Krippendorff K. 1980. *Q*: an interpretation of the information theoretical *Q*-measures. *In*: Trappl R, Klir G J, Pichler F (eds.), *Progress in Cybernetics and Systems Research* (Vol. VIII, pp. 63-67). New York: Hemisphere.

Krippendorff K. 2009a. Information of interactions in complex systems. *International Journal of General Systems*, 38 (6): 669-680.

Krippendorff K. 2009b. Ross Ashby's information theory: a bit of history, some solutions to problems, and what we face today. *International Journal of General Systems*, 38 (2): 189-212.

Laafia I. 2002. Employment in high tech and knowledge intensive sectors in the EU continued to grow. *Statistics in Focus*: *Science and Technology*, *Theme* 9 (4): http: //www. eds-destatis. de/en/downloads/sif/ns _ 02 _ 04. pdf.

Lengyel B, Leydesdorff L. 2011. Regional innovation systems in Hungary: the failing synergy at the national level. *Regional Studies*, 45 (5): 677-693.

Leydesdorff L. 2003. The mutual information of university-industry-government relations: an indicator of the Triple Helix dynamics. *Scientometrics*, 58 (2): 445-467.

Leydesdorff L, Fritsch M. 2006. Measuring the knowledge base of regional innovation systems in Germany in terms of a Triple Helix dynamics. *Research Policy*, 35 (10): 1538-1553.

Leydesdorff L, Ivanova I A. 2014. Mutual redundancies in inter-human communication systems: steps towards a calculus of processing meaning. *Journal of the American Society for Information Science and Technology*, 65 (2): 386-399.

Leydesdorff L, Strand Ø. 2013. The Swedish system of innovation: regional synergies in a knowledge-based economy. *Journal of the American Society for Information Science and Technology*, 64 (9): 1890-1902.

Leydesdorff L, Sun Y. 2009. National and international dimensions of the Triple Helix in Japan: university-industry-government versus international co-authorship relations. *Journal of the American Society for Information Science and Technology*, 60 (4): 778-788.

Leydesdorff L, Dolfsma W, van der Panne G. 2006. Measuring the knowledge base of an economy in terms of Triple-Helix relations among "technology, organization, and territory". *Research Policy*, 35 (2): 181-199.

McGill W J. 1954. Multivariate information transmission. *Psychometrika*, 19 (2): 97-116.

Park H W, Hong H D, Leydesdorff L. 2005. A comparison of the knowledge-based innovation systems in the economies of South Korea and the Netherlands using Triple Helix indicators. *Scientometrics*, 65 (1): 3-27.

Park H W, Leydesdorff L. 2010. Longitudinal trends in networks of university-industry-government relations in South Korea: the role of programmatic incentives. *Research Policy*, 39 (5): 640-649.

Perevodchikov E, Uvarov A, Leydesdorff L. 2013. Measuring synergy in the Russian innovation system. Paper presented at the 12th International Conference about the Triple Helix of University-Industry-Government Relations, London, UK.

Ribeiro S P, Menghinello S, de Backere K. 2010. *The OECD ORBIS Database*: *Responding to the Need for Firm-level Micro-data in the OECD*. Paris: OECD Publishing.

Schaaper M. 2009. *Measuring China's Innovation System*: *National Specificities and International Comparisons*. Paris: OECD Publishing.

Shannon C E. 1948. A mathematical theory of communication. *Bell System Technical Journal*, 27: 379-423, 623-656.

Storper M. 1997. *The Regional World—Territorial Development in a Global Economy*. New York: Guilford Press.

Strand Ø，Leydesdorff L. 2013. Where is synergy in the Norwegian innovation system indicated? Triple Helix relations among technology，organization，and geography. *Technological Forecasting and Social Change*，80（3）：471-484.

Sun Y，Negishi M. 2010. Measuring the relationships among university，industry and other sectors in Japan's national innovation system：a comparison of new approaches with mutual information indicators. *Scientometrics*，82（3）：677-685.

Theil H. 1972. *Statistical Decomposition Analysis*. Amsterdam/London：North-Holland.

Tsujishita T. 1995. On triple mutual information. *Advances in Applied Mathematics*，16（3）：269-274.

Ye F Y，Yu S S，Leydesdorff L. 2013. The Triple Helix of university-industry-government relations at the country level，and its dynamic evolution under the pressures of globalization. *Journal of the American Society for Information Science and Technology*，64（11）：2317-2325.

# 1-3 Exploring the Deterministic Factors That Influence the Publication and Citation Process

**Liu Yuxian**①

## Abstract

We locate the position of scientometrics in the knowledge system. Scientometrics belongs to formal science as well as empirical science. Scientometrics could explore the deterministic factors that influence the publication and citation process. We present the main idea of Liu Yuxian's dissertation, which won the 2011 Emerald/EFMD Outstanding Doctoral Research Award in the information science category. We expound the implications of this research to the field of scientometrics.

**Keywords**: deterministic factor; publication; academic citation process

## 1 Writing for the Special Conference on Celebration of Professor Ronald Rousseau's Collaboration with Chinese Colleagues for 15 Years

Professor Ronald Rousseau has two Chinese students, Professor Liang Liming and Liu Yuxian. Both of his Chinese students won the Emerald/EFMD Outstanding Doctoral Research Award in the information science category. Professor Liang Liming won this prestigious award in 2007, in which year she was 58 and already a famous professor. At that time I just began to pursue my doctoral program and said to her: "Congratulations for being elected to receive this prize, an achievement which I do not expect to follow." She answered me that we were both lucky to have a great

① Library of Tongji University, Siping Street 1239, Shanghai, 200092 (China).

supervisor and that under his guidance, also my doctoral thesis would be excellent. I never took her encouraging words serious because I was just working as a librarian with limited means...I even could not write in decent English! However, after four years hard work, I too won the Emerald/EFMD Outstanding Doctoral Research Award in 2011.

On this treasured occasion, I would like to express my sincere thanks to Professor Ronald Rousseau. When pursuing my doctoral degree, I acted as a kid who pointed to the brightest star in the sky, and said to my dad: "I want that brightest star..." I just had a strong ambition and didn't notice whether there was a ladder leading to the star or not. Even if there was just one vague fulcrum in the intellectual landscape of human being to support my idea, Professor Ronald Rousseau tried hard to construct a ladder and put it in the right place for me to pick up that brightest star.

After I finished the first version of my thesis, its main ideas were not accepted by the field. Articles based on my thesis were rejected by journals, sometimes more than once. Sometimes, I thought I had to give up. But Professor Ronald Rousseau wrote to me: the more I read your thesis, the more I know how deep you have reflected on the scientific investigations you had done. It is now my responsibility, as your thesis advisor, to convince colleagues of the value of your ideas.

Then he invited Professor Glänzel and Professor Leydesdorff to be jury members for my thesis defense. Professor Glänzel and Professor Leydesdorff gave me their valuable comments and finally made my ideas permeate into the vein of development of the fields of scientometrics and informetrics. My scientific ideas then have struck roots in our field and my thesis got its final shape.

After the defense, Professor Ronald Rousseau and I published the main ideas in *JASIST*, the Journal of Documentation, Aslib Proceedings, and Information Processing & Management.

I also want to take this chance to thank all my friends who helped me overcome the obstacles I encountered. Indeed, I didn't have any social or economic support to pursue my doctoral degree. Although I obtained the diploma from the University of Antwerp, Belgium, I studied most of the time on my own in my spare time in Shanghai, China. The absence of daily practice in expressing my thoughts in English and the lack of proper academic circumstances made the process full of difficulties. For this, my son awarded me the "Hard Work Nobel Prize" ...Luckily, I also had friends supporting me such as Jiang Guohua, Wu Yishan and Liang Liming.

I also wish to share with you a story about a dream and how this dream came true. When I was a teenager I announced to the world that I wanted to study abroad and obtain a doctoral degree. I was just a little girl who lived in a poor rural region of North China. Born in an illiterate family, I had nothing real at hand besides ambition inspired by a novel, *Second Hand Clasp*, which described the life of a Chinese woman who studied abroad and became a famous physical scientist. Actually, I saw no road that could bring my dream to reality. In retrospect it seems magic that my dream of obtaining a doctoral degree abroad came true thirty years after I announced it, at a time when it was really an impossibility. Then what makes the impossible possible? Time is elapsing when rivers flow to the east, and hope is always hiding in a corner. And even when one cannot see the corner, one should not be misled to think that hope is not there: just work hard and wait. Believe that all dreams will come true one day. With this belief, I will concentrate all my strengths to go ahead.

## 2 Formal Science, Empirical Science and Metrics of Science

The formal sciences are those branches of knowledge that deal with formal conceptual systems. They presuppose no knowledge of contingent facts and do not describe the real world, so that they do not involve empirical procedures. Because of their non-empirical nature, the formal sciences are normally construed by outlining a set of axioms and definitions from which other statements (theorems) are deduced. In other words, theories in formal sciences contain no synthetic statements: all their statements are analytical. They are not concerned with the validity of theories based on observations in the real world, but instead with the properties of formal systems based on definitions and rules. Formal sciences are in content and validity independent of any empirical procedure. Knowledge as studied in logic, mathematics, theoretical computer science, information theory, system theory, decision theory, statistics, and some aspects of linguistics belongs to the formal sciences (Bunge, 1985).

Empirical sciences include not only the natural sciences but also the social sciences and the humanities. Natural sciences study natural phenomena, while the social sciences and the humanities investigate human behavior and societies. All empirical knowledge must be based on observable phenomena, and can be tested for their validity by other researchers working under the same conditions.

Formal sciences are vital to the empirical sciences. Major advances in the formal sciences have often led to major advances in the empirical sciences (and vice versa). The formal sciences are essential in the formulation of hypotheses, theories and laws, and thus in discovering and describing how things work (natural sciences) as well as how people think and act (social sciences).

Using mathematical methods to study the quantitative aspects of scientific documents, in printed or electronic form, is nowadays developed as a scientific discipline. It has some resemblance to the formal sciences, yet it is an empirical science. According to the carriers and communication style, the study of quantitative aspects of documents has evolved to a set of related subfields: informetrics, bibliometrics, scientometrics, cybermetrics and webometrics. Their relations are shown in Figure 1.

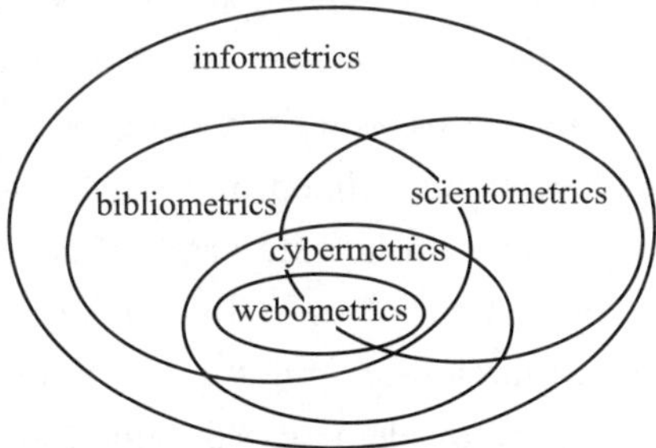


Figure 1 Relationships between the library and information science fields of informetrics/bibliometrics/scientometrics/webometrics

Notes: Sizes of the overlapping ellipses are made for the sake of clarity only and do not reflect quantitative relations (Björneborn and Ingwersen, 2004)

Among these metrics fields, "scientometrics" is the study of the quantitative aspects of science as a discipline or economic activity. Indicators are formally defined, their statistical characteristics are analyzed and the results are used to measure science. In this sense scientometrics has some similarity to a formal science.

However, science does not evolve according to our defined formal indicators: it evolves at its own special rhythm internal laws steering the direction of its evolution. Science itself is an object that can be observed, giving abundant empirical data—such as publications and citations—that can be used to sum up the regularities hidden in the evolution of science.

Publications are final products of scientific investigations and once the final result is published, it enters into the scientific communication system. It may be read and inspire further investigations on the phenomenon under study. When the results of the new investigation are published, older publications are acknowledged as citations. So we may say that citations are produced as the result of interactions (inspiration) of different ideas on phenomena under study. Publications and citations are two basic elements in the metrics of science. Can we use them as instruments to observe the evolution of science?

# 3 Is the Publication and Citation Process Determined by Contingent or by Deterministic Factors?

In the history of citation analysis, scientometrics is a multifaceted approach that can be used to depict the structure and development of science (van Raan, 2008). From a scientometric perspective the development of science is approximated by the citation process. However, this process is usually described as a stochastic process.

Figure 2, taken from Glänzel and Schoepflin (1994), shows the citation process as a stochastic process. It represents an infinite array of cells indexed in succession by the non-negative integers. The content of the $i$-th cell is denoted by $x_i$ and the finite content of all cells by $x$. As a function of time this is written as $x(t)$. The fractions $y_i = x_i/x(i \geqslant 0)$ denote the share of elements contained in the $i$-th cell. The change of content is guided by the following postulates:

(1) Substance may not enter the system from the external environment nor leak out into the external environment;

(2) Substance may be transferred only in one direction from the $i$-th cell to the $(i+1)$-th at a rate $f_i$.

Consequently, $x(t) = x(0) > 0$ at any time $t > 0$.

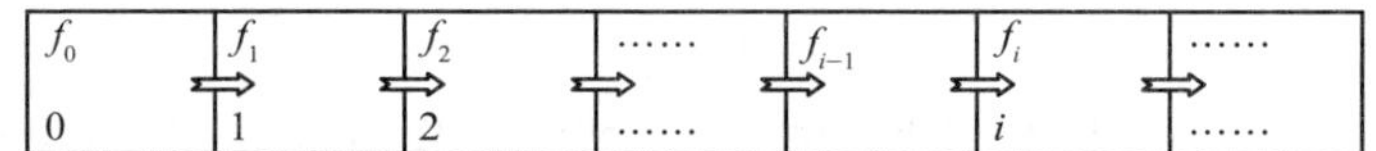


Figure 2 A stochastic citation process

Translating this general framework into an informetric one the word "substance" becomes "publication" and "$i$-th cell" becomes "$i$ citations". Then the postulates mean that no articles will leave the system nor will further articles join in; and articles cannot lose citations. Time 0 is the date of publication. At that time all articles are uncited. The transfer rate $f_i$ expresses that the chance of being cited is proportional to the number of citations already received. The ratio $y_i$ is interpreted as probabilities with which a paper is contained in the $i$-th cell. The stochastic process is then formed by the change of the content of the cells, i. e., by the change of the number of citations received by the given set of articles. If $x(t)$ is the stochastic variable, denoting the number of citations received by this set of articles, then $P[x(t) = i] = y_i$ is the probability of $i$ citations that an article in this set has received exactly in a period of length $t$. Assuming $x(t) = 1$ leads to the citation process $x(t)$ of an arbitrary publication. Here $x(0) = x(t)$ denotes the finite content of all cells. If the total is 1 then we have a probability distribution, hence as a function of $t$, a stochastic process.

Yet, stochastic processes are determined by contingent factors. Now we ask the question: is there a deterministic factor determining the publication and citation process?

Liu and Rousseau (2013) mapped publication and citation into thc scientific investigation. Science develops by the investigation of specific phenomena. Observations and the ideas about the phenomenon articulated in the publication inspire the new thinking of scientists on this phenomenon. The scientists then began to do new investigations and formulate their results in a new publication.

In the same sense, each publication is expected to interact with preceding ones, by incorporating into its own line of reasoning arguments developed in other publications; and each new publication, due to the claims it makes to new knowledge, invites reactions and hence further publications.

Science evolves through the interactions of different scientific ideas about phenomena under investigation. We may say that these scientific ideas are diffused in an abstract intellectual landscape. During this diffusion process, these scientific ideas interact with each other. This, in turn, changes the field and the intensity of the diffusion process.

Actual scientific activities determine the publication and citation process; in this way the publication and citation process represents the scientific investigation process. In this sense, the publication and citation process can describe the evolution of science.

## 4 Empirical Studies

We used splines to describe diffusion process as a spline-fitted curve that can represent changing information. All information about change resides in the original data. Splines not only reflect this change, but keep this information. By integration and the use of differential calculus we can revive and reproduce the original information (Liu and Rousseau, 2011, 2012a).

Liu and Rousseau (2014) illustrate how different interactions of scientific ideas influence the citation diffusion process by the characteristics of the citation diffusion process of the Nobel Prize winning articles in physics written by some Chinese Americans (Ashkin, et al., 1986; Aubert et al., 1974; Kao and Hockham, 1966; Lee and Yang, 1956; Tsui et al., 1982).

The article by Lee and Yang and the one by Ting et al. both questioned the common sense of that era. Yet, their ideas were shown to be right and were acknowledged by a Nobel Prize within a few years after their articles were published. So Lee and Yang's articles about non-conservation of parity gave rise to a strong argument between their ideas and the knowledge system of their time. Similarly, it was believed by physicists that there were only three types of subatomic particles. Yet Samuel C. C. Ting and his collaborators found a fourth type of subatomic particles, so it challenged the old knowledge system. Hence we may say that a fierce conflict between new insights and old knowledge system stirred up a strong argument leading to concave curve. However, when the new ideas were confirmed to be right and accepted by the academic system a Nobel Prize came soon. Later, the rate of increase becomes smaller as it refers to accepted knowledge. This is the reason why the curve becomes convex at the end (Figure 3).

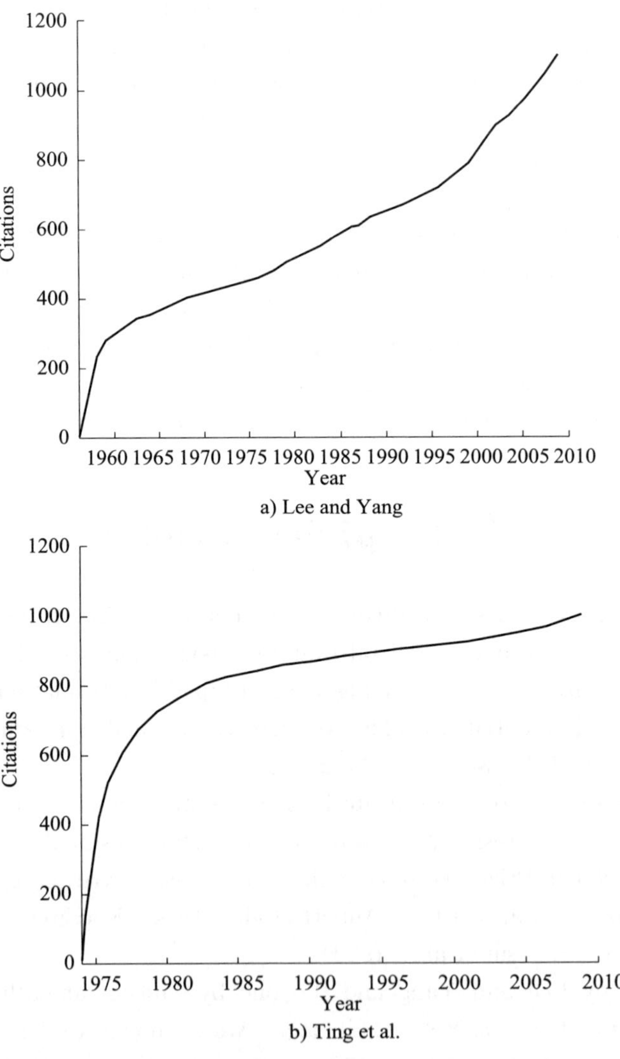


Figure 3 The citation diffusion process of Lee and Yang and Ting et al.

Figure 4 shows the citation diffusion process of Tsui and Chu's Nobel article. The scientific value of their work was not recognized at the beginning. Some aspects were even thought to be impossible. Yet, their ideas inspired scientists to begin an intense search for the truth leading finally to their acceptance. Such a process is described by a convex or linear citation curve (Figure 4).

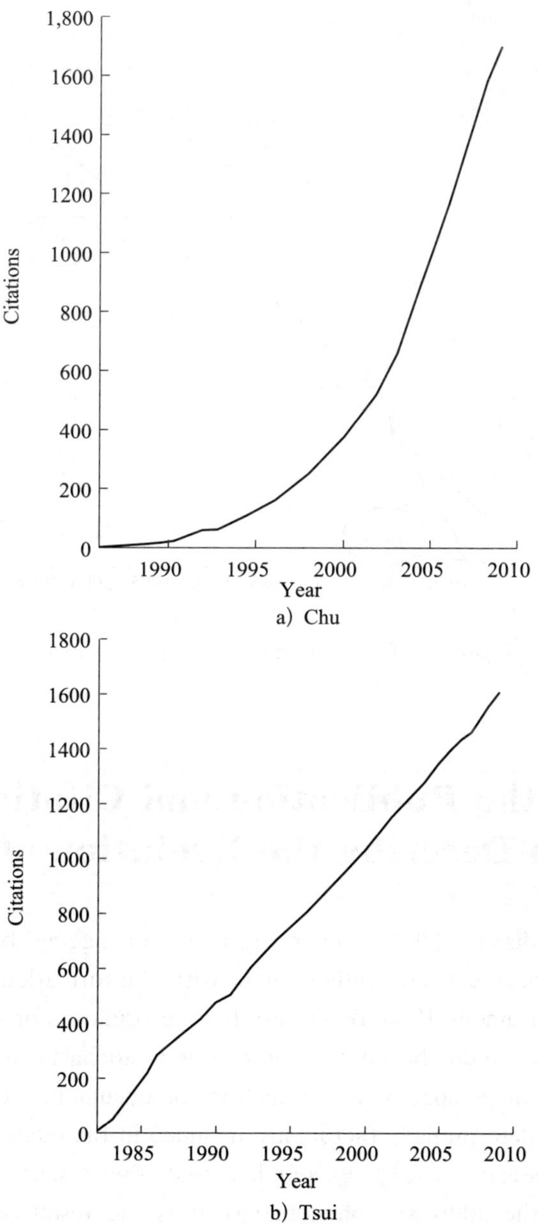


a) Chu

b) Tsui

Figure 4 The citation diffusion process of Tsui and Chu

Liu and Rousseau (2012b) illustrated how the development of science influences the diffusion process. Figure 5 shows that Kao's case starts with an academic phase, which is largely convex, that is followed by a concave technology dominated one. The inflexion points mark the turning points of different phases. Extrema correspond to breakthroughs.

We can see that how scientific investigations are reflected in the publication and citation process.

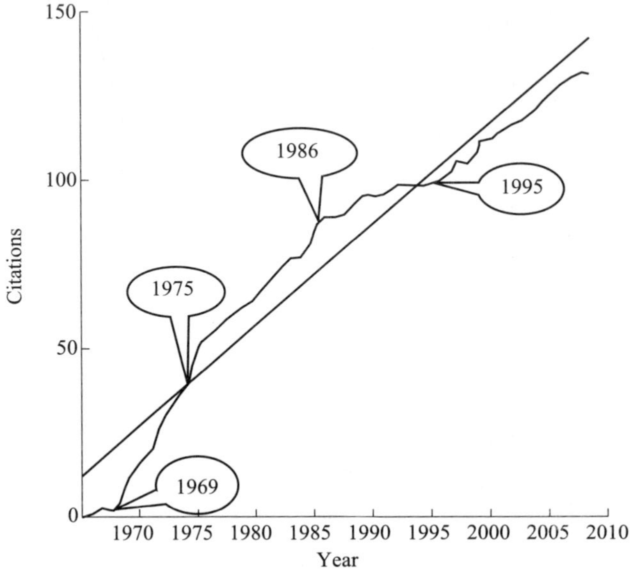


Figure 5 The citation diffusion process of Kao

## 5 How the Publication and Citation Process Can Describe the Evolution of Science

Line and Sandison (1974) regard growing or ageing by citation as more concerned with document use rather than with "information" or "knowledge" contained in the document. If we don't care how information or knowledge is used in the citation process, then the citation process is a stochastic process. But when we consider how ideas or perspectives contained in the document are used in the citation diffusion process, deterministic factors are included in the citation process.

Liu and Rousseau (2013) expounded that the essence of citation is the interestingness of the addressed phenomena; it is the result of the interaction of different ideas regarding the objects of study. In this sense the number of citations is an instrument to measure to what extent different ideas and perspectives of a phenomenon aroused arguments in their interaction. Citation analysis can bear the commitment to measure how science evolves between interactions among different ideas (Leydesdorff, 1998; Leydesdorff and Wouters, 1999)

In this framework citation analysis is not only based on accidents, but can also lead to further investigations on the evolution of science. When there are intensive arguments regarding certain phenomena, the interestingness of these phenomena will increase. The underlying reasons will give us a monitor to check how new insights and the old knowledge system interact and how science develops in the interaction among different scientific ideas. When the arguments become heated, the phenomenon becomes a hot topic, even more increasing its interestingness. Possibly a breakthrough is about to occur, promoting further investigations about the phenomenon. In this sense we may define the notion "citation" in essence as the historical remains left by the interaction of different ideas on a phenomenon, and define the number of citations as the extent to which the scientific ideas aroused arguments between the interactions of different ideas on this phenomenon.

A citation is produced when a new insight reacts on preceding knowledge. When the new insight and the preceding knowledge are coherent, the extent of the arising argument may not be very fierce. When a new reaction on the ideas and perspectives formulated on the article occurs, the number of citations will increase, and science develops according to the same paradigm (Kuhn, 1962). When the paradigm reaches maturity, the ideas or perspectives become common knowledge and are absorbed in current practice and handbooks. The number of citations will not increase anymore. When the old and the new points of view are not coherent, perhaps even contradicting each other, there will be fierce arguments and the number of citations will grow intensively. During these arguments, new ideas or perspectives proposed in the original article will be clarified. These ideas may lead to new insights, and when accepted a new paradigm will be established.

It is still possible that scientists react to and even support wrong ideas and perspectives. The reasons behind this also reflect the interestingness of the phenomenon and the ideas and perspectives deduced from it. Also in such cases we see an increase in citations.

In this sense we can describe scientific evolution via the citation diffusion process and further on to reveal more regularities in the evolution of science.

## Acknowledgement

Work of the authors is supported by NSFC with grant number 71173154.

## References


Ashkin A, Dziedzic J M, Bjorkholm J E, et al. 1986. Observation of a single-beam gradient force optical trap for dielectric particles. *Optics Letters*, 11: 288-290.

Aubert J J, Becker U, Biggs P J, et al. 1974. Experimental observation of a heavy particle. *Physical Review Letters*, 33 (23): 1404-1406.

Björneborn L, Ingwersen P. 2004. Toward a basic framework for webometrics. *Journal of the American Society for Information Science and Technology*, 55 (14): 1216-1227.

Bunge M A. 1998. *Philosophy of Science*: *From Problem to Theory*. New Jersey: Transaction Publishers.

Glänzel W, Schoepflin U. 1994. A stochastic model for the ageing of scientific literature. *Scientometrics*, 30 (1): 49-64.

Kao K C, Hockham G A. 1966. Dielectric-fibre surface waveguides for optical frequencies. *Proceedings of the Institution of Electrical Engineers*, 113: 1151-1158.

Kuhn T. 1962. *The Structure of Scientific Revolutions*. Chicago: University of Chicago Press.

Lee T D, Yang C N. 1956. Question of parity conservation in weak interactions. *Physical Review B*, 104: 254-258.

Leydesdorff L. 1998. Theories of citation. *Scientometrics*, 43: 5-25.

Leydesdorff L, Wouters P. 1999. Between texts and contexts: advances in theories of citation (a rejoinder). *Scientometrics*, 44: 169-182.

Line M B, Sandison A. 1974. Obsolescence and changes in the use of literature with time. *Journal of Documentation*, 30 (3): 283-350.

Liu Y, Rousseau R. 2011. Splines can recover dynamic information contained in discrete data. *In*: Noyons E, Ngulube P, Leta J (eds.), *Proceedings of the ISSI* 2011 *Conference* (pp. 1022-1024). Durban: University of Zululand.

Liu Y, Rousseau R. 2012a. A continuous description of discrete data points in informetrics: using spline functions. *ASLIB Proceedings*, 64 (2): 193-200.

Liu Y, Rousseau R. 2012b. Towards a representation of diffusion and interaction of scientific ideas: the case of fiber optics communication. *Information Processing & Management*, 48 (4): 791-801.

Liu Y, Rousseau R. 2013. Interestingness and the essence of citation. *Journal of Documentation*, 69 (4): 580-589.

Liu Y, Rousseau R. 2014. Citation analysis and the development of science: a case study using articles by some Nobel Prize winners. (Accepted by *Journal of the American Society for Information Science and Technology*)

Tsui D C, Stormer H I, Gossard A C. 1982. Two-dimensional magnetotransport in the extreme quantum limit. *Physical Review Letters*, 48: 1559-1562.

van Raan A F J. 2008. Scaling rules in the science system: influence of field-specific citation characteristics on the impact of research groups. *Journal of the American Society for Information Science and Technology*, 59: 565-576.

# 1-4 合作网络中的合作关系链接预测

于 琦[①]

科研合作是指由同一个国家或不同国家的科研人员组成的团队共同完成科研任务。普遍认为，科研人员之间的合作具有协同作用，团队成员相互合作而产生的科研成果远远超过由科研人员个人产生成果的总和。然而，组建并管理好一支优秀的团队并不是一件容易的工作。科研工作人员最关心的问题之一就是如何找到一个合适的合作对象。提前预测应该怎样进行合作是非常困难的，因此领域专家经常不确定应该与谁进行合作。

如果我们拥有专家的语义信息，那么事情会变得容易很多。例如，如果某一科研人员了解其他研究人员在其所属研究领域的研究状况，他就会很容易地知道自己的合作对象应该是谁。然而，专家的语义信息是很难获取的。

给定一个文献集合，我们可以很容易地建立起一个合著网络，该网络中结点表示科研人员，连线表示合作关系。合著网络中作者之间的拓扑特征可以预测未来作者之间是否会产生合作关系。换句话说，如果我们能够以一定的准确度预测两个作者之间是否存在新的连接，那么这些新的连接将为未来的科研合作提供很好的建议。

本文从过去的合著网络中提取作者之间的拓扑结构特征，采用监督式机器学习算法对这些拓扑结构特征进行建模。而后这些模型就可以用来预测未来的网络中两个作者之间是否能够产生连接关系。我们利用冠心病研究领域的合作网络对上述模型进行了测试，结果表明，未来合著关系的存在与过去合著网络的拓扑结构是密切相关的，监督式学习方法能够帮助我们找出这种相关性，从而进行合作关系链接预测。

## 1 方法和数据

### 1.1 拓扑结构特征

定义 $G=\langle V,E\rangle$ 为图，结点 $v_i\in V$，连线（$v_i$，$v_j$）$\in E$，$1\leqslant i$，$j\leqslant|V|$。可以据此计算出许多不同的网络拓扑结构特征，这些拓扑结构特征与未来结点之间的

① 于琦，山西医科大学，yuqi351@gmail. com。

连接出现的概率相关。某一对结点的所有拓扑结构特征构成该结点对的特征向量。

Borner 等（2007），以及 Liben-Nowell 和 Kleinberg（2007）在他们的文章中提出了多种网络拓扑结构。本文选取了可广泛应用于各种网络的 8 种拓扑结构特征进行链接预测。

1）基于邻居的拓扑特征

Common neighbors. Common neighbors 是指 $v_i$ 和 $v_j$ 的共同 neighbor 数量，即 $|\Gamma(v_i)\cap\Gamma(v_j)|$。$\Gamma(v_i)$ 表示 $v_i$ 的所有 neighbor。

Jaccard's coefficient. Jaccard's coefficient 是 common neighbor 的标准化，即 $|\Gamma(v_i)\cap\Gamma(v_j)|/|\Gamma(v_i)\cup\Gamma(v_j)|$。

Preferential attachment. Preferential attachment 指的是 $v_i$ 与 $v_j$ neighbor 数量的乘积，即 $|\Gamma(v_i)|g|\Gamma(v_j)|$。

2）基于路径的拓扑特征

$\text{Katz}^\beta$。$\text{Katz}^\beta$ 是一种优化后的 shortest path，即 $\sum_{s=1}^{\infty}\beta^s\cdot \text{paths}_{ij}^s$. $\text{paths}_{ij}^s$ 表示连接 $v_i$ 和 $v_j$ 的长为 $s$ 的路径数量。它将两结点之间的所有路径都计算在内，并且赋予长度较短的路径更高的权重。参数 $\beta\in[0,1]$。

PropFlow。PropFlow 将每条边 flow 比例的乘积作为权重赋予每条路径（Lichtenwalter et al.，2010）。

Shortest path count。shortest path count 是指 $v_i$ 和 $v_j$ 之间最短路径的数量。

3）基于中心度的拓扑特征

Rooted PageRank。Rooted PageRank 是 $y$ 在随机 walk 中的固定分布：以 $\alpha$ 的概率跳至 $x$，以 $1-\alpha$ 的概率跳向该结点的其他 neighbor。

我们使用跨平台软件 Lpmade（Lichtenwalter and Chawla，2011）计算合著网络拓扑结构特征。

本文中的参数设置为 $\beta=0.05$，$\alpha=0.15$。这些都是目前使用比较广泛的参数值（Katz，1953；Leroy et al.，2010）。

## 1.2 预测模型

我们以拓扑结构特征建立方程对两作者之间产生合著关系的概率进行建模。本文中，我们选取了 logistic 回归（LR）和支持向量机（SVM）作为预测模型。LR 是应用最为广泛的一种分类模型，而 SVM 是最近兴起的一种分类模型算法。

1）LR

对于每个训练作者对（$v_k1, v_k2$），定义 $x_k$ 为（$d+1$）维向量，该向量包括一个常量以及作者对之间的 $d$ 个拓扑结构特征。$y_k$ 用来标示上述作者对之间在将来是否会出现合著关系（$y_k=1$ 表示出现，$y_k=0$ 表示没有出现），它服从概率为 $p_k$ 的二项分布，概率 $p_k$ 定义如下：

$$p_k=\frac{e^{x_k\beta}}{e^{x_k\beta}+1}$$

$\beta$ 是一个与常量和拓扑结构特征对应的权重矢量，我们利用标准 MLE（极大似然估计）计算 $\beta$，使如下公式达到最大值：

$$L = \prod_k P_k^{y_k}(1 - p_k)$$

2）SVM

SVM 的基本思想是：包含 $n$ 个特征的向量可以映射到 $n$ 维空间的一个点（每个维度对应一个特征）。因此，我们的作者对可以对应于这个空间的点集合。每个点拥有自己的 0/1 标识。该方法的目标是将这些点分成两组，使得拥有同样标识的点分在同一个组。这可以由线性分割器来实现，本文采用的即是这种。为了减小误差，线性分割器的选择要使得其两侧的空白达到最大值。我们采用序列最小优化算法，普遍认为这一算法与线性 SVM 结合很好。

我们使用 Weka（Waikato Environment for Knowledge Analysis）实现 LR 和 SVM（Hall et al.，2009）。

我们使用分层十倍交叉检验来预测上述模型的准确度。

## 1.3 模型评估

我们采用三种指标评估模型的准确性：准确（precision）率、召回（recall）率，以及 AUC。

准确率是指真阳性预测在所有的阳性预测里所占的比例。它用来评估模型拟合数据的程度。

召回率是指真阳性预测在所有阳性标识中所占的比例。它意味着该模型预测未来合作关系的能力如何。

AUC，即 ROC 曲线下的面积，是指该模型辨别产生新的合作关系的作者对与没有产生新的关系的作者对的能力。

## 1.4 特征选取

我们使用 Wrapper 方法从特征向量中选取最有效的特征。Wrapper 方法使用 subset 评估器生成特征向量的所有子集，然后它会使用一种分类算法（如本文中的 LR 和 SVM）生成分类器，从而确定哪个特征子集采用这个分类算法最为有效。为了找到特征子集，评估器将使用诸如 random search、breadth first search、depth first search、hybrid search 的搜索技术。本文选取 breadth first 搜索技术。

## 1.5 数据来源

我们选取 Web of Science（WOS）作为数据来源构建合著网络。WOS 拥有出版物的大量信息，如作者、出版物、题名、参考文献等。我们选取冠心病为研究领域。数据包括 2007～2012 年的 118 850 篇文献、224 663 个作者。通过在该网络中测试我们的方法，我们希望找到哪些拓扑结构特征最为有效，能够帮助我们辨认出那些尚未发生的合作关系。

# 2 结　果

我们考虑两个时间段的网络：$T_1$ = ［2007－2009］及 $T_2$ = ［2010－2012］。$T_1$ 阶段提取拓扑结构特征，$T_2$ 阶段标识两作者在 $T_2$ 阶段是否出现新的连接。下面的情况很可能发生：某些作者只在 $T_1$ 阶段出现，在 $T_2$ 阶段没有产出；某些作者从 $T_2$ 阶段才开始有产出。所以我们将作者限制为在两个阶段都有产出的作者。最终得到 55 813 个作者。

然后，我们又将作者集限制为文章产量为 5 篇以上，并且我们只考虑 2-hop 的作者对，即两作者之间有不少于 1 个连接的合著者。根据上述限制，我们找到所有在第二阶段产生新连接的作者对，用以作为阳性作者对。第二阶段中共发现 166 538 个新连接，占所有可能连接的 1.6%。然后，我们随机选取了与阳性作者对相同数量的阴性作者对，使得阳性作者对与阴性作者对数量平衡。

为了查明我们的模型在预测高产作者与低产作者时的区别，我们设置了阈值将作者分为高产作者组（发表 10～20 篇及以上文章的作者）和低产作者组（发表 5 篇以上 10 篇以下文章的作者），见表 1。

**表 1　作者集合**

| 作者类型 | 作者数 | 新的合著关系数 | 所有可能合著关系数 |
|---|---|---|---|
| 论文数≥ 5 | 13 927 | 166 538 | 9 933 944 |
| 论文数≥ 10 | 5 433 | 132 546 | 7 082 693 |
| 10>论文数≥5 | 8 494 | 70 736 | 4 845 167 |
| 论文数≥20 | 1 670 | 85 417 | 3 796 520 |
| 20>论文数≥5 | 12 257 | 123 326 | 7 847 478 |

## 2.1　模型的总体准确性

我们使用 10 倍交叉检验评估了我们的方法在上述数据集上的准确性。

我们首先比较了 LR 模型与 SVM 模型对于这个数据集（发表 5 篇以上论文的作者）的测试结果。如表 2、图 1 和图 2 所示，两模型在三个准确率指标上的得分都很高。LR 模型在召回率与 AUC 两个指标上略强于 SVM 模型（0.708 vs. 0.707，0.781 vs. 0.779），而 SVM 模型在准确率上高于 LR 模型（0.730 vs. 0.729）。因此，我们可以说两个模型对于我们的数据集拟合程度都很好，它们能够预测至少 70.8% 的在未来将出现的连接，而且在辨别那些产生新连接的作者对与那些没产生新连接的作者对时的能力都很强。

**表 2　LR 与 SV 的测试结果**（论文 5 篇以上作者）

| 学习模型 | 准确率 | 召回率 |
|---|---|---|
| LR | 0.729 | 0.708 |
| SVM | 0.730 | 0.707 |

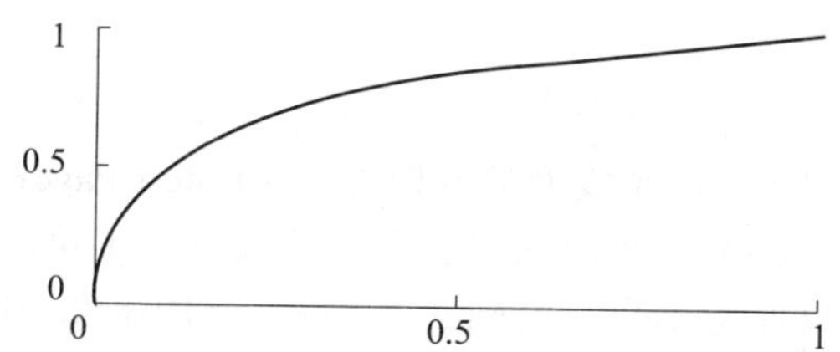


图 1 LR 模型的 ROC 曲线，AUC = 0.781（论文 5 篇以上作者）

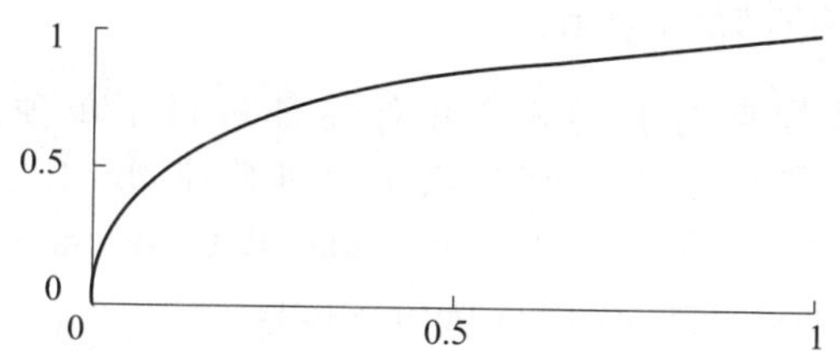


图 2 SVM 模型的 ROC 曲线，AUC = 0.779（论文 5 篇以上作者）

然后我们比较了两个模型预测高产作者集与低产作者集的测试结果。如图 3 和图 4 所示，对于两个模型而言，高产作者的链接预测并没有比低产作者的链接预测显得容易（甚至有点难），这一结果与 Sun 等（2011）的研究是不一致的。

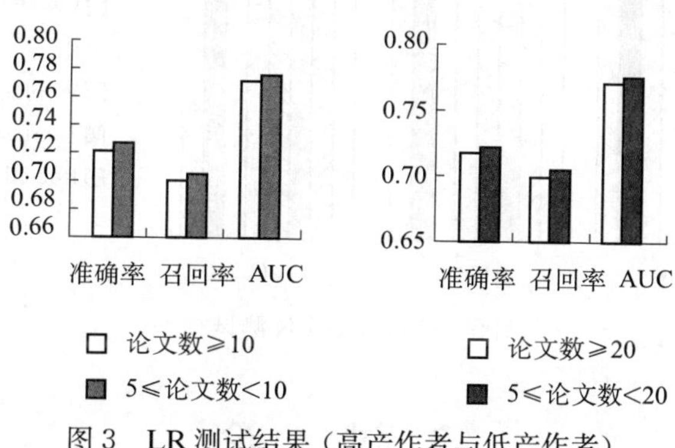


图 3 LR 测试结果（高产作者与低产作者）

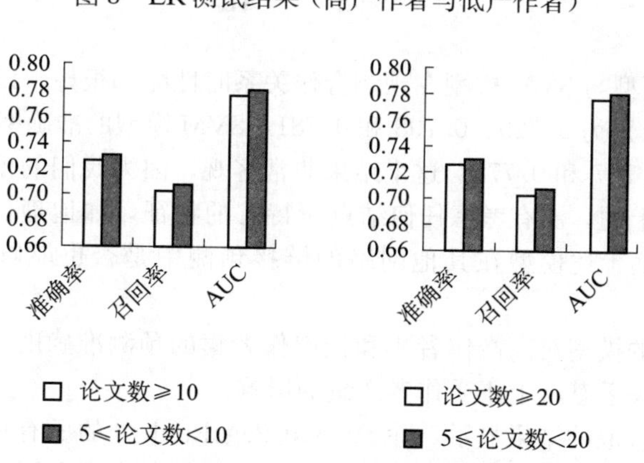


图 4 SVM 测试结果（高产作者与低产作者）

## 2.2 特征选取

特征提取的结果显示，两个模型都选取除了 Rooted PageRank 以外的 7 个拓扑特征作为最有效的拓扑特征。我们利用这 7 个特征进行了重新测试，发现对于 LR 模型而言，其测试结果保持不变（准确率 0.729，召回率 0.707，AUC 0.781），而对于 SVM 模型而言，准确率和召回率保持不变，AUC 由 0.779 降至 0.777。

## 2.3 单独的拓扑特征测试

我们还利用 LR 模型对上述 8 个拓扑特征进行了单独测试。如图 5 所示，Adamic/Adar 在三个指标上均达到最大值（准确率 0.73，召回率 0.702，AUC 0.765）。基于邻居的拓扑特征（如 Adamic/Adar 和 Common Neighbor）的准确率要高于基于路径的拓扑特征（如 shortest path count）。

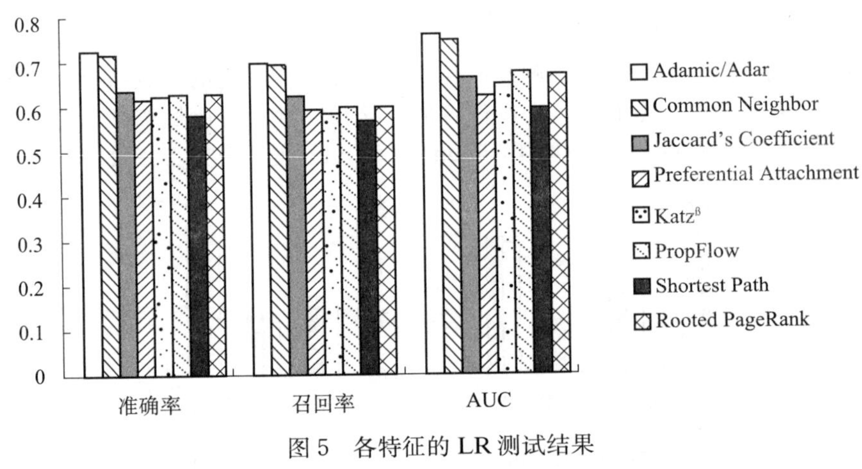


图 5 各特征的 LR 测试结果

# 3 讨 论

首先，LR 模型与 SVM 模型在预测合作关系时性能都很好。LR 模型的准确率、召回率和 AUC 分别为 0.729、0.708 和 0.781，SVM 模型的准确率、召回率和 AUC 分别为 0.730，0.707 和 0.779。这个结果非常客观，因为我们的模型仅仅是建立在网络的拓扑结构上的，没有考虑任何结点所特有的特征，如地理位置、单位隶属及研究主题等。然而上述模型在其他网络的链接预测中是否也能够成功需要进一步验证。

其次，我们的模型对高产作者集和低产作者集的预测准确度一样。这意味着两作者之间的合作关系基本上不受作者产量的影响。

再次，特征选取的结果显示，除 Rooted PageRank 外的所有拓扑结构特征在合作关系预测中都是有效的。原因可能因为合著网络是一个无向网络（尽管可以看做是双向网络），而 Rooted PageRank 算法是建立在有向网络的基础上的。去除 Rooted PageRank 后，两个预测模型的测试结果基本上没有变化。

最后，单独的拓扑特征测试结果显示，基于邻居的拓扑特征在合作关系预测中的表现要强于基于路径的拓扑特征。这表明，作者主要是依靠自己的邻居（合著者）来建立合作关系。

总体而言，我们的模型能够准确有效地在合著网络中进行链接预测。由于该方法仅仅是依靠网络中的拓扑特征以及监督式学习方法，所以很容易应用到其他的网络链接预测中。

## 4 结　论

本文使用监督式机器学习方法，利用合著网络中的拓扑结构特征建立链接预测模型。这些模型可以用于建立尚未实现的合作关系，从而帮助我们组建和管理强大的科研团队。此外，我们得到了大量有关哪些拓扑结构特征最为有效的信息，这些信息可以用来建立专家档案。

本研究有许多研究方向。其中的一个重要方向就是在异构网络中测试链接预测问题，异构网络中的结点可以为多种（如作者和论文），链接也为多种（撰写/被撰写、引用/被引用）。由于异构网络能够为作者对提供更多的拓扑结构特征，所以有可能会产生更好的链接预测效果。

## 致谢

本研究为国家自然科学基金资助项目“基于知识网络的科技创新团队成员合作研究——以生物医学为例（71103114）”，以及国家自然科学基金资助项目“论文合著和专利合作视角的生物医学领域科技合作研究（71240006）”的研究成果之一。

## 参考文献

Bilgic M，Namata G M，Getoor L. 2007. Combining collective classification and link prediction. *In*：*Seventh IEEE International Conference on Data Mining Workshops*（ICDMW 2007），Omaha，Nebraska：381-386.

Borner K，Sanyal S，Vespignani A. 2007. Network science. *Annual Review of Information Science and Technology*，41：537-607.

Chao W，Satuluri V，Parthasarathy S. 2007. Local probabilistic models for link prediction. *In*：*Seventh IEEE International Conference on Data Mining*. Omaha NE，USA：322-331.

Clauset A，Moore C，Newman M E J. 2008. Hierarchical structure and the prediction of missing links in networks. *Nature*，453（7191）：98-101.

Hall M, Frank E, Holmes G, et al. 2009. The WEKA data mining software: an update. *SIGKDD Explorations*, 11 (1): 10-18.

Huang Z, Li X, Chen H. 2005. Link prediction approach to collaborative filtering. *In*: *Proceedings of the 5th ACM/IEEE-CS Joint Conference on Digital Libraries*. Denver, CO, USA: 141-142.

Kashima H, Abe N. 2006. A parameterized probabilistic model of network evolution for supervised link prediction. *In*: *Proceedings of the Sixth International Conference on Data Mining*, Hong Kong, China: 340-349.

Katz L. 1953. A new status index derived fromsociometric analysis. *Psychometrika*, 18 (1): 39-43.

Leroy V, Cambazoglu B B, Bonchi F. 2010. Cold start link prediction. *In*: *Proceedings of the 16th ACM SIGKDD International Conference on Knowledge Discovery and Data Mining*. Washington, DC, USA: 393-402.

Liben-Nowell D, Kleinberg J. 2007. The link-prediction problem for social networks. *Journal of the American Society for Information Science and Technology*, 58 (7): 1019-1031.

Lichtenwalter R N, Chawla N V. 2011. LPmade: link prediction made easy. *Journal of Machine Learning Research*, 12: 2489-2492.

Lichtenwalter R N, Lussier J T, Chawla N V. 2010. New perspectives and methods in link prediction. *In*: *Proceedings of the 16th ACM SIGKDD International Conference on Knowledge Discovery and Data Mining*. Washington, DC, USA: 243-252.

Lü L, Jin C-H, Zhou T. 2009. Similarity index based on local paths for link prediction of complex networks. *Physical Review E*, 80 (4): 046122.

Lü L, Zhou T. 2010. Link prediction in weighted networks: the role of weak ties. *EPL*, 89: 18001.

Newman M E J. 2003. The structure and function of complex networks. *Siam Review*, 45 (2): 167-256.

Pavlov M, Ichise R. 2007. Finding experts by link prediction in co-authorship networks. *In*: *Proceedings of the Workshop on Finding Experts on the Web with Semantics*. Busan, South Korea: 42-55.

Popescul A, Popescul R, Ungar L. 2003. Statistical relational learning for link prediction. *In*: *Proceedings of the Workshop on Learning Statistical Models from Relational Data at the International Joint Conference on Artificial Intelligence*. Acapulco, Mexico: 143-144.

Redner S. 2008. Networks—teasing out the missing links. *Nature*, 453 (7191): 47-48.

Sun Y Z, Barber R, Gupta M, et al. 2011. Co-author relationship prediction in heterogeneous bibliographic networks. The advances in social networks analysis and mining (ASONAM). *In*: *2011 International Conference on Advances in Social Network Analysis and Mining*. Taiwan, China: 121-128.

# 二、科学计量评价实践研究

# 2-1 Scientific Papers on Renewable Energy Published by the People's Republic of China and Spain, 2003-2012: a Bibliometric Analysis

Elías Sanz-Casado, Antonio Serrano-López, Daniela de Filippo, María Luisa Lascurain-Sánchez①

## Abstract

In light of the growing environmental concern that has driven research on the use of renewable energy in many countries, the present paper studies Spanish and Chinese scientific papers published on the subject. Bibliometric techniques were used to analyse the papers listed in the Web of Science in 2003-2012, focusing on factors such as output, collaboration, impact and visibility. The findings show a substantial rise in both countries' output: activity has increased steeply especially in China, which has climbed to second place in the country ranking after the United States. While international collaboration has intensified in both countries, partnership tends to be regional in scale. The impact findings show that Chinese papers are cited more intensely by US institutions, while Spanish papers are more frequently cited by Chinese researchers. No differences are observed in the profiles of the citing publications.

**Keywords**: renewable energy; sustainability; bibliometric analysis; China; Spain

## 1 Introduction

Humanity may have never had to rise to a challenge as daunting and dramatic as the one it is facing in the early twenty-first century. What distinguishes this challenge

① Laboratory of Metric Studies on Information (LEMI). Department of Library Science and Documentation. University Carlos Ⅲ of Madrid. C/Madrid 126, Getafe 28903, Madrid, Spain; Research Institute for Higher Education and Science; Associated Unit IEDCYT-LEMI.

from past dilemmas is that it is directly related to the ongoing depletion of the planet's resources and the change in the environmental conditions on which the species depends for its survival.

The analyses conducted of various future scenarios call for decisions and measures that will entail substantial changes in human behaviour. In its latest report, the International Energy Agency (IEA, 2013), for instance, warns that "Global greenhouse gas emissions are increasing rapidly and, in May 2013, carbon dioxide ($CO_2$) levels in the atmosphere exceeded 400 parts per million for the first time in several millennia". These developments will accelerate climate change, and therefore "the long-term average temperature increase is more likely to be between 3.6℃ and 5.3℃ (compared with pre-industrial levels), with most of the increase occurring this century" (IEA, 2013).

The energy industry is the major source of the greenhouse gas emissions inducing climate change, specifically as a result of its use of fossil fuels (coal, oil, gas), the raw materials for over 80% of the energy produced. The present economic crisis and its effects on a sizeable number of countries, along with the Fukushima nuclear plant accident, have altered many nations' energy agenda in terms of the *Kyoto Protocol* targets. Those targets included bringing greenhouse gas emissions in 2008-2012 down by 1.8% from the 1990 levels with a view to preventing the Earth's temperature from rising by more than 2℃. That target has not been reached. Rather, emissions are rising in most countries for a number of reasons. One is the growing use of coal as a source of energy production due to its lower price, and another the revocation or reduction of many countries' nuclear programmes in the wake of the Japanese accident and the concomitant rise in the use of fossil fuels (World Energy Council, 2013).

Energy is an obvious imperative for comprehensive development and as such its consumption will continue to rise. According to the most conservative forecasts, world energy demand is expected to grow by over one third by 2035, with China, India and the Middle East accounting for 60% of that increase (IEA, 2012).

A pressing need has consequently arisen to seek other sources of energy to sustainably meet countries' development demands while ensuring protection of the environment and the conservation of resources for future generations. Science and technology are essential to the development of renewable and sustainable energy (Kajikawa et al., 2008). The new economical patterns in most of the developed countries are closely linked to the incorporation of sustainability criteria to the production sectors based on scientific and technological knowledge (Sanz-Casado et al., 2013).

For several years, many countries have been investing a good deal of effort in researching and developing renewable energy to replace fossil fuel-derived sources. One of the European Union's (EU) targets for 2020 calls for meeting 20% of the energy demand with renewable sources (Eurostat, 2012). Spain overshot that target as early as 2010, when nearly 30% of the energy consumed was sourced from renewables (Eurostat, 2012).

The Chinese Government has also invested heavily in research on renewable energy, and in the middle of the first decade of this century launched a series of policies and strategies for this type of energy. Since 2009, energy efficiency, $CO_2$ emissions reduction and renewable energy have constituted a separate item on the Chinese national budget (Duan, 2011).

Drawing from bibliometric techniques, the present study aims to analyse the research on renewable energy published by Chinese and Spanish institutions, focusing on output and impact trends in the last ten years (2003-2012).

## 2 Materials and Methods

The data for this study were taken from the records in the Web of Science databases (SCI, SSCI and the proceedings databases) accessible on the Web of Knowledge platform. The entries were filtered both chronologically (for the period 2003-2012) and geographically (at least one Spanish or Chinese institution in the author affiliations).

The subject area was delimited in the manner designed by the authors for earlier studies on wind and solar energy, which is based on the selection of specific terms for identifying output in these fields (Sanz-Casado et al., 2013; Ingwersen et al., 2013). The main field chosen, "renewable energy generation", covers three sub-fields, "alternative energy", "green energy" and "energy policy".

The records retrieved were downloaded to a labelled text file. A series of scripts were designed in Perl language to process the texts of the records and include them in an MySQL relational database. The one-dimensional indicators obtained in this study are listed below.

(1) Worldwide scientific output on renewable energy contained in the Web of Science;

(2) Chinese and Spanish output in renewable energy compared to worldwide trends;

(3) Subject areas covered by Chinese and Spanish institutions researching renewable energy;

(4) Most productive Chinese and Spanish institutions in the field of renewable energy;

(5) International collaboration in Chinese and Spanish research;

(6) Citing countries for Chinese and Spanish papers on renewable energy;

(7) Citing institutions for Chinese and Spanish papers on renewable energy.

One multi-dimensional indicator was studied: characteristics of citations of Spanish and Chinese papers on renewable energy (principal component analysis).

# 3 Results and Discussion

## 3.1 Renewable Energy Research: Output

World scientific output on renewable energy is broken down in Table 1. The period studied was divided into two 5-year sub-periods to better illustrate country trends. While the United States was world leader in both sub-periods, its relative weight actually declined slightly over the decade studied, while China leapt from the fourth place in the first 5 years to second in the latter sub-period. More striking than that, however, was the increase in its output, which grew fivefold in absolute terms and climbed to 17.1% of world production. Prior studies have reported similarly steep upturns in the number of papers published on renewable energy by Chinese institutions (Sanz et al., 2013).

Rather a different pattern was observed in Spanish from seventh to ninth place over the period, even though its output grew 2.7-fold and it maintained its 3.8 per cent share of worldwide production.

**Table 1 World output on renewable energy**

| 2003-2007 | | | 2008-2012 | | |
|---|---|---|---|---|---|
| Country | Number of papers | Percentage | Country | Number of papers | Percentage |
| USA | 6 102 | 21.4 | USA | 15 782 | 20.6 |
| Japan | 2 942 | 10.3 | **China** | **13 093** | **17.1** |
| Germany | 2 700 | 9.5 | Germany | 5 616 | 7.3 |
| **China** | **2 437** | **8.5** | Japan | 4 782 | 6.2 |
| United Kingdom | 1 758 | 6.2 | South Korea | 4 354 | 5.7 |
| France | 1 112 | 3.9 | United Kingdom | 3 876 | 5.1 |
| **Spain** | **1 073** | **3.8** | India | 2 995 | 3.9 |
| India | 1 020 | 3.6 | **Spain** | **2 911** | **3.8** |
| Canada | 929 | 3.2 | France | 2 540 | 3.3 |
| Italy | 928 | 3.2 | Canada | 2 471 | 3.2 |

Continued

| 2003-2007 | | | 2008-2012 | | |
|---|---|---|---|---|---|
| Country | Number of papers | Percentage | Country | Number of papers | Percentage |
| Australia | 855 | 3.0 | Italy | 2 383 | 3.1 |
| Netherlands | 838 | 2.9 | Australia | 2 008 | 2.6 |
| South Korea | 766 | 2.7 | Turkey | 1 525 | 2.0 |
| Turkey | 633 | 2.2 | Netherlands | 1 414 | 1.8 |
| Switzerland | 632 | 2.2 | Switzerland | 1 333 | 1.7 |
| Sweden | 542 | 1.9 | Sweden | 1 072 | 1.4 |
| Greece | 496 | 1.7 | Singapore | 1 058 | 1.4 |
| Denmark | 467 | 1.6 | Denmark | 1 001 | 1.3 |
| Russia | 436 | 1.5 | Brazil | 899 | 1.2 |
| Brazil | 383 | 1.3 | | | |

Chinese Scientific output and Spanish scientific output are compared to worldwide output in Figure 1. The curves plot the index numbers that reflect the variation in scientific production with respect to the base year (2003) in the two countries and the world overall. As the figure shows, Spanish output varied very much in line with the world as a whole in the 10 years analysed, while Chinese production diverged significantly from the general pattern, with a several-fold increase over the base value.

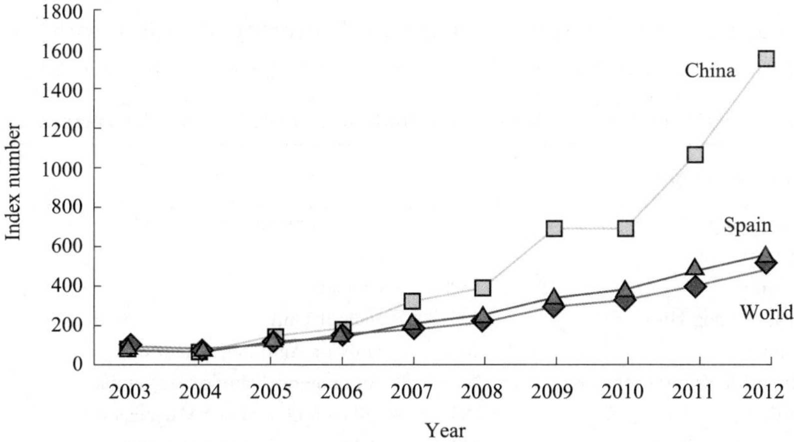


Figure 1 Chinese, Spanish and world output, 2003-2012

The Chinese and Spanish research institutions studying renewable energy addressed similar subject areas. Eight of the ten categories with the highest output in each country concurred and many had also been observed in prior studies on renewable energy research (Romo-Fernández, et al., 2011). That notwithstanding, the subjects at the top of the two lists differed. "Materials science", the area of greatest interest for Chinese institutions, ranked fourth among Spanish researchers. The category that headed the Spanish list, in turn, "energy and fuels", was fifth in importance for Chinese institutions (Table 2).

**Table 2 Chinese and Spanish publications on renewable energy: major areas of study**

| China | | Spain | |
|---|---|---|---|
| WOS category | Total papers | WOS category | Total papers |
| Materials science | 4990 | Energy and Fuels | 1329 |
| Engineering | 4227 | Engineering | 1151 |
| Chemistry | 3780 | Physics | 881 |
| Physics | 3635 | Materials Science | 756 |
| Energy and Fuels | 3274 | Chemistry | 749 |
| Science and technology—other topics | 1472 | Environmental sciences and ecology | 364 |
| Environmental sciences and ecology | 720 | Science and technology—other topics | 237 |
| Computer Science | 647 | Thermodynamics | 144 |
| Construction and building technology | 544 | Computer science | 138 |
| Polymer science | 505 | Geology | 130 |

Most of the most productive Chinese and Spanish institutions working in renewable energy are universities, although China's Academy of Sciences (CAS) indisputably prevailed in that country, putting out five times more papers than the second ranking institution, Tsinghua University. In Spain, the opposite is true: the most productive institution is the Politechnical University of Madrid with 330 papers, followed by the Spanish National Research Council (CSIC) with 256 (Table 3).

**Table 3 Most productive Chinese and Spanish institutions in renewable energy research**

| Chinese Institutions | Number of papers | Spanish Institutions | Number of papers |
|---|---|---|---|
| Chinese Academy of Sciences | 2741 | Polytechnic univ of Madrid | 330 |
| Tsinghua Univ | 533 | CSIC | 256 |
| Zhejiang Univ | 462 | Ciemat | 206 |
| Shanghai Jiao Tong Univ | 455 | JaumeI Univ | 161 |
| Peking Univ | 405 | Univ of Autonoma Madrid | 145 |
| North China Electric Power Univ | 361 | Autonomous Univ of Barcelona | 133 |
| Nankal Univ | 329 | Politechnic Univ of Valencia | 125 |
| Dalian Univ of Technology | 314 | Politechnic Univ of Catalonia | 116 |
| South China Univ of Technology | 314 | Complutense Univ of Madrid | 110 |

## 3.2 Collaboration in Chinese and Spanish Renewable Energy Research

Chinese and Spanish research partnering with centres in other countries in the time period analysed is depicted in Figure 2. As the figure shows, the number of papers that Chinese institutions co-authored with institutions in other countries rose exponentially beginning in 2007. That growth concurred with the Chinese government's launch of a series of energy policies and strategies designed to encourage Chinese energy stakeholders to develop renewable energy and new energy technologies through scientific and technological innovation and active international cooperation (Duan, 2011).

In Spain, collaboration grew much more slowly. Percentage-wise, however (Figure 3), Spanish institutions exhibited greater capacity to partner with foreign institutions than their Chinese counterparts, with 34% of the papers involving international co-authorship, compared to 17% for Chinese research.

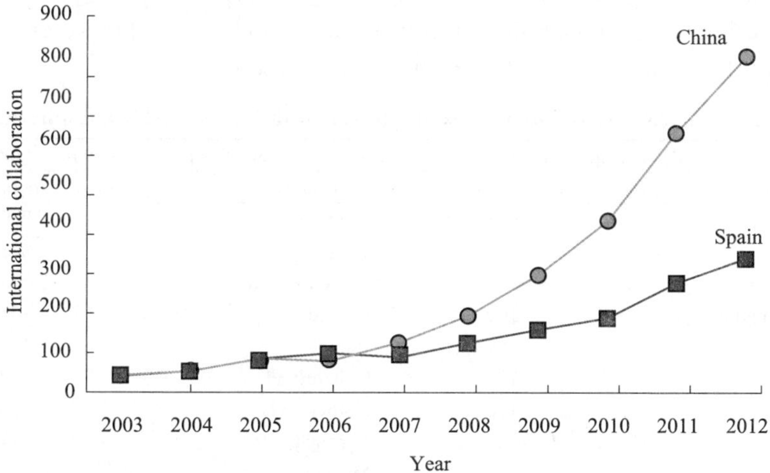


Figure 2 Number of Chinese and Spanish papers co-authored with institutions in other countries

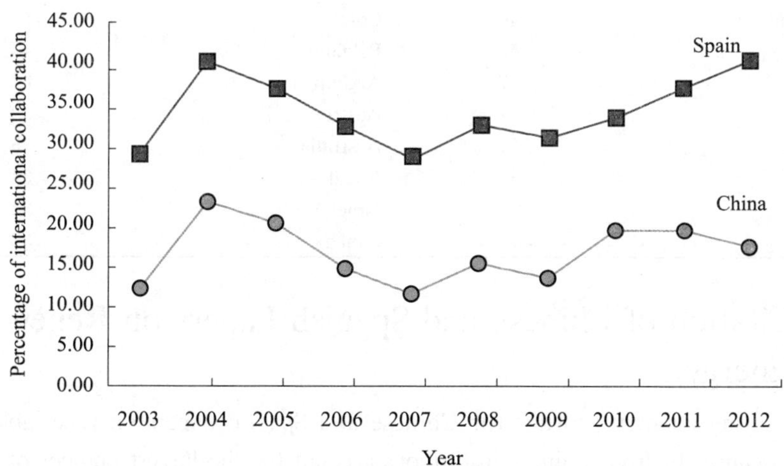


Figure 3 International partnering by Chinese and Spanish institutions (in per cent)

The countries with which Chinese and Spanish researchers collaborate and the intensity of their partnering are listed in Table 4. United States is the country of choice for Chinese institutions, followed by Japan, Australia, United Kingdom and Singapore. Spain, in turn, partners most intensely with Germany, followed by the United States, United Kingdom, France and Italy. A certain regional and cultural component can be detected in these preferences. China, for instance, in addition to collaborating with certain European countries (United Kingdom, Germany and Sweden), routinely partners with Japan, Australia, Singapore, and South Korea, whereas Spain tends to favour cooperation with the European Union and Latin America (Mexico, Argentina, Brazil) (Table 4, Figure 4).

Partnering between Spanish and Chinese institutions lies at the lower end of the scale: 15th for Spanish and 13th for Chinese institutions.

**Table 4 Nationality of research centres partnering with Chinese and Spanish institutions**

| Countries partnering with China | | Countries partnering with Spain | |
|---|---|---|---|
| Country | Number of papers w/collab. | Country | Number of papers w/collab. |
| USA | 948 | Germany | 286 |
| Japan | 345 | USA | 203 |
| Australia | 203 | United Kingdom | 189 |
| United Kingdom | 190 | France | 164 |
| Singapore | 184 | Italy | 108 |
| Germany | 180 | Switzerland | 82 |
| Canada | 170 | Netherlands | 73 |
| Sweden | 135 | Portugal | 71 |
| South Korea | 109 | Mexico | 59 |
| Switzerland | 75 | Denmark | 56 |
| France | 68 | Sweden | 48 |
| Denmark | 49 | Japan | 48 |
| Netherlands | 46 | China | 45 |
| Spain | 45 | Belgium | 43 |
| Italy | 27 | Argentina | 42 |
| Portugal | 22 | Austria | 38 |
| Belgium | 21 | Australia | 35 |
| Finland | 18 | Brazil | 34 |
| India | 17 | Israel | 33 |
| | | Canada | 33 |

## 3.3 Citation of Chinese and Spanish Papers on Renewable Energy

The citing country profiles for Chinese and Spanish papers on renewable energy vary significantly. Firstly, Chinese institutions account for the largest number of citations received by Spanish papers, whereas Chinese papers are most frequently cited by US

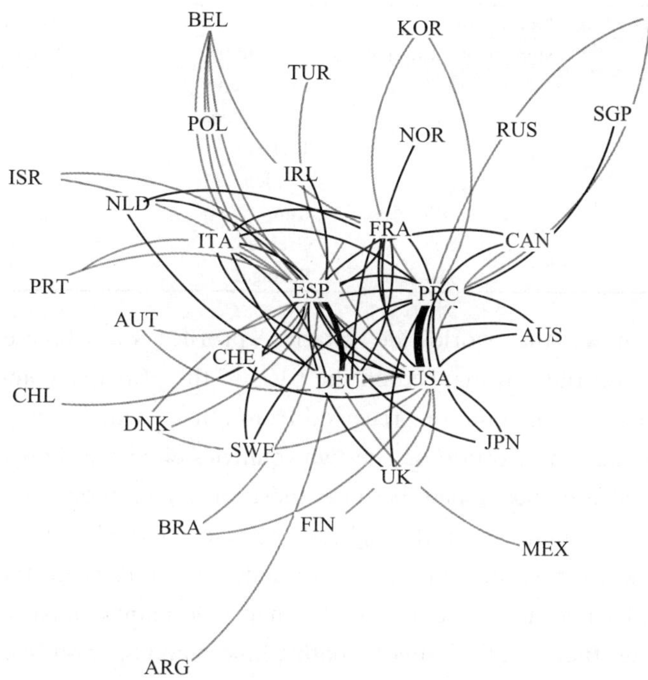


Figure 4 Renewable energy research co-authorship network for Spain and China (>44 joint papers)

researchers. Spain ranks thirteenth on that list. The regional component is likewise visible in this respect: Chinese papers receive many citations from the Asian region (South Korea, Japan, India), whereas Spanish output is cited more intensely in the European Union (Germany, United Kingdom, France, Italy) (Table 5).

**Table 5 Countries citing Chinese and Spanish publications**

| Countries citing Chinese papers | | Countries citing Spanish papers | |
|---|---|---|---|
| Country | Number of citing papers | Country | Number of citing papers |
| China | 34 017 | Spain | 5 934 |
| USA | 11 139 | China | 5 557 |
| South Korea | 4 804 | USA | 5 421 |
| Japan | 4 141 | Germany | 2 401 |
| Germany | 3 323 | United Kingdom | 2 056 |
| India | 3 263 | France | 1 759 |
| United Kingdom | 2 596 | Japan | 1 734 |
| France | 2 021 | Italy | 1 627 |
| Austria | 1 908 | South Korea | 1 470 |
| Spain | 1 818 | India | 1 228 |
| Canada | 1 786 | Canada | 1 123 |
| Singapore | 1 750 | Austria | 977 |

Continued

| Countries citing Chinese papers | | Countries citing Spanish papers | |
|---|---|---|---|
| Country | Number of citing papers | Country | Number of citing papers |
| Italy | 1716 | Switzerland | 880 |
| Iran | 984 | Netherlands | 649 |
| Turkey | 878 | Turkey | 618 |
| Switzerland | 826 | Sweden | 581 |
| Netherlands | 806 | Iran | 570 |
| Sweden | 805 | Greece | 559 |
| Malaysia | 780 | Portugal | 529 |

Figure 5 shows the worldwide citation profile for Chinese and Spanish institutions. This profile was mapped on the basis of the principal component analysis dipicted in Figure 4 of the citations received from other countries by papers authored by Chinese and Spanish institutions. The two countries cited are shown in bold.

Spain is centrally positioned because most of its citations come from other countries, whereas China lies on the right of the map, very close to China as a citing country, because Chinese institutions are responsible for most of the citations. The citing countries located at the centre of the map (Germany, Japan, South Korea, France) distribute their citations among both Chinese and Spanish institutions.

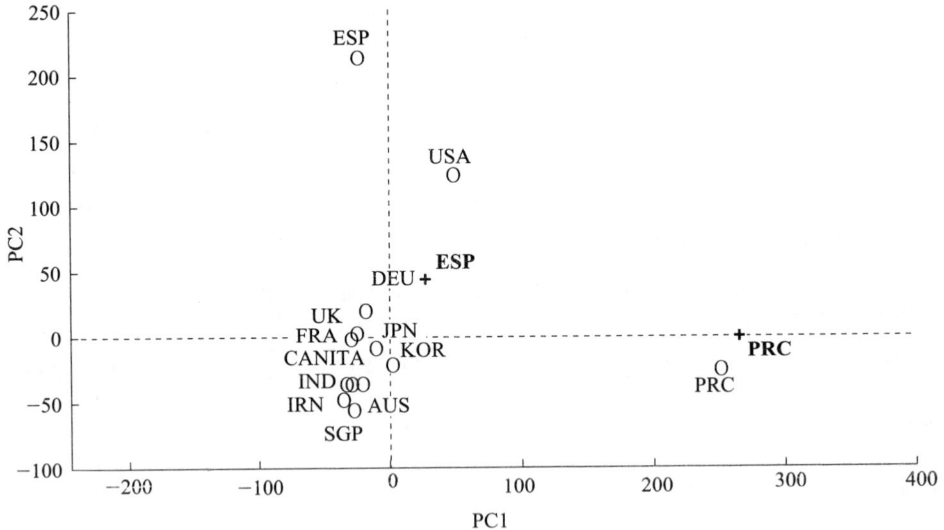


Figure 5 Principal component analysis of countries citing Chinese and Spanish papers on renewable energy

Another factor that merits analysis is the identity of the institutions that cite Chinese and Spanish papers on renewable energy. As a review of the names listed in Table 6 shows, the Chinese Academy of Sciences headed both countries' list of citing institutions.

Chinese papers are cited essentially by institutions in China and other countries in the area. The Spanish papers are cited by a wider variety of countries, and although many of the citing institutions are Spanish, a sizeable proportion is located in other EU nations and a significant number are in Asian ones.

**Table 6 Citing institutions for Chinese and Spanish papers on renewable energy**

| Institutions citing Chinese papers | | Institutions citing Spanish papers | |
|---|---|---|---|
| Institution | Number of citing papers | Institution | Number of citing papers |
| Chinese Academy of Sciences | 7603 | Chinese Academy of Sciences | 1042 |
| Peking University | 1158 | Spanish National Research Council | 635 |
| Tsinghua University | 1112 | Polytechnic University of Madrid | 310 |
| Zhejiang University | 1076 | Autonomous University of Madrid | 301 |
| Jilin University | 1014 | Barcelona University | 271 |
| Nanyang Technological University | 980 | Complutense University of Madrid | 254 |
| Nanjing University | 860 | National University of Singapore | 253 |
| Shanghai Jiao Tong University | 847 | National Center for Scientific Research | 247 |
| "National" Taiwan University | 782 | Tsinghua University | 243 |
| University of Science and Technology of China | 715 | Imperial College of Science, Technology and Medicine University of London | 241 |
| Fudan University | 698 | Polytechnic University of Valencia | 240 |
| Dalian University of Technology | 696 | Zhejiang University | 234 |
| National University of Singapore | 687 | Nanyang Technological University | 234 |
| South China University of Technology | 631 | University of Erlangen-Nurnberg | 233 |
| Nankai University | 616 | Ecole Polytech Fed Lausanne | 227 |
| Wuhan University | 578 | "National" Taiwan University | 217 |
| City University of Hong Kong | 539 | Osaka University | 215 |
| Beijing Normal University | 523 | James I University | 213 |
| East China University of Science and Technology | 519 | Granada University | 209 |
| Tianjin University | 518 | Almeria University | 198 |

注：①CSIC是西班牙高等科研理事会的简称，西语是 Consejo Superior de Investigaaiones Cientificas (CSIC)，Spanish Nativnal Research Center 是英译名。
②CNRS 是法国国家科学研究中心，法文是 Center National de la Recherche Scientifique, National Center for Scientific Research 是英译名。
③Universitat Jaume I 是西班牙一所大学，英文名是 James I University。

Finally, this study aimed to determine the quality of the citations based on the visibility of the citing journals, defined on the grounds of the quartile in which they are ranked. Table 7 shows the absolute number and percentage of citations found in journals in each quartile. The similarity in the percentages for the two countries is an indication that the quality of the citing sources is essentially the same.

**Table 7 Number and percentage of papers cited by journals in each quartile**

| China | | | Spain | | |
|---|---|---|---|---|---|
| Quartile | Total citations | Percentage | Quartile | Total citations | Percentage |
| 1Q | 54 208 | 34. 46 | 1Q | 25 545 | 33. 74 |
| 2Q | 47 221 | 30. 02 | 2Q | 22 640 | 29. 91 |
| 3Q | 30 590 | 19. 45 | 3Q | 15 237 | 20. 13 |
| 4Q | 25 266 | 16. 06 | 4Q | 12 279 | 16. 22 |
| Total | 157 285 | 100 | Total | 75 701 | 100 |

## 4 Conclusions

The conclusions drawn from the above findings denote, firstly, the steep rise in Spanish and Chinese scientific output in the area of renewable energy over the last 10 years. Chinese institutions have been especially active and their publication rate is much higher than recorded for the world as a whole. By the end of the period studied, they accounted for 17% of world output, second only to the United States.

Close similarity was observed between the two countries in terms of the areas researched, for eight of the ten most intensely studied subjects concur. Nonetheless, "materials science" is the area of choice for Chinese researchers, while their Spanish colleagues prefer "energy and fuels".

In both countries, most of the most productive institutions are universities. In China, however, the Chinese Academy of Sciences plays a particularly preponderant role, publishing five times more papers than the second listed Chinese research institution.

International collaboration is another important factor. In absolute terms, Chinese institutions engage in more joint research with foreign institutions than their Spanish counterparts. Percentage-wise, however, the latter partner is more intensely, at twice the rate recorded for Chinese institutions (34% compared to 17%).

Regional and cultural factors are observed in the identity of partnering countries. While both countries collaborate routinely with institutions in the United States, China conducts research with many Asian countries and Spain works in conjunction with European Union and Latin American nations.

Collaboration between Chinese and Spanish institutions in this area of research is not especially significant.

Chinese papers receive more citations from US institutions and Spanish institutions than from institutions of any other country. Like partnering, citations exhibit a regional component: several of the countries that cite Chinese research most frequently are Asian, while Spanish papers receive more European citations.

A closer analysis of the origin of citations shows that Chinese papers are heavily cited by institutions in China, whereas a more significant proportion of the citations received by Spanish papers come from foreign institutions.

No differences are observed in the profiles of the citing journals. Measured as the percentage breakdown of citations by the quartile of the citing journal, the visibility of the two countries' research is similar.

## Acknowledgement

This research was funded by the Spanish Ministry of Economy and Competitiveness under the project CSO2010-21759-C02-01 titled "Análisis de lascapacidadescientíficas y tecnológicas de la eco-economía en España partir de indicadorescuantitativos y cualitativos de I+D+i".

## References

Duan L. 2011. Analysis of the relationship between international cooperation and scientific publications in energy R&D in China. *Applied Energy*, 88: 4229-4238.

Eurostat. 2012. http://epp.eurostat.ec.europa.eu/portal/page/portal/eurostat/home/.

Ingwersen P, Larsen B, Garcia-Zorita C, et al. 2013. Contribution and influence of conference papers to citation impact in seven conference and journal-driven sub-fields of energy research 2005-11. *In*: *Proceedings of the ISSI Conference*. Vienna, July 2013. http://www.issi2013.org/Images/ISSI_Proceedings_Volume_I.pdf.

International Energy Agency (IEA). 2012. *World Energy Outlook* 2012. Paris: International Energy Agency (IEA).

International Energy Agency (IEA). 2013. *Redrawing the Energy-Climate Map. World Energy Outlook Special Report*. *Paris*: International Energy Agency (IEA).

Kajikawa Y, Yoshikawa J, Takeda Y, et al. 2008. Tracking emerging technologies in energy research: toward a roadmap for sustainable energy. *Technological Forecasting and Social Change*, 75 (6): 771-782.

Romo-Fernández L M, López-Pujalte C, Guerrero Bote V P, et al. 2011. Analysis of Europe's scientific production on renewable energies. *Renewable Energy*, 36: 2529-2537.

Sanz-Casado E, García-Zorita J C, Serrano-López A E, et al. 2012. Renewable energy research 1995-2009: a case study of wind power research in EU, Spain, Germany and Denmark. *Scientometrics*, 95 (1): 197-224.

World Energy Council (WEC). 2013. *World Energy Issues Monitor*. London: World Energy Council.

# 2-2 Two Elements of Evaluative Bibliometrics in Sweden: The National Indicator and the Bibliometric Unit of Stockholm University

**Per Ahlgren**[①]

## Abstract

This paper treats two elements of evaluative bibliometrics in Sweden: the national indicator and the bibliometric unit of Stockholm University. The national indicator, proposed in the year 2007, is used yearly to distribute 5% of the research funds across Swedish higher education institutions. The indicator has two components, a normalized article production score and the well-known mean normalized citation score. Both these components are normalized with respect to field (subject). The production score is obtained by dividing the number of articles, for a given institution and a given subject, by the estimated mean number of articles per (Nordic) author within the subject. Computer simulation indicates, however, that mean estimates generated from empirical data might be bad approximations of true population means. The bibliometric unit of Stockholm University was established in September 2007 as a part of the university library. The unit provides the main university administration and other decision-makers within the university with analyses of the university's scientific publishing. Publication production, the impact of publications and collaboration in terms of co-publishing are analyzed by the bibliometric unit, and the aggregation level is normally department. The unit also coordinates data collection and data delivery regarding producers of university rankings, like Times Higher Education/Thomson Reuters.

**Keywords**: evaluative bibliometrics; national indicator; Stockholm University; Sweden

---

① Stockholm University Library, SE-106 91 Stockholm, Sweden per. ahlgren@sub. su. se.

# 1 Introduction

In recent years, bibliometrics, both in Sweden and in several other countries, is increasingly being used as a tool for evaluating research. This is related to the circumstance that demands on universities, as well as their subunits, to demonstrate accountability are growing stronger. The Swedish Research Council uses a national bibliometric indicator in order to annually measure the performance of Swedish higher education institutions (HEIs). The indicator was proposed in the year 2007 by the Swedish bibliometrician Ulf Sandström, in a public document on resource allocation with respect to research and postgraduate studies at Swedish HEIs (Resursutredningen, 2007)

Several Swedish HEIs, like Stockholm University (SU), have staff that analyses the research of the institutions bibliometrically. Bibliometric units at the Swedish HEIs normally belong to the libraries. However, in a few cases the units belong to the administration. Uppsala University, for instance, is such a case. The following HEIs have the most developed bibliometric units (alphabetical order):

(1) Karolinska Institute (Stockholm);

(2) Royal Institute of Technology (Stockholm);

(3) SU;

(4) University of Gothenburg.

Since spring semester year 2007, there is a loosely composed group of (foremost) library staff of the HEIs that are interested in bibliometrics. Members of the group organize, one time per semester, a meeting, where invited speakers give talks on bibliometric themes or on themes related to bibliometrics.

# 2 The National Indicator

As stated above, the Swedish Research Council annually uses a bibliometric indicator for the measurement of the performance of Swedish HEIs. The indicator has two components, both of them normalized with respect to subject: (a) a citation impact component, and (b) a production component. Component (a) is the well-known "mean normalized citation score" (MNCS), used, for example, by the Centre for Science and Technology Studies (CWTS) in its Leiden Ranking 2013. ①

---

① http://www.leidenranking.com.

Component (b) is a "normalized article production score" (NAPS). NAPS is an estimate of what the number of articles, for a given HEI and a given subject, corresponds to in terms of number of normal productive authors within the subject, where the authors are affiliated to the Nordic countries (Denmark, Finland, Iceland, Norway and Sweden). Here, the subjects are macro-classes of Web of Science journals, obtained by cluster analysis based on citations between the journals. For both MNCS and NAPS, address fractionalization is used: a given HEI is assigned an article after the share of unique addresses occurring in the article that the HEI stand for.

To obtain the NAPS, the number of articles for a given HEI in the macro-class is divided by the estimated average number of articles per (Nordic) author within the macro-class. The resulting value expresses what the number of articles corresponds to in terms of number of normal productive (Nordic) authors within the macro-class. The reason that estimation is applied is that the number of authors, within the macro-class, that have produced zero articles is unknown.

The estimated average number of articles per author is the Waring reference value. The expression "Waring" is used in the name of the reference value, since the involved production distributions are assumed to approximate the Waring distribution, a two-parameter ($\alpha$ and $N$) discrete distribution. In Braun et al. (1990), where the Waring distribution is related to bibliometrics, the following recursive definition of the distribution is given ($i = 1, 2, \cdots$):

$$p_0 = \alpha/(\alpha + N)$$

$$p_i = p_{i-1} \cdot (N + i - 1)/(N + i + \alpha) \tag{1}$$

The Waring reference value is obtained by weighted least squares regression with the number of articles as $x$ values and truncated means as $y$ values. The first truncated mean, corresponding to the $x$ value 1, is the mean number of articles per author, where the authors have produced at least one article. The second truncated mean, corresponding to the $x$ value 2, is the mean number of articles per author, where the authors have produced at least two articles. And so on, up to about 12 articles. The weights used in the regression are such that the first points in the series influence the regression line most. This is reasonable, since these points have considerably more underlying observations compared to the last points in the series.

The Waring reference value is taken to be the intercept of the regression. Clearly, then, the series is extrapolated to $x = 0$. In Figure 1, the extrapolation is visualized. The figure is not based on empirical data. Instead, computer simulation was used to generate an approximate Waring distribution with 10 million observations. $\alpha$ and $N$ were set to 4 and 5, respectively. The variable values of the distribution are 0, 1, ⋯, 100. The involved computer program performs a weighted least squares regression in the way indicated in the preceding paragraph. The weights $w_i(i=1, \cdots, 12)$ of the regression were defined as follows (Telcs et al., 1985):

$$w_i = \frac{D(i)}{\sqrt{n(i)}} \tag{2}$$

where $D(i)$ is the standard deviation of the truncated distribution of all observations $\geqslant i$, and $n(i)$ is the number of observations in this distribution. Thus, $w_i$ gives the estimated standard error of the $i$-th truncated mean. In this case, where a huge number of observations are used, the Waring reference value constitutes a very good approximation of the mean of the generated distribution, 1.68 and 1.66, respectively.

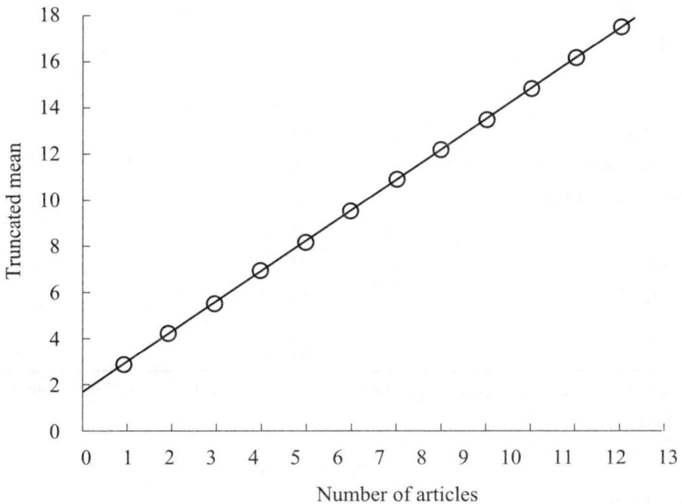


Figure 1 Regression line extrapolated to $x=0$

Notes: This yields the Waring reference value 1.68 (population mean = 1.66)

Now, to obtain the indicator value for a given HEI, the following is done:

(1) Calculation of the products of MNCS and NAPS for each macro-class.

(2) Summation of the products of the macro-classes.

Once the indicator values are obtained, the value for a HEI is divided by the sum of the indicator values over all the considered HEIs. The share of the sum each HEI accounts for is then obtained, and research funds are distributed according to these shares. However, only 5% of the funds are distributed on the basis of the shares.

Table 1, where the data is taken from Resursutredningen (2007), reports two distributions for 16 Swedish HEIs. In the rightmost column, the actual shares (in percentage) of the research funds for the HEIs at the year 2007 are given. The second column gives the shares that would have been obtained if the national indicator had been used for research fund allocation. We can see winners and losers. An example of the former is Karolinska Institutet, while Umeå University exemplifies the latter.

**Table 1 Hypothetical and actual distribution of research funds with regard to 16 Swedish HEIs**

| HEI | Hypothetical distribution (MNCS×NAPS) | Actual distribution year 2007 |
|---|---|---|
| Lund University | 15. 22 | 14. 33 |
| Uppsala University | 15. 19 | 13. 16 |
| Karolinska Institutet | 13. 35 | 11. 17 |
| University of Gothenburg | 10. 30 | 12. 53 |
| Stockholm University | 9. 38 | 8. 30 |
| Royal Institute of Technology | 7. 67 | 4. 74 |
| Swedish University of Agricultural Sciences | 7. 51 | 6. 70 |
| Chalmers | 6. 39 | 4. 46 |
| Umeå University | 5. 63 | 9. 21 |
| Linköping University | 5. 00 | 5. 88 |
| Luleå University of Technology | 1. 35 | 2. 25 |
| Stockholm School of Economics | 0. 84 | 0. 57 |
| Öbro University | 0. 65 | 1. 81 |
| Karlstad University | 0. 63 | 1. 55 |
| Mid Sweden University | 0. 55 | 1. 76 |
| Växjö University | 0. 36 | 1. 59 |
| Total | 100. 00 | 100. 00 |

Notes: Shares are given in Percentage. The rows are sorted descending after the values of the hypothetical distribution.

## Critical Points

Not surprisingly, critical points have been raised regarding the national indicator. The Swedish Research Council, the unit that is responsible for the generation of the indicator values, points to the fact that some of the macro-classes are small in number of authors, which leads to instable calculations of the Waring reference values. The council further states that the choice of subject classification matters: small adjustments of the subject borders can give large effects in the calculated Waring reference value (Fröberg et al. , 2010). Another concern has to do with the humanities. For this area, the MNCS is set to 1. One reason for this is that the reference citation values used for field normalization normally are very low for humanistic fields. When this is the case, high normalized citation values for articles with only one or two citations are obtained. In practice then, there is no impact/quality component for the humanities.

We end this subsection by reporting the outcome of a computer simulation on the differences between Waring reference values and corresponding population means. A computer program was written that iteratively creates approximate Waring distributions and performs corresponding weighted least squares regressions (number of iterations = 10 000). In each iteration $I$, an approximate Waring distribution, with randomly selected values on the parameters $\alpha$ ($\alpha > 2$) and $N$, and with 5000 observations, is generated. The variable values of the distribution are 0, 1, ⋯, 100. The difference between the Waring reference value, which is the intercept in the least squares regression of $I$, and the mean of the distribution of $I$ is calculated. Also a normalized difference is calculated: the difference in question divided by the mean of the distribution of $I$.

The histogram of Figure 2 shows the distribution of absolute differences (10 000 observations). It is clear from the figure that the intercept in several cases deviates from the distribution mean by 0.2 or more (regardless of sign).

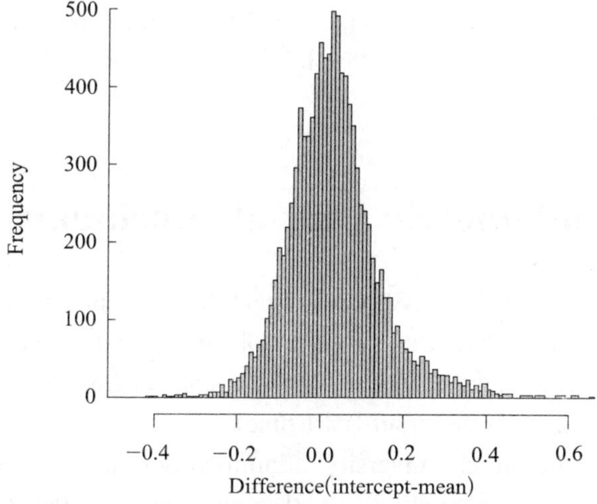


Figure 2 Distribution of differences (intercept-mean) with randomly selected values on parameters $\alpha$ and $N$ (sample size = 5000, number of iterations = 10 000)

In Figure 3, the normalized differences are plotted against the distribution means. Larger normalized differences tend to occur at lower means. This is a concern, since calculated Waring reference values for the various macro-classes in many cases are fairly small (<1) (Resursutredningen, 2007). Moreover, it should be kept in mind that the computer simulation scenario we treat in this subsection is a best-case scenario, where the generated distributions are very good approximations of the Waring distribution. In the light of this, one might wonder how well the calculated Waring reference values for the macro-classes, values that concern distributions that might deviate considerably from the Waring distribution, approximate the true population means.

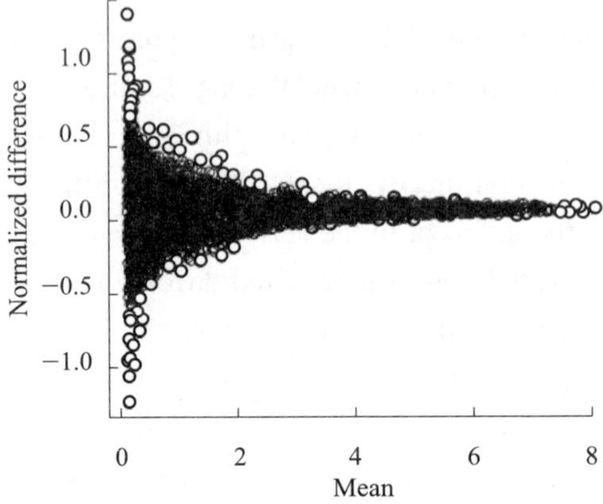


Figure 3 Normalized differences (difference-mean) plotted against means (sample size = 5000, number of iterations = 10 000)

## 3 The Bibliometric Unit at Stockholm University

The bibliometric unit at SU (BSU) was established in September, 2007 as a part of the university library. Kåre Bremer, professor of botany and former vice-chancellor of SU, initiated BSU. At the time of writing, BSU has one bibliometrician (full time; 20% research in the post) and one librarian (part time).

BSU provides the main university administration and other decision-makers within the university with analyses of the university's scientific publishing. The following are analyzed by BSU:

(1) Publication production;

(2) The impact of publications;

(3) Collaboration (in terms of co-publishing).

The aggregation level in the analyses is normally department. The results of the analyses are reported in publications, published on the website of BSU. In Figure 4, part of the start page of the English version of the site is shown.

BSU applies, one time per year, the Norwegian weighting model (Schneider, 2009; Sivertsen, 2010) for analysis of publication activity to all four faculties (Faculty of Humanities, Faculty of Law, Faculty of Science, and Faculty of Social Sciences) of SU. The model, which is designed for all scientific areas, combines

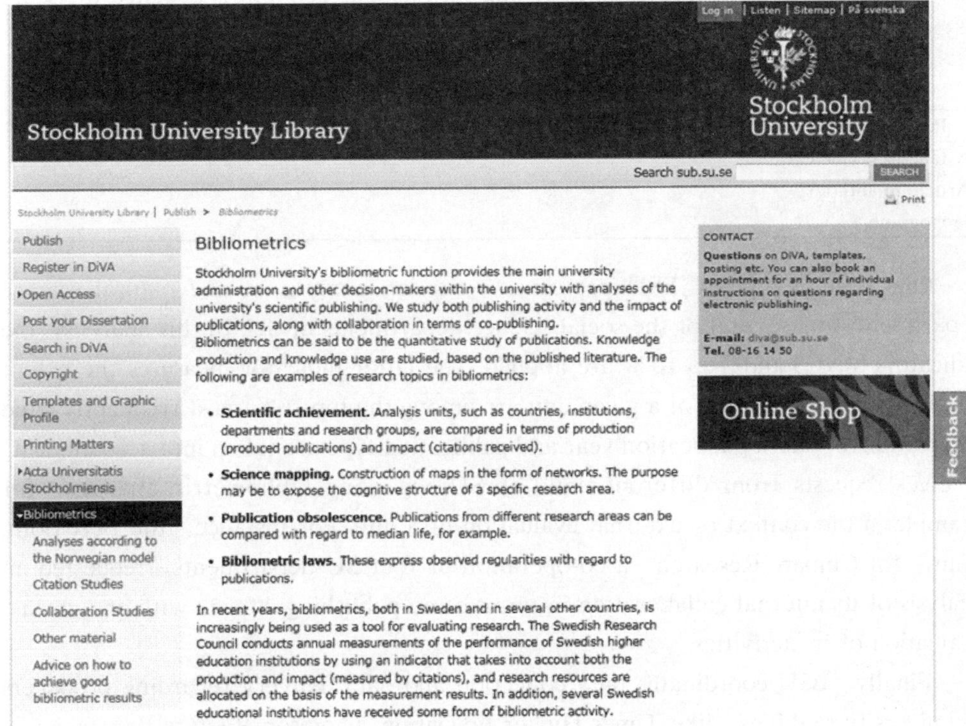


Figure 4 Part of the start page of the English version of the BSU website

production and impact. The weight of a publication is a function of its type and the level of its publishing channel. Three types are taken into consideration—article in periodical, article in anthology and monograph—whereas the considered channels are journals, publishers and series. The model makes use of three levels: 0 (non-scientific channel), 1 (scientific channel) and 2 (scientific channel with extra large prestige).

Author fractionalization is used in the Norwegian model: a unit (e. g., a department) is assigned a publication after the share of authors that have declared affiliation to the unit. Each analyzed unit is assigned a publication score, an indicator of both output and impact in scientific publishing. Such a score is the sum over the scores of the involved publications, where the score for a publication is the product of the author share (with respect to the analyzed unit) and the weight of the publication. Table 2 gives the weighting scheme of the Norwegian model (a publication such that is publishing channel has been assigned to level 0 has the weight 0). For instance, an article with three authors and published in *Journal of the American Society for Information Science and Technology*, a level 2 journal, and such that exactly one of the three authors belongs to the analyzed unit is assigned the score $1/3 \times 3 = 1$. ①

**Table 2 The weighting scheme of the Norwegian model**

| Publication type | Level 1 | Level 2 |
|---|---|---|
| Article in periodical | 1 | 3 |
| Article in anthology | 0.7 | 1 |
| Monograph | 5 | 8 |

One time per year, BSU performs a citation analysis of the natural science departments and several of the social science departments of SU. In this analysis, the indicators MNCS and Top 10% are applied. The latter indicator measures the extent to which the publications of a given unit are among the top 10% most frequently cited publications, where publication year and publication type are taken into account. BSU receives requests from different units of SU regarding bibliometric analyses, for example in the context of external evaluation of a unit. For instance, the Bert Bolin Centre for Climate Research, a co-operation of four SU departments, requested an analysis of its internal collaboration in terms of co-publishing, faced with an external evaluation of its activities.

Finally, BSU coordinates data collection and data delivery regarding producers of university rankings, like Times Higher Education/Thomson Reuters.

## 4 Concluding Remarks

In this work we have dealt with two elements of evaluative bibliometrics in Sweden: the national indicator and the bibliometric unit of SU. With respect to the national indicator, we focused on its production component, the NAPS, and the computer simulation outcome reported in the subsection. "Critical points" indicates that mean estimates generated from empirical data might be bad approximations of true population means. The standpoint of the present author is that the government should consider replacing NAPS by a more reliable production indicator.

The present author takes the position that it is time for SU to take the next step with regard to bibliometrics. The author has proposed to the university management that a scientometric centre should be established at the Department of Computer and

① In cases where an author has declared affiliation to more than one unit, the contribution of the author to the author share is weighted down according to the number of units the author has stated. In the example, we assume that the only author that belongs to the analyzed unit has not stated any other unit.

System Sciences. According to the proposal, the potential centre should have at least two senior researchers, at least two PhD students and one senior administrator. At the bibliometric unit of SU, research plays a minor role. At the potential centre, however, research would be a very important activity.

## References

Braun T, Gläzel W, Schubert A. 1990. Publication productivity: from frequency-distributions to scientometric indicators. *Journal of Information Science*, 16 (1): 37-44.

Fröerg J, Gunnarsson A, Karlsson S. 2010. *Kan Man Använda Waring Metoden för att Uppskatta Antalet Forskare?* Stockholm: Vetenskapsrådet.

Resursutredningen. 2007. *Resurser för Kvalitet*. Stockholm: Fritze.

Schneider J W. 2009. An outline of the bibliometric indicator used for performance-based funding of research institutions in Norway. *European Political Science*, 8 (3): 364-378.

Sivertsen G. 2010. A performance indicator based on complete data for the scientific publication output at research institutions. *ISSI Newsletter*, 6 (1): 22-28.

Telcs A, Glänzel W, Schubert A. 1985. Characterization and statistical test using truncated expectations for a class of skew distributions. *Mathematical Social Sciences*, 10 (2): 169-178.

# 2-3 Analysis of the Development of ISSI and Scientometrics and Informetrics: A View of ISSI Conference Papers

**Zhao Rongying**①, **Zhao Yuehua**②

## Abstract

In order to reflect the development and evolution of scientometrics and informetrics, this paper selected the ISSI conference papers from CPCI-S and CPCI-SSH databases as the sample dataset. Statistical and visualization software were used to analyse the dataset from aspects of spatial and temporal distribution, core papers, core authors, international collaboration and the evolution of conference topics and focuses. It concludes that: (1) The overall number of conference papers shows a growing trend, and the research force of scientometrics and informetrics is located mainly in European region, such as Spain, the Netherlands, the United Kingdom and Germany. Furthermore, the host of ISSI conference will encourage participation of the local researchers. (2) Researches on citation analysis, bibliometrics analysis, bibliometric indicators, Science Citation Index, and impact factor receive the most concern. Themes of the conference papers were under the guidance of ISSI conference topics, and in turn will help decide the future ISSI conference topics. (3) China is the only country which shows growth both in number and percentage of conference papers, but the gap still exists between China and the world average in the influence of ISSI conference papers.

---

① Wuhan University, the Center for the Studies of Information Resources, Research Center for China Science Evaluation, Luojia Hill, 430072, Wuhan (China), zhaorongying@126. com.

② University of Wisconsin-Milwaukee, School of Information Studies, P. O. Box 413, 53201WI Milwaukee (USA), yuehua@ uwm. edu.

## Conference Topic

The development of the International Society for Scientometrics and Informetrics (ISSI), ISSI's international conferences and related international conferences (Topic 1)

# 1 Introduction

In 1987, the International Society for Scientometrics and Informetrics (ISSI) held the 1st International Conference on Bibliometrics and Theoretical Aspects of Information Retrieval at Belgium, and then published the proceeding of *Informetrics*, which attracted huge attention from academia. After that, the ISSI has been holding its biennial conference all over the world. The title of the conference has been confirmed as International Conference of the International Society for Scientometrics and Informetrics since 1995. The foundation of the international society and the determination of the conference title have promoted the acceptance of scientometrics and informetrics for international academia.

As the organizers of ISSI 2013 conference presented, the ISSI conference will provide an international open forum for scientists, research managers, authorities and information professionals to debate the current status and advancements of informetrics and scientometrics theories and their deployment (detailed information can be get in this portal: http://www.issi2013.org/about.html). Aleixandre-Benavent et al. (2009) proposed that participating into such ISSI conference will access the scientists to the international forum, and exploit and enhance their reputations. For one thing, it is an investment for their research, since it allows them to extend their network of collaborators to include other peers; and for another, conferences are the ideal forum for discussing research that is in progress and in new trends.

In sight of the above, the aim of the present study is to perform a systematic review and analysis of conference papers derived from past ISSI conferences in order to explore the development of scientometrics and informetrics all around the world from a view of international conference.

## Overview of Past ISSI Conferences

ISSI conferences havebeen successfully held 13 sessions. Previous ISSI conferences took place in Belgium (1987), Canada (1989), India (1991), Germany (1993), USA (1995), Israel (1997), Mexico (1999), Australia (2001), China (2003),

Sweden (2005), Spain (2007), Brazil (2009), and South Africa (2011) (detailed information can be get in this portal: http: //www. issi-society. info/). And the 14th conference will be hosted by the University of Vienna and the Austrian Institute of Technology (AIT) in Vienna, Austria. Specific information[①] detailed in Table 1.

**Table 1 Overview of past ISSI conferences**

| Item | Conference title | Date | Place | Themes |
|---|---|---|---|---|
| 1st | International Conference on Bibliometrics and Theoretical Aspects of Information Retrieval | 25-28 August, 1987 | Belgium | ①Discussion of Several Major Laws; ②Application of Citation Analysis |
| 2nd | International Conference on Bibliometrics, Scientometrics and Informetrics | 5-7 July, 1989 | Canada | ①Definitionof Metrics Boundary; ②Promotion of Several Major Laws |
| 3rd | International Conference on Informetrics | 9-12 August, 1991 | India | ①Applicationof Statistical Methods in Informetrics; ②Application of Mathematical Methods |
| 4th | International Conference on Bibliometrics, Informetrics and Scientometrics | 11-15 September, 1993 | Germany | ①Relationship Between Three Metrics Research; ② Application of Citation Analysis |
| 5th | International Conference of the International Society for Scientometrics and Informetrics | 7-10 June, 1995 | USA | ① Discussion of Journal Evaluation; ②Development of Several Major Laws |
| 6th | The same as above | 16-19 June, 1997 | Israel | ①Application of Citation Analysis; ②Study of Document Obsolescence and Scattering; ③Data Compression; ④R&D Management, etc. |
| 7th | The same as above | 5-8 July, 1999 | Mexico | ①Evaluation of Scientific Journals; ②Content Analysis; ③Citation Analysis and Mathematical Models; ④Law Distribution, etc. |
| 8th | The same as above | 16-20 July, 2001 | Australia | ①Dynamics of Scientific Fields; ②Mathematical Modelling of Informetric Laws; ③Citer Motivation; Evaluation of Scientists' Research Performance; ④Development of Indicators; ⑤Mapping and Visualisation of Knowledge; ⑥Publication Productivity and Research Cooperation; ⑦Library Management; ⑧Economic Factors in Information Production and Dissemination; ⑨Science Policy Analysis and Forecasting |

① The conference information of 8th to 13th stemmed from the website of ISSI (http: //www. issi-society. info/), whereas other information cited from *Informetrics*, published by Qiu (2007) in Chinese.

Continued

| Item | Conference title | Date | Place | Themes |
|---|---|---|---|---|
| 9th | The same as above | 25-29 August, 2003 | China | ①Mathematical Modelling of Informetric Laws; ②Research Evaluation and Methodology of University Ranking; ③Citation Analysis and Database; ④Quantitative Analysis of Technology Innovation (Patent); ⑤Network Information Retrieval, etc. |
| 10th | The same as above | 24-28 July, 2005 | Sweden | ①Dynamics of Scientific Fields; ②Interdisciplinarity; ③History of Bibliometrics and Scientometrics; ④Mathematical Modelling of Informetric Laws; ⑤Citer and Linker Motivation; ⑥Webometrics; ⑦Evaluation of Science Research Performance; ⑧Development of Indicators; ⑨Mapping and Visualisation of Knowledge; ⑩Publication Productivity and Research Cooperation; ⑪Collection Management; ⑫Economic and Social Factors in Information Production and Dissemination; ⑬Science Policy Analysis and Forecasting |
| 11th | The same as above | 25-27 June, 2007 | Spain | ①Dynamics of Scientific Fields; ②Interdisciplinarity; ③History of Bibliometrics and Scientometrics; ④Mathematical Modelling of Informetric Laws; ⑤Citer and Linker Motivation; ⑥Webometrics; ⑦Evaluation of Science Research Performance; ⑧Development of Indicators; ⑨Mapping and Visualisation of Knowledge; ⑩Publication Productivity and Research Cooperation; ⑪Collection Management; ⑫Economic and Social Factors in Information Production and Dissemination; ⑬Science Policy Analysis and Forecasting |
| 12th | The same as above | 14-17 July, 2009 | Brazil | ①Methods and Techniques; ②Citation Analysis; ③Indicators; ④Webometrics; ⑤Journals; ⑥Open Access and Electronic Publications; ⑦Productivity and Publications; ⑧Collaboration; ⑨Research Policy; ⑩Patent Analysis; ⑪Bibliographic Database |

Continued

| Item | Conference title | Date | Place | Themes |
|---|---|---|---|---|
| 13th | The same as above | 4-7 July, 2011 | South Africa | ①Theory; ②Methods and Techniques; ③Citation and Co-citation Analysis; ④Indicators; ⑤Webometrics; ⑥Mapping and Visualization; ⑦Research Policy; ⑧Productivity and Publications; ⑨Journals, Databases and Electronic Publications; ⑩Collaboration; ⑪Country Level Studies; ⑫Patent Analysis |
| 14th | The same as above | 15-19 July, 2013 | Austria | ①Scientometrics Indicators; ②Data Sources for Scientometric Studies; ③Science Policy and Research Evaluation; ④Research Fronts and Emerging Issues; ⑤Technology and Innovation; ⑥Collaboration Studies and Network Analysis; ⑦Webometrics; ⑧Visualisation and Science Mapping; ⑨ Management and Measurement of Bibliometric Data; ⑩Open Access and Scientometrics; ⑪Modeling the Science System, Science Dynamics and Complex System Science; ⑫Scientometrics in the History of Science; ⑬Sociological and Philosophical Issues and Applications; ⑭Bibliometrics in Library and Information Science |

Brookes B. C. once proposed to complement the term "Informetrics" to the title of second ISSI conference at the first conference in 1987 (Tague-Sutcliffe, 1992). And this proposal received generally agreement and support from participating scholars. Subsequently, Brookes' view was to some extent accepted. Finally, at the 5th conference held in Chicago, the title of the conference determined to be International Conference of the International Society for Scientometrics and Informetrics. And bibliometrics has been excluded from the title but still included by the conference subjects. This change of title meant the research object of metrics has been extended to border scope of science and information instead of merely literatures.

It can be seen in Table 1 that after the 26-year course of ISSI conferences, the conferences have formed a stable pattern: conferences were mainly hold from June to September each year, lasting for 3-5 days, and organizers were mostly well-known colleges and universities or research institutions in the conference places. Meanwhile, these institutions are also the main force of scientometrics and informetrics research.

# 2 Methodology

## 2.1 Data Collection and Process

All the data was drawn from CPCI-S (Conference Proceedings Citation Index-Science) and CPCI-SSH (Conference Proceedings Citation Index-Social Science and Humanities), available through the Web of Knowledge. We retrieved the conference papers from databases with the conference field "informetrics", the time span is from 1996 to present, which is the full coverage of the database. Then results were refined by the field of "conference title", which returned 901 ISSI conference papers. Because some full papers and posters presented at the conference were later published in journals, excluding such journal articles by the term of "publishing source", we finally got 726 ISSI conference papers as our sample dataset. The dataset consists of the conference papers derived from six sessions of conferences: namely the 7th, 8th, 10th, 11th, 12th and 13th ISSI conferences. The data was retrieved on November 22, 2012.

## 2.2 Method of Data Analysis

In order to achieve a systematic review, statistical analysis and visualization analysis are performed by Microsoft Excel 2010 and CiteSpace Ⅱ. CiteSpace Ⅱ is a java application developed by Dr. Chen Chaomei for analysing and visualizing emerging trends and changes in scientific literatures.

# 3 Results

## 3.1 Timing Distribution of ISSI Conference Papers

Figure 1 depicts the timing distribution of bibliographic records of six sessions of conference papers. Utilizing "Create Citation Report" of Web of Knowledge platform sorts citation frequency of 901 pieces of conference papers in descending orders, and counts the citation number of each session. Figure 1 shows the trend of change during the time span.

According to Figure 1, we can divide the development into two phases. Firstly, during the period of 7th to 11th, the number of conference papers increased significantly, in 7th, the number of papers was 51 and in 11th the number was 170,

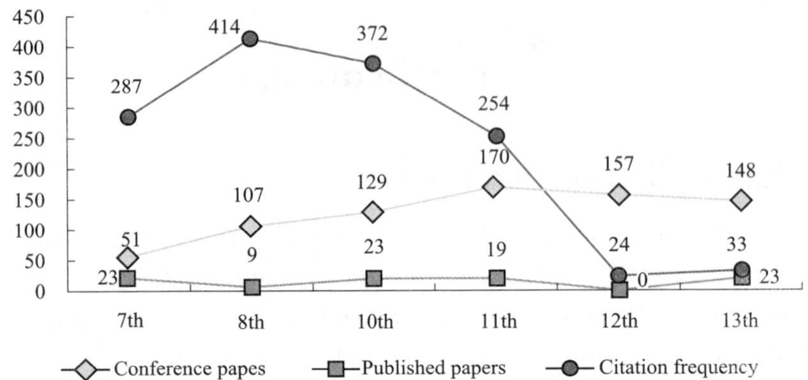


Figure 1 Timing distribution of conference papers

which is the peak of all sessions. As for citation frequency, it always stayed at a high level and reached the peak 414 in 8th, suggesting that ISSI conference had aroused more and more concern all over the world and a number of research achievements were presented on the conference. Secondly, during 11th to 13th, the number of papers began to decline continuously with the speed of roughly 10 papers per year. What is more, the citation frequency drops sharply, in 11th, the number of citations was 254 and in 12th the number was 0, followed by a small rise in 13th.

The number of conference papers published after presenting on the conference kept stable. Out of all the presentations at six conferences, 139 (18.24%) were later published and then picked up by the CPCI-S and CPCI-SSH databases. 130 (17.06%) of them were published on *Scientometrics*, 2 of them were published on *Journal of Informetrics*, 2 of them were published on *Research Evaluation*, and each of *Information Processing & Management*, *Journal of the American Society for Information Science*, *National Medical Journal of India*, *Publishing Research Quarterly* and *Research Policy* published 1 paper derived from ISSI conference. The statistical result indicates that most (94.96%) published conference paper was accepted by journals of scientometrics or informetrics field. Besides, journals from the field of research evaluation and policy, information science, medicine also published ISSI conference papers, suggesting that the convening of ISSI conference has promoted the exchange of knowledge with related disciplines.

## 3.2 Spatial Distribution of ISSI Conference Papers

CiteSpace Ⅱ can intuitively display the spatial distribution of papers. On the graphical user interface of CiteSpace Ⅱ, we selected "country" as the type of network nodes. "Time slicing value" was set to 1999-2012 and "per slice " was 2, namely the entire time interval of 1999-2012 was divided into seven 2-year slices for processing. In each time slice, the top 30 most occurred items were selected out for constructing the network.

Table 2 clearly specifies the top ranked countries with most publications. The frequency distribution of papers in Table 2 indicate that Spain and USA are the largest contributors both publishing 65 pieces of ISSI conference papers. India's publication counts are 59 times and ranks third, followed by China and Brazil respectively with 58 and 51 times, ranking the fourth and fifth in terms of publication counts.

**Table 2 The top 10 most productive countries**

| Country | 7th | 8th | 10th | 11th | 12th | 13th | Total |
|---|---|---|---|---|---|---|---|
| Spain | 2 | 2 | 10 | 30 | 14 | 7 | 65 |
| USA | 3 | 6 | 15 | 19 | 11 | 11 | 65 |
| India | 11 | 24 | 10 | 6 | 2 | 5 | 59 |
| China | 1 | 3 | 7 | 14 | 10 | 23 | 58 |
| Brazil | 1 | 0 | 2 | 6 | 30 | 12 | 51 |
| Netherlands | 3 | 6 | 6 | 12 | 8 | 12 | 47 |
| England | 1 | 5 | 11 | 10 | 9 | 10 | 46 |
| Germany | 3 | 6 | 7 | 4 | 9 | 8 | 37 |
| Belgium | 4 | 3 | 6 | 7 | 5 | 11 | 36 |
| Canada | 0 | 3 | 8 | 6 | 7 | 5 | 29 |

Figure 2 employs Google Earth to vividly display the spatial distribution network developed by CiteSpace Ⅱ. Each node on the Google map represents an institute presented papers on the conference, and the more vivid the colour of the node, on behalf of the more recent publishing time. ISSI conference papers were contributed by 50 countries spread over the entire globe, while the major contribution of the total output mainly came from European countries, such as Spain, the Netherlands, England and Germany. Just like Figure 2 shows, European region is the densest area of nodes, which means European countries are the core position of international scientometrics and informetrics research.

The distribution of publication frequencies according to country of origin is given in Table 2, and Figure 3 vividly shows the trend of percentage changes. From Figure 3, we can draw three patterns of changes: (1) Remaining stable, such as the Netherlands. (2) Keeping an upward trend. China is the only one exhibiting this characteristic. (3) Achieve the peak at one session but subsequently decline, which is the most type, especially to India, Spain and Brazil.

Comparing the changes of the number of conference papers with the changes of ISSI conference locations, it suggests Spain and Brazil, respectively, reached the publication peak when they held the conference. Furthermore, 8 of the top 10 productive countries once hosted one session of ISSI conference. All these phenomena reveal that the organization of ISSI conference will not only promote the enthusiasm of local researchers to participate in the conference, but also provide opportunities for the country's researchers to embark on the world stage.

Figure 2 Spatial distribution of conference papers (mainly shows European region)

In Table 2 and Figure 3, only China keeps upward trend. From only 1 presentation in 7th to the largest number, 23 presentations in 13th, the sharp rise of contribution shows substantially increased number of researchers engaged in the study of scientometrics and informetrics in China, and the English ability of Chinese scholars has much improved. On the other hand, presenting papers in such top conferences also shows the research outcomes of Chinese scholars gradually gained international approval.

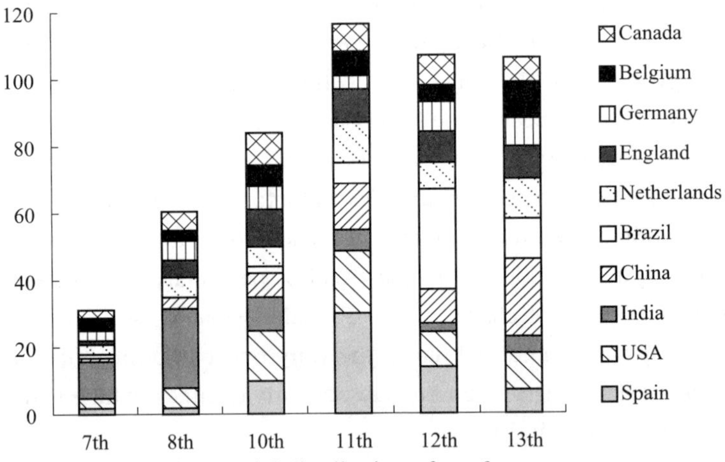


Figure 3 Spatial distribution of conference papers

## 3.3 Important Papers of ISSI Conferences

Citation analysis is designed to reflect the impact of academic papers; highly cited papers reflect the impact of countries, regions, institutions, journals or personnel (Liang, 2012). As of November 30, 2012, CPCI-S and CPCI-SSH databases contain 901

ISSI conference papers (including conference papers published in the journal after the conferences), cited by 1, 501 papers, and the total citation times is 2012. Top 10 highly cited papers are shown in Table 3.

**Table 3 The top 10 highly cited papers**

| Title | Author | Source Title | Conference | Frequency |
|---|---|---|---|---|
| Journal Impact Measures in Bibliometric Research | Glänzel, W; Moed, HF | *Scientometrics* | 8th | 146 |
| Advantages and Limitations in the Use of Impact Factor Measures for the Assessment of Research Performance in a Peripheral Country | Bordons, M; Fernandez, MT; Gomez, I | *Scientometrics* | 8th | 78 |
| A Bibliometric Study of Reference Literature in the Sciences and Social Sciences | Glänzel, W; Schoepflin, U | *Information Processing and Management* | 5th | 65 |
| How Much Is a Collaboration Worth? A Calibrated Bibliometric Model | Katz, JS; Hicks, D | *Scientometrics* | 6th | 61 |
| Explaining Australia's Increased Share of ISI Publications—the Effects of a Funding Formula Based on Publication Counts | Butler, L | *Research Policy* | 8th | 60 |
| Towards Appropriate Indicators of Journal Impact | Moed, HF; van Leeuwen, TN; Reedijk, J | *Scientometrics* | 7th | 53 |
| Tracking and Predicting Growth Areas in Science | Small, H | *Scientometrics* | 10th | 47 |
| Benchmarking Scientific Output in the Social Sciences and Humanities: The Limits of Existing Databases | Archambault, E; Vignola-Gagne, Etienne; Cote, G; Lariviere, V; Gingras, Y | *Scientometrics* | 10th | 44 |
| Development and Application of Journal Impact Measures in the Dutch Science System | van Leeuwen, TN; Moed, HF | *Scientometrics* | 8th | 38 |
| Selecting Scientific Excellence Through Committee Peer Review—A Citation Analysis of Publications Previously Published to Approval or Rejection of Post-doctoral Research Fellowship Applicants | Bornmann, L; Daniel, Hans-Dieter | *Scientometrics* | 10th | 37 |

In terms of citation frequency, Wolfgang Glänzel and Henk F. Moed's (Glänzel and Moed, 2002) paper takes the first place with the most citations of 146 times, suggesting that it is the most influential paper presented in the ISSI conferences. This paper was presented in the 8th ISSI conference and then published on *Scientometrics*, cited by 146 times, much higher than the other conference papers (Table 3). The paper *Journal Impact Measures in Bibliometric Research* mainly discussed both strengths and flaws of the impact factor, and statistical reliability of journal citation measures. Analysis found that the citing papers of this conference paper focused on the themes of impact factor and journal evaluation, which means that this paper played an important inspiration for the follow-up study of the topic.

Maria Bordons, Maria Teresa Fernandez and Isabel Gomez's paper *Advantages and Limitations in the Use of Impact Factor Measures for the Assessment of Research Performance in a Peripheral Country* (Bordons et al., 2002) has been cited 78 times since its publication in 8th ISSI conference and then on *Scientometrics*, ranks second in Table 3. In this paper, the authors pointed out several limitations of impact factor when applied to the analysis of peripheral countries, and displayed the usefulness of the impact factor measures in macro-, meso- and micro-analyses.

*A Bibliometric Study of Reference Literature in the Sciences and Social Sciences* (Glänzel and Schoepflin, 1999) occupied the third place on the list of citation frequency, received 65 citation times. This paper was co-authored by Wolfgang Glänzel and UrsSchoepflin, and published on *Information Processing & Management* in 1999. In this study, the author indicated that the model of information transfer from scientific literature to scientific (journal) literature assumed by standard bibliometrics requires substantial revision before valid results can be expected through its application to social science areas.

As can be seen, top 3 highly cited conference papers concentrated on the topic of journal evaluation indicator, suggesting it is one of the most hot spot in this field.

## 3.4 The Impact of ISSI Conference Papers from China

The number of outcomes can show the research strength of a given country, whereas the citation frequency can reflect its scientific influence. 58 China's conference papers was presented on the ISSI conferences, according to the six sessions of ISSI conference papers in data set 2, and 8 of them were published on journals after the conference. Although 12 papers have been cited and the sum of citation times is 40, the citation frequency of per paper is 0.61. Thus, comparing with the overall citation frequency of per paper (2.23), influence of conference papers published by Chinese scholars still has a certain gap between the world average levels. Besides, 8 of 12 papers which have been cited were the fruit by co-authoring with foreign researchers. And 7 of them were collaboration with Professor Ronald Rousseau. It means cooperation with well-known experts and scholars from abroad have an important promotion to China's scientific impact.

Dr. Liu Yuxian, comes from Tongji University Library in China, collaborated with Professor Ronald Rousseau to present a paper *Hirsch-type Indices and Library Management: The Case of Tongji University Library* (Liu and Rousseau, 2007) on 11th ISSI conference. This paper was cited for 15 times, and lists the first place among all conference papers from China. Liu and Rousseau made a comparison between the properties of different Hirsch-type indices, and used Tongji University Library as the case study.

## 3.5 The Publication Source of Important ISSI Conference Papers

Analysis of the citation frequency of ISSI conference papers found that only 1 of 63 highly cited papers which was cited over 10 times was cited from the conference proceeding, whereas other 62 papers received citation when they were published on journals. Furthermore, 1815 (90.2%) of 2012 citation times all derived from journals. It shows journals perform higher cited rate and influence compared with conference proceedings.

In summary, a large number of high-level papers were intensively provided to journals by the convention of ISSI conferences. At the same time, top journals in this field, such as *Scientometrics* and *Journal of Informetrics* published ISSI conference papers expanded the international impact of ISSI conferences. Thus, collaboration of international conferences and important international journals can jointly promote the rapid development of scientometrics and informetrics.

## 3.6 Important Authors of ISSI Conferences

### 3.6.1 Most productive authors

Data set 2 is cleaned through merging the different abbreviated forms of the same author. As of November 30, 2012, CPCI-S and CPCI-SSC databases included the ISSI conference papers of a total of 982 authors.

Table 4 shows the most productive authors of ISSI conference: Ronald Rousseau, the present incumbent of ISSI, Mike Thelwall, expert of link analysis research, T. N. (Thed) van Leeuwen, an important contributor of impact factor research, Wolfgang Glänzel, the founder of SCI database, Grant Lewison, a British scientist of informetrics, and Peter Ingwersen, the pioneer of webometrics, both rank in the forefront. These experts presented their latest scientific production in the ISSI conferences, so that they can discuss with conference participants and receive useful comments and suggestions.

**Table 4 The most productive authors**

| Output | Author | Output | Author |
|---|---|---|---|
| 20 | Rousseau，R | 11 | Lewison，G |
| 19 | Thelwall，M | 11 | Gupta，BM |
| 15 | Leta，J | 10 | Bordons，M |
| 13 | van Leeuwen，TN | 10 | Ingwersen，P |
| 12 | Glänzel，W | 10 | Archambault，E |
| 11 | Wilson，CS | | |

Statistics of paper counts showed an increasingly high position of Chinese scientists. Liang Liming and Liu Zeyuan，both Chinese scientists of scientometrics，respectively presented 8 papers and 5 papers in the ISSI conferences，and separately ranked 15 and 28 on the production list.

### 3. 6. 2 Highly cited authors

The citation and reference of scientific literature illustrates the inheritance and utilization of knowledge and represents the development of science. Thus，the cited extent by others is a measure of the paper's academic value and influence，meanwhile reflects the impact and position of author in certain field（Zhao and Xu，2010）. As for ISSI conferences，Professor Wolfgang Glänzel，Price Award winner in 1999，showed the citation frequency up to 359 times（Figure 4），and about 18 times per paper. And other Price Award winners，such as Henk F. Moed，Peter Vinkler and Henry Small，also performed high impact from the perspective of citation frequency. When it comes to Chinese authors，Dr. Liu Yuxian was the most cited author attributed to two ISSI conference papers collaborated with Professor Ronald Rousseau.

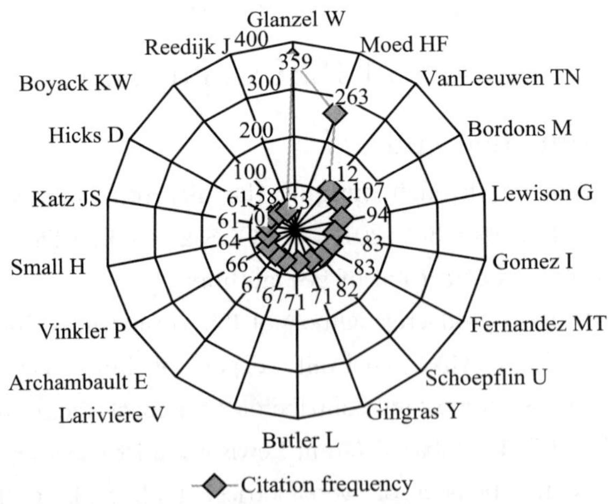


Figure 4 Highly cited authors in ISSI conferences

## 3.7 International Scientific Collaboration of ISSI Conferences

As we have mentioned at the first of this study, the core aim of ISSI conferences is to provide an open forum for scientists all over the world to promote the international collaboration. CiteSpace Ⅱ was employed to visualize the collaboration network of ISSI conference papers.

### 3.7.1 Macroscopic international collaboration

A comprehensive network consisting of nodes on behalf of the collaborating countries and authors between 1999 and 2012 was shown in Figure 5. If researchers from different countries co-authored published papers on ISSI conferences, these countries would be connected in the visualization map.

Figure 5 demonstrates that two research groups formed through ISSI conferences: the first one was consisted of the United Kingdom, Spain, the Netherlands and India; the United States, Germany, China, Belgium, and France composed of another research cooperative group. Countries in the groups show the larger node in the knowledge map, which means the higher amount of publications. Therefore, these countries not only played the core role involved in ISSI conferences, but also international cooperation.

### 3.7.2 Microscopic international collaboration

From microscopic view, scientific collaboration was completed by researchers. In terms of international collaboration network, the author nodes connecting different country nodes are the key roles between these countries. And high centrality nodes in the network show outstanding contribution for international cooperation. Table 5 is made up of successive presidents and members of ISSI boards (detailed information can be get in this portal: http://www.issi-society.info/board.html), and their centrality in the international collaboration network. Information in Table 5 indicates 6 of 17 scholars in successive ISSI boards that showed high centrality in the network, which means members of ISSI board did strengthen the international cooperation in the process of conferences.

**Table 5 Successive ISSI boards**

| Item | President | Centrality | Secretary /Treasurer | Centrality | Members | Centrality |
|---|---|---|---|---|---|---|
| 1st | Hildrun Kretschmer | 0.26 | | | | |
| 2nd | Michael E. D. Koenig | | | | | |
| 3rd | Bluma Peritz | | | | | |
| 4th | César Macías-Chapula | | | | | |
| 5th | Mari Davis | 0.04 | | | | |
| 6th | Henry Small | | | | | |
| 7th | Ronald Rousseau | 0.12 | | | | |

Continued

| Item | President | Centrality | Secretary /Treasurer | Centrality | Members | Centrality |
|---|---|---|---|---|---|---|
| 8th | Ronald Rousseau | 0. 12 | Wolfgang Glänzel | 0. 19 | Aparna Basu,<br>Leo Egghe,<br>Peter Ingwersen,<br>Grant Lewison,<br>Henk Moed,<br>Ed Noyons, | 0. 14<br>0. 17 |
| 9th | Ronald Rousseau | 0. 12 | Wolfgang Glänzel | 0. 19 | Judit Bar-Ilan,<br>Aparna Basu,<br>Peter Ingwersen,<br>Grant Lewison,<br>Martin Meyer,<br>Olle Persson | 0. 14<br>0. 17 |

Present incumbent Professor Ronald Rousseau played a crucial role in international cooperation between China and Belgium (Figure 5). He has served as the present incumbent of ISSI since 2003. And from then on, he visited universities and research institutions in China twice a year, and paid close attention to the progress of the construction and research work in China (Liang, 2012).

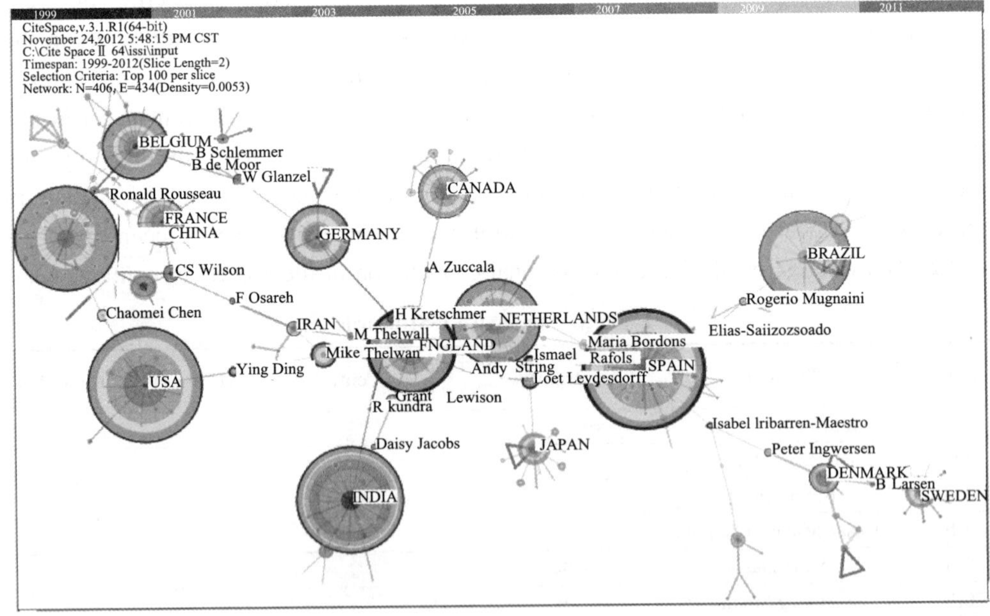


Figure 5 International collaboration network in ISSI conferences

## 3. 8 The Evolution of ISSI Conference Themes and Hotspots

ISSI conferences convene every two years, and conference papers are contributed by the researchers all over the world. Hence, the evolution of ISSI conference themes and hotspots indicates the development of scientometrics and informetrics. In CiteSpace Ⅱ, hotspot terms were contracted from the fields of title, abstract, descriptors and identifiers. Chen (2006) stated a time-zone view consists of an array of vertical strips as time zones, and the time zones are arranged chronologically from left to right.

We chose "term" as the node of the network, and the term type was "noun phrases". Figure 6 shows a network of terms with 122 nodes and 175 links. Figure 6 describes the evolution of conference hotspots in the time-zone view. Square shapes in the map indicate terms from title, abstract, descriptors and identifiers. The node size in the map represents the overall occurrence frequency of terms, and the link between different nodes shows the co-occurrence relation. The temporal distribution of terms corresponds to six periods: (1) 1999-2001, (2) 2001-2003, (3) 2005-2007, (4) 2007-2009, (5) 2009-2011, and (6) 2011 to present.

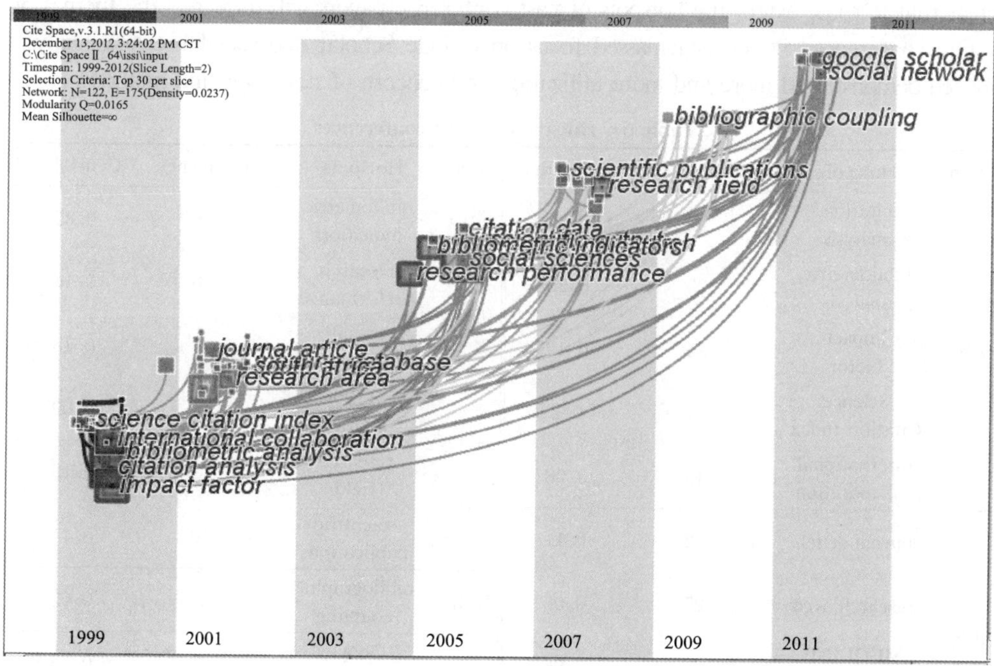


Figure 6 A time-zone view of the noun phrases network

Details of the research focuses of the six sessions' ISSI conference papers during 1999 to 2011 are listed in Table 6. 7th conference displayed most hotspots including citation analysis, bibliometric analysis, impact factor, Science Citation Index, and international collaboration, thus the main emphases were put on bibliometrics and citation analysis during this period. Then, journal article, research area, MEDLINE database and South Africa sprang up in the 8th conference. Furthermore, this session firstly set research area as a separate theme of the conference (Table 6), and the term "research area" exactly became one of hotspots. It suggests the arrangement of conference themes guides the study direction. In 10th conference, four hotspots emerged: bibliometric indicators, research performance, social sciences and citation data, which showed this phase focused on metric analysis of research performance evaluation and bibliometric indicators was still research concentration. Until 2007, increased researchers paid attention to research field and scientific publications. Particularly, themes related to the research evaluation were set in 11th conference, such as Evaluation of Science Research Performance and Publication Productivity (Table 1). This evolution illustrated scientific evaluation function of metrics study received more and more attention. Bibliographic coupling as an innovative hot topic arisen in 12th conference and coupling relationship turned into a new focus of citation analysis. With the emergency of a variety of new theories and methods, Google Scholar and social network analysis attracted augmented focus in 2011, and social network analysis has been settled as an individual theme in 14th conference. It is thus clear that to some extent the hotspots of past conference papers will influence the themes of future conferences. Besides, a massed focus on Google Scholar and social network analysis indeed demonstrated more and more utilization and concern of new tools and methods.

**Table 6 Hotspots of ISSI conferences**

| Item | Hotspots | Frequency | Centrality | Item | Hotspots | Frequency | Centrality |
|---|---|---|---|---|---|---|---|
| 7th | citation analysis | 29 | 0.3 | 10th | bibliometric indicators | 15 | 0.22 |
| | bibliometric analysis | 25 | 0.25 | | research performance | 20 | 0.16 |
| | impact factor | 24 | 0.19 | | social sciences | 29 | 0.15 |
| | Science Citation Index | 17 | 0.2 | | citation data | 10 | 0.09 |
| | international collaboration | 19 | 0.08 | 11th | research field | 15 | 0.13 |
| 8th | journal article | 12 | 0.11 | | scientific publications | 10 | 0.11 |
| | research area | 25 | 0.06 | 12th | bibliographic coupling | 8 | 0.02 |
| | MEDLINE database | 5 | 0.06 | 13th | Google Scholar | 7 | 0.03 |
| | South Africa | 7 | 0.02 | | social network analysis | 8 | 0.03 |

By comparison of the research focus of eachsession of conferences，citation analysis，bibliometrics analysis，bibliometric indicators，Science Citation Index and impact factor gained highest attention and have always been the most active research topics. In addition，Science Citation Index，MEDLINE database，and Google Scholar successively became research focuses demonstrated studies of citation database persistently run through the evolution of scientometrics and informetrics.

## 4 Conclusions

This paper aims to depict the development of ISSI and scientometrics and informetrics by presenting the results of a systematic and comprehensive review of ISSI conference papers published in six sessions of conferences：namely the 7th，8th，10th，11th，12th and 13th ISSI conference. For ISSI conferences，the overall number of papers grows，and the number of conference papers published on journals remained relatively stable. The contributing research community of ISSI conference papers distribute over various countries. Spain and USA are the most productive countries，and European countries，such as the Netherlands，England and Germany，are the main force of scientometrics and informetrics research. "Journal impact measures in bibliometric research" is the most influential paper from ISSI conferences. The most productive authors are Ronald Rousseau and Mike Thelwall，whereas，in terms of impact，Wolfgang Glänzel takes up the first place. Analysis of international scientific collaboration reveals two research groups that formed through ISSI conferences，and from microscopic view，members of ISSI board are pivotal nodes in the international cooperation network. The evolution of conference hotspots is described in the time-zone view，and the result shows that citation analysis，bibliometrics analysis，bibliometric indicators，Science Citation Index and impact factor have always been the hottest research topics. Furthermore，our analysis reveals that only China remained growing in conference papers；nevertheless，influence of their fruits still has a certain gap with the world averages. Besides，collaboration with celebrated foreign scholars will promote the impact of Chinese studies.

As an exploratory study，this research has limitations. We have only collected bibliographic records of ISSI conference papers from CPCI-S and CPCI-SSH databases. Also，these databases did not begin to index ISSI conference proceedings until the 7th conference，and also did not include 9th conference，so the research missed papers of seven sessions of conferences. Future studies may carry out a broader study based on the whole records of ISSI conference papers to complement the preliminary results of the current study. In addition，qualitative methods also should be used in the future for in-depth examination with other aspects such as the collaboration patterns of researchers.

## Acknowledgement

This paper is supported by Major Program of National Social Science Foundation in China（11&ZD152）. We wish to express our gratitude to Professor Ronald Rousseau, Chen Bikun and Yu Houqiang for their useful comments and suggestions.

## References

Aleixandre-Benavent R, Gonzalez-Alcaide G, Miguel-Dasit A, et al. 2009. Full-text publications in peer-reviewed journals derived from presentations at three ISSI conferences. *Scientometrics*, 80: 407-418.

Bordons M, Fernandez M T, Gomez I. 2002. Advantages and limitations in the use of impact factor measures for the assessment of research performance in a peripheral country. *Scientometrics*, 53: 195-206.

Chen C. 2006. CiteSpace Ⅱ: detecting and visualizing emerging trends and transient patterns in scientific literature. *Journal of the American Society for Information Science and Technology*, 57: 359-377.

Glänzel W, Moed H F. 2002. Journal impact measures in bibliometric research. *Scientometrics*, 53: 171-193.

Glänzel W, Schoepflin U. 1999. A bibliometric study of reference literature in the sciences and social sciences. *Information Processing and Management*, 35: 31-44.

ISSI. 2012. *Past conferences*. http://www.issi-society.info/ [2012-12-02].

ISSI. 2012. *Society's board*. http://www.issi-society.info/board.html [2012-12-22].

ISSI. 2012. *About ISSI* 2013. http://www.issi2013.org/about.html [2012-12-22].

Liang L. 2012. Development of scientometrics and informetrics in mainland China and in Taiwan. *Library and Information Service*, 56: 5-12.

Liu Y, Rousseau R. 2007. Hirsch-type indices and library management: the case of Tongji University Library. *In*: *Proceedings of ISSI 2007*: *11th International Conference of the International Society for Scientometrics and Informetrics.* Madrid, Spain: 514-522.

Ma H, Lü H. 2012. Knowledge mapping analysis of information science in China based on Chinese Social Science Citation Index. *Journal of the China Society for Scientific and Technical Information*, 31: 470-478.

Qiu J. 2007. *Informetrics*. Wu Han: Wuhan University Press.

Tague-Sutcliffe J. 1992. An introduction to informetrics. *Information Processing and Management*, 28: 1-3.

Zhao R, Xu L. 2010. The knowledge map of the evolution and research frontiers of the bibliometrics. *Journal of Library Science in China*, 36: 60-68.

# 2-4 Evaluation Index System for Academic Papers of Humanities and Social Sciences

Ren Quan'e①, Gong Xuemei②

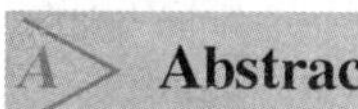


Abstract

An index system for evaluating academic papers is constructed and verified based on the empirical analysis of papers that has gained the 6th Chinese Academy of Social Sciences Award for Outstanding Achievements. Some new index, such as Paper Discipline Impact Factor (PDIF), Discipline Average Cited Rate per Paper (DACRP) and Discipline Average Downloaded Rate per Paper (DADRP) have been put forward in this paper. The empirical research results show that the ranking of papers calculated by this evaluation index system is in conformity with the awards determined by peer review in general, but still needs to be verified and improved in practice.


**Keywords**: academic papers; humanities and social sciences; evaluation index system; AHP; Delphi method

## 1 Introduction

Scientific achievements of humanities and social sciences in complicated forms cannot create economic and social value instantly as natural sciences. When evaluating these achievements, some rules can be operated artificially. Nevertheless, the academia is still keen on awards in the mainland in China because awards are directly related to interests and resource allocation, such as professional titles, bonuses and the honors brought by awards.

---


① Research Center for Bibliometrics and Science Evaluation, Chinese Academy of Social Sciences, Beijing, 100732, China. e-mail: renqe@cass.org.cn.

② College of Information Science and Technology, Drexel University, Philadelphia, PA 19104, USA. e-mail: xuemei.gong@drexel.edu.

Thus it can be seen that the achievement evaluation would be manufactured, once the evaluation results are tied up to benefit distribution. The achievements cannot be evaluated fairly only by experts' judgments, but also by objective indicators and comprehensive evaluation system. Citation analysis is an easy method to measure the importance of research, but not always effective in some disciplines, such as regional studies and local history (Finkenstaedt, 1990). And there is significant correlation between the citation counts and the rankings by 2001 UK Research Assessment Exercise (RAE) (Norris and Oppenheim, 2003). However, the importance of a paper to the articles that cite it cannot be simply assessed by counting the number of citations (Hanney et al., 2005). Moed, Luwel and Nederhof adopted bibliometric method which is much more than conducting citation analysis to eliminate the influence of subjective factors (Moed et al., 2002). Recently, Zhang et al. (2010) established a paper assessment system based on disciplinary benchmarks, utilizing total number of papers, order of authors, impact of journals, citation count, *h*-index, *e*-index, *a*-index and *m*-quotient as the assessment criteria. The research on paper evaluation tends to construct a more comprehensive evaluation system. This study aims to design a set of objective evaluation index system, consistent with the subjective judgments, in order to provide the basis for awarding, based on empirical analysis of the papers awarded by Chinese academy of social sciences.

## 2 Relative Citation Index and Downloaded Index

The normalization of citation indicators such as the so-called "crown-indicator" (CPP/FCSm) of the Leiden Vniversity's Center for Science and Technology Studies (CWTS), and the MNCR of the Flemish evaluation center ECOOM in Louvain (Glänzel et al., 2009) have been well-established or used. These indicators aggregate the numerator and the denominator separately, and then normalize by dividing the two means as a ratio. However, the new crown indicator (like the old one) is based on using (arithmetic) averages of highly skewed citation distributions.

As shown in Table 2, in this evaluation index system, the weights of relative citation index and relative downloaded index are greater, directly influencing the evaluation results. Relative citation/downloaded index of a paper, which is called paper's disciplinary impact factor (PDIF), can be defined as the ratio between citation/downloaded rate to average cited/downloaded rate per paper in the discipline the paper belongs to.

Calculation formula is as follows:

$$\mathrm{PDIF}_a = \mathrm{CIT}_a / \frac{1}{n}\sum_{k=1}^{n}\mathrm{CIT}_k$$

where $\mathrm{PDIF}_a$ is paper $a$'s disciplinary impact factor; $\mathrm{CIT}_a$ is the absolute citation/downloaded rate of paper $a$ ; $n$ is the total number of papers in the discipline to which paper $a$ belongs; and $\frac{1}{n}\sum_{k=1}^{n}\mathrm{CIT}_k$ is the average cited/downloaded rate per paper in that discipline, which is called Discipline Average Cited Rate per Paper (DACRP) and Discipline Average Downloaded Rate per Paper (DADRP).

Therefore, it's critical to find an appropriate method to calculate Discipline Average Cited Rate per Paper (DACRP) and Discipline Average Downloaded Rate per Paper (DADRP) in each discipline. In this scheme the focus is not only on (relative) citation rates, but also on the top-cited papers.

In this paper, three calculation methods are discussed based on statistics on utilization of papers that were published in 2003 and 2004: the first one is to consider all papers as a whole; DACRP/DADRP is the ratio between general cited/downloaded rate of all papers in a discipline and the number of the papers. The latter two methods focus on the papers with high quality, just considering the top 20% and top 30% of core paper zone which refers to the collection of papers cited at least once and in descending order of citation. These two methods accord with the skewed distribution of citation of papers as only a small proportion is frequently cited and downloaded among all the papers in each discipline. Table 1 shows the DACRPs and DADRPs calculated by those three methods respectively.

As can be seen from Table1, DACRP/DADRP calculated with number, citation and downloading of all the papers sorted by subject cannot reflect the differences between the disciplines, reducing the comparability of papers in different discipline. Take "reform of economic system" and "Chinese ancient history" as an example, their overall DACRPs in 2004 are 2.21 and 2.16 respectively, reflecting little differences. However, the distributions of paper citation are significantly different. There were 23, 202 papers published in "reform of economic system", with the highest cited rate being 792 and 37.57% of all papers having been cited; in the field of "Chinese ancient history", 3, 119 papers were published, with the highest citation frequency being 33 and 64.16% of the papers having been cited. This difference shows that the papers in the area of "reform of economic system" have greater influence, wider range of readers, more papers of high quality, but also more low-quality papers; in the field of "Chinese ancient history", a certain kind of expertise is needed for readers to understand the papers, so its influence is smaller, the total number of papers is less, but it has a relatively high proportion of high-quality papers. Thus, DACRPs/DADRPs based on all papers cannot eliminate the differences between disciplines, and then cannot be used to calculate the citation index and downloaded index.

**Table 1 DACRPs and DADRPs calculated by three methods respectively**

| Number | Discipline | DACRP | | | | | | DADRP | | | | | |
|---|---|---|---|---|---|---|---|---|---|---|---|---|---|
| | | Overall | | 20% | | 30% | | Overall | | 20% | | 30% | |
| | | 2003 | 2004 | 2003 | 2004 | 2003 | 2004 | 2003 | 2004 | 2003 | 2004 | 2003 | 2004 |
| 1 | Chinese politics and international politics | 1.36 | 1.34 | 16.83 | 15.1 | 13.07 | 11.85 | 45.59 | 61.64 | 288.99 | 351.67 | 253.34 | 311.98 |
| 2 | Finance | 2.72 | 2.63 | 23.95 | 21.00 | 18.24 | 16.15 | 52.05 | 76.06 | 262.14 | 336.15 | 227.70 | 296.42 |
| 3 | Reform of economic system | 1.90 | 2.21 | 19.70 | 19.77 | 14.98 | 15.07 | 42.27 | 63.92 | 235.71 | 302.20 | 203.63 | 261.84 |
| 4 | Economic theory and history of economic thought | 5.05 | 5.31 | 32.86 | 28.19 | 24.67 | 21.63 | — | 156.19 | — | 468.98 | — | 400.04 |
| 5 | Archeology | 1.01 | 1.05 | 9.08 | 7.38 | 7.21 | 6.04 | 50.54 | 66.54 | 176.51 | 176.29 | 157.01 | 161.21 |
| 6 | World history | 2.03 | 1.79 | 9.26 | 7.95 | 7.61 | 6.65 | — | 135.79 | — | 295.20 | — | 272.02 |
| 7 | World literature | 2.27 | — | 10.76 | — | 8.67 | — | 179.17 | — | 399.40 | — | 367.99 | — |
| 8 | Finance and taxation | 1.32 | — | 16.13 | — | 12.63 | — | 30.84 | — | 181.73 | — | 158.76 | — |
| 9 | Industrial economy | — | 0.79 | — | 13.95 | — | 10.86 | — | 28.85 | — | 257.46 | — | 215.51 |
| 10 | Agricultural economy | 1.87 | — | 22.37 | — | 17.04 | — | 27.99 | — | 208.13 | — | 174.62 | — |
| 11 | Investment | — | 2.38 | — | 20.10 | — | 15.71 | — | 70.62 | — | 324.79 | — | 288.23 |
| 12 | Information economy and post economy | 0.71 | — | 12.26 | — | 9.58 | — | 21.94 | — | 169.50 | — | 145.57 | — |
| 13 | Administrative law and local law | — | 2.69 | — | 21.50 | — | 17.98 | — | 77.46 | — | 328.25 | — | 299.49 |
| 14 | Administration and state administration | 2.34 | — | 25.09 | — | 19.16 | — | 56.20 | — | 320.98 | — | 273.60 | — |
| 15 | Politics | 5.30 | — | 23.33 | — | 19.70 | — | 155.44 | — | 376.62 | — | 345.98 | — |
| 16 | Demography and family planning | — | 3.00 | — | 22.47 | — | 17.62 | — | 93.24 | — | 441.64 | — | 368.37 |
| 17 | Sociology and statistics | — | 2.74 | — | 23.04 | — | 18.09 | — | 101.45 | — | 492.12 | — | 420.01 |
| 18 | Chinese ancient history | — | 2.16 | — | 8.47 | — | 7.06 | — | 106.60 | — | 212.10 | — | 188.14 |
| 19 | Chinese modern history | 1.74 | — | 9.81 | — | 7.99 | — | 80.74 | — | 248.60 | — | 213.14 | — |
| 20 | History of China | 1.93 | — | 9.98 | — | 8.07 | — | 91.95 | — | 249.57 | — | 219.75 | |
| 21 | Chinese literature | 0.82 | 0.93 | 7.67 | 6.98 | 6.21 | 5.71 | 55.86 | 74.42 | 241.71 | 340.45 | 236.05 | 298.04 |
| 22 | Chinese language | 4.44 | — | 26.68 | — | 20.20 | — | 136.50 | — | 477.83 | — | 395.77 | — |
| 23 | Religion | — | 0.93 | — | 6.31 | — | 5.34 | — | 67.77 | — | 317.28 | — | 274.64 |
| 24 | Philosophy | — | 2.53 | — | 11.26 | — | 9.00 | — | 134.89 | — | 303.58 | — | 271.64 |
| 25 | Travel | — | 3.33 | — | 23.03 | — | 17.88 | — | 109.11 | — | 478.65 | — | 399.90 |
| Standard deviation | | 1.42 | 1.15 | 7.78 | 7.18 | 5.80 | 5.46 | 49.72 | 33.16 | 90.89 | 91.32 | 79.65 | 73.77 |

As we chose the award-winning papers with high quality as the sample, DACRPs/DADRPs based on top 20% and top 30% of the discipline core paper zone could better reflect the differences between disciplines. For example, in the fields of "reform of economic system" and "Chinese ancient history", their DACRPs in 2004 based on top 20% are 19. 77 and 8. 47 respectively, and 15. 07 and 7. 06 based on top 30%, which can better differentiate these two disciplines. Standard deviation of DACRPs/DADRPs can be used to denote the degree of difference among disciplines. According to the standard deviation of DACRPs based on all papers, the degrees of difference in 2003 and 2004 are 1. 42 and 1. 15 respectively, showing little difference among disciplines. They increase to 5. 80 and 5. 46 respectively when considering the top 30% of core paper zone, showing obvious difference, and increase to 7. 78 and 7. 18 respectively when only considering the top 20% of core paper zone, showing more significant difference. And it is much the same in DADRPs. The standard deviation of DADRPs based on core-paper zone is much larger than that based on all papers, indicating higher degree of difference among disciplines. And the standard deviation of DADRPs based on a smaller proportion of core paper zone shows there is more significant difference. We can draw a conclusion that compared to the total papers, core paper zone can better reflect the characteristics of disciplines, and the smaller proportion of the core area, the more obvious the difference is between disciplines.

It still needs further study on the proportions of the core paper zone through relationship between the proportions and the evaluation results. In the evaluation results based on the overall papers, the papers awarded the first or second prizes place on the first half of the list, in the latter part of the list are all papers awarded the third prizes. But when coming to the individual, the first prize paper drops behind several second prize papers, and some third prize papers place before the first and second prize papers. In the evaluation result based on top 20% of the core paper zone, the first prize paper rises to the ranking while all second prizes dropping down, even sliding into the latter part of the ranking. When the proportion rises to top 30%, the ranking of the first prize is lower, with all second prizes higher. Thus, with the proportion of the core paper zone increases, the first prize will gradually fall with the second prizes rising in the ranking. Furthermore, the evaluation results sorted by disciplines based on either top 20% or top 30% of the core paper zones are in accordance with awards.

As there is only one paper awarded the first prize, the papers awarded the second prize can be considered as the representatives of best papers which should be elevated to a higher rank. Therefore, this paper chose the top 30% of the core paper zone to calculate the DACRP and DADRP.

# 3 Defining an Evaluation Index System for Papers in Humanities and Social Sciences

## 3.1 Delphi Method

Delphi method is a structured technique for forecasting based on a panel of independent experts (Landeta, 2006). It can avoid the misjudgment caused by the incomplete knowledge of a single expert (Skulmoski et al., 2007). There are two basic types of Delphi method, normal Delphi method and indirect survey method. By the normal Delphi method, the consensus of opinion of a group of experts can be obtained by a series of questionnaires with controlled opinion feedback (Dalkey and Helmer, 1963), as the experts will revise their earlier judgments according to the feedback of last questionnaires. Indirect survey method, which is a kind of assumption analysis method, mainly depends on the sufficient information of the questionnaires without several rounds of questionnaires and feedback. Considering the period of time the experts need to spend, we adopted the indirect survey method.

We chose famous editors, specialists from multiple disciplines in humanities and social sciences and experts on management of scientific research to be the group of experts, as experts with different academic backgrounds and working experiences hold different views towards the problem of evaluating papers. Famous editors were chosen randomly from the editors in charge of editorial offices of core journals in humanities and social sciences; the specialists, who were chosen from the list of conscientious specialists published by National Planning Office of Philosophy and Social Science, are researchers with research area in different disciplines in humanities and social sciences; and the experts on management of scientific research, who were chosen from the National Planning Office of Philosophy and Social Science and the scientific research administrations in famous universities, are the administrative staff taking charge of research management, research performance evaluation, research policy making and fund allocation. The weights of the editors, specialists and experts are 0.1, 0.6 and 0.3 respectively. In this survey, 85 questionnaires were sent out and 62 available questionnaires were retrieved. In the questionnaire, some indexes and choices for each one were listed for the experts to choose. The choices for the commonly used indexes included: very important, important and not important. And the choices for some new indexes included: should be included, can be included with light weight, can be considered in the future as lack of the statistics now, and should not be considered. Then the evaluation criteria and indicators were defined according to the statistics of questionnaires.

## 3. 2 Evaluation Criteria and Indicators

Papers in humanities and social sciences can be evaluated from three aspects, including level of innovation, research norms and academic values. However, once these evaluation elements are transformed to evaluation criteria and decomposed into indicators, the borderline between criteria will be indefinable, and the indicators will be interrelated with each other. Therefore, the first-class indicators are adjusted according to the second-class indicators properly.

### 3. 2. 1 Academic value

Cited rate, considered as the main index of academic value, is an international bibliometric evaluation method for innovativeness and academic level of papers. In that case, academic value, which becomes the aggregative indicator to judge both the research value and innovativeness of papers, has the same implication as innovativeness.

Citation index and downloaded index are set as the second-class indicators of academic value, because the citation and downloading are two mainforms of communication in academia, implying that the paper has taken effect in the academic community and the knowledge has been transferred between the entities. Meantime, citation index and downloaded index can be divided into absolute citation, relative citation, absolute downloading and relative downloading. Absolute citation/downloading refers to the cited/downloaded rate of a paper, while relative citation/downloading of a paper which aims to eliminate differences among disciplines can be calculated by the ratio between absolute citation/downloading and average cited/downloaded rate per paper in the discipline the paper belongs to. For multi-disciplinary paper, the average cited/downloaded rate is equal to the mean value of average cited/downloaded rate in all disciplines it belongs to.

### 3. 2. 2 Innovativeness and normalization

As the innovativeness and comprehensiveness of papers are judged mainly by novelty search departments, editors and experts, the second-class indicators of them contain novelty retrieval, digest, journal level and the award level, etc. Therefore, in accordance with the principle of operability, the indicator of innovativeness and completeness, which merges those indicators, is divided into journal index and digest index. The journal index of a paper refers to the impact factors of journals the paper publishes on; the digest index is determined by paper's digest such as full text copy or paper abstract.

### 3. 2. 3 Social value

Theinnovation which is regarded as the basic requirement of research achievements is also the most obvious feature of academic papers. Normally, innovative research is more likely to create academic and social value. Social value of papers in humanities and social sciences is not an indicator as important as academic value and innovation, and more subjective than academic values, but still embodies the quality of papers. Therefore, social value composes the evaluation index system with academic value and innovation together at a lower weight.

Social value can be decomposed into social echo and adoption by executive branch. However, because it's too difficult to obtain data on adoption, only the social echo is adopted to represent the social value in this research. Internet information search tools can be used to measure the social echo whose value is the volume of relevant data retrieved by search engine. For Chinese research papers, we use search engine Baidu to value the spreading scope and social impact in cyberspace.

## 3.3 Evaluation Index System

Analytic Hierarchy Process (AHP) is a structured technique for dealing with complex decisions, integrating both quantitative and qualitative aspects (Vidal et al., 2010). It was proposed by Saaty (1977) in the 1970s, and has been extensively studied and refined from then on. The AHP can be applied to decision situations, including evaluation, ranking and selection (Forman and Gass, 2001). Using this approach, decision-makers decompose the complex problem into several levels and some elements, and then numerical priority for each decision alternative can be calculated according to criteria and the weights of all elements.

After determining the evaluation indexes, we still need to assign the weight of each index according to their relative importance to the evaluation objective, thus constructing the evaluation index system. First of all, according to the target, identify the related indicators at all levels by principal component analysis, and then apply AHP to establish hierarchical structure diagram of evaluation indexes (Figure 1).

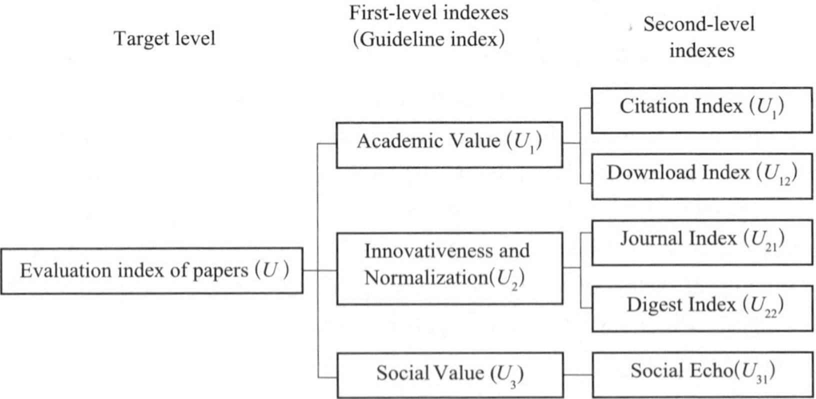


Figure 1 Hierarchical structure diagram of paper evaluation index system

Based on the hierarchical structure diagram above, compare every two indicators that have the same upper indicator and assign the relative weights according to "1-9" scale judgment method, constructing comparison matrix. Finally, the weights of each index is calculated by software yaahp 5. 0.

In the questionnaires, the options on the importance of each index are set. After analyzing and summarizing the feedback of experts, comparison matrix can be constructed to calculate the weight of each index. Table 2 shows the evaluation index system of papers.

**Table 2　Evaluation index system of papers**

| First-class indexes | Weights | Second-class indexes | Weights | Third-class indexes | Weights | Synthetical weights |
|---|---|---|---|---|---|---|
| Academic value | 0. 643 4 | Citation index | 0. 873 3 | Absolute citation | 0. 166 7 | 0. 093 7 |
| | | | | Relative citation | 0. 833 3 | 0. 468 2 |
| | | Download index | 0. 126 7 | Absolute downloading | 0. 366 7 | 0. 029 9 |
| | | | | Relative downloading | 0. 633 3 | 0. 051 6 |
| Innovativeness and completeness | 0. 207 4 | Journal index | 0. 250 0 | | | 0. 051 9 |
| | | Digest index | 0. 750 0 | Class A digest | 0. 400 | 0. 062 2 |
| | | | | Class B digest | 0. 300 | 0. 046 7 |
| | | | | Class C digest | 0. 200 | 0. 031 1 |
| | | | | Class D digest | 0. 100 | 0. 015 6 |
| Social value | 0. 149 2 | Social echo | 1. 000 0 | | | 0. 149 2 |

The digest index of a paper is calculated based on the level of publications that digest the paper. A-class digest refers to 4 abstracting journals, including *Xinhua Digest*, *China Social Science Digest*, *Journal of China Renda*, and *Chinese Journal of College of Liberal Arts Digest*. B -class digest refers to a variety of newspapers which published comprehensive social sciences abstracts, such as *Digest News*. C-class digest refers to the special column of digest in various subject journals, such as *Review of Economic Research* and *Academics in China*. D-class digest refers to various informally published digests serials, such as *Reference Information for Social Science Research*.

# 4 Empirical Analysis

## 4. 1　Objects and Data Sources

A total of 52 journal papers gained 6th Chinese Academy of Social Sciences Award for Outstanding Achievements, but only 49 papers are included in CNKI (China National Knowledge Infrastructure), among which, 43 papers were published between 2003-2004 and only 6 papers were published between 2000 and 2002. So this paper selected 43 papers published in 2003 or 2004 as the sample to ensure the comparability among the papers on citation and downloading. According to the classification in CNKI database, these papers can be classified into 25 disciplines, including finance, economic system reform, politics, world history and archeology, etc.

Research data came from CNKI database, search engine Baidu and *A Guide to the Humanities & Social Science Core Journals in China* (2008 *Edition*) (Jiang et al., 2009). The data on academic value was obtained from CNIK database where we can get the downloaded rate and cited rate of each paper, thus calculating the citation index and downloaded index in each discipline, which will be discussed below. The value of journal index of a paper directly adopted the Journal Impact Factor of the journal on which the paper published, which came from *A Guide to the Humanities & Social Science Core Journals in China* (2008 *Edition*). Value of digest index was calculated according to the digested condition of each paper which can be obtained from *Humanities & Social Sciences Digest Database*. Social echo index of a paper was represented by the retrieval result through the search engine Baidu with the paper title as keywords; the larger volume of retrieval result generally reflected that the theme of the paper attracts more attention and social echo.

## 4.2 Results

Then we evaluated the selected 43 papers with above evaluation index system. It must be noted that the scores of all papers on each index need to be normalized before calculating the integrated scores of each paper. Finally, the papers are ranked according to their integrated scores (Table 3).

According to Table 3, the evaluation results of award-winning papers are consistent with the awards in general. The ranking list has some exceptions, answering for the phenomenon of subjective judgments and artificial balance in the humanities and social sciences achievements awarding.

(1) In the overall ranking list, the correlation between the ranking and awards is significant and reasonable. Most of the papers getting the first and second prizes are listed in the top ten; nearly all papers getting the third prize are listed in the latter part of the ranking list. But there are also some exceptions which may be relevant to the discipline characteristics and other factors: two papers getting the second prize are listed in 24 and 26, but a third prize paper get the first place. When putting this evaluation system into practice, it's available to pick out the papers that have abnormal data on indicators and consider subjective evaluation independently.

(2) Sorting papers by the disciplines, the ranking of papers in each discipline is exactly the same as the awards. This shows that it's necessary to calculate the citation index and downloaded index in different disciplines respectively, especially research the DACRP and DADRP in core paper zone.

**Table 3 The results of evaluating papers**

| Number | Institutes | Awards | Disciplines | Absolute citation | Relative citation | Absolute download | Relative downloading | Journal index | Digest index | Social echo | Synthetical score |
|---|---|---|---|---|---|---|---|---|---|---|---|
| 1 | Institute of Law | 3rd prize | Administrative law and local law | 1 | 1 | 1 | 1 | 0. 354 5 | 0. 666 7 | 0. 467 3 | 0. 835 2 |
| 2 | Finance Institute | 2nd prize | Finance and taxation | 0. 261 1 | 0. 371 8 | 0. 170 6 | 0. 350 4 | 0. 142 3 | 1 | 0. 116 0 | 0. 401 9 |
| 3 | World Economy and Politics Institute | 3rd prize | Finance | 0. 362 8 | 0. 357 6 | 0. 430 3 | 0. 575 6 | 0. 347 4 | 0. 666 7 | 0. 205 5 | 0. 396 4 |
| 4 | Economy Institute | 2nd prize | Economic reform | 0. 230 1 | 0. 276 2 | 0. 341 8 | 0. 517 5 | 1 | 0. 666 7 | 0. 219 8 | 0. 376 1 |
| 5 | Rural Development Institute | 3rd prize | Finance | 0. 292 0 | 0. 325 2 | 0. 456 4 | 0. 459 8 | 1 | 0. 666 7 | 0. 016 3 | 0. 375 0 |
| 6 | Institute of Population and Labor Economics | 3rd prize | Demography and family planning | 0. 451 3 | 0. 460 5 | 0. 726 6 | 0. 583 0 | 0. 220 8 | 0 | 0. 095 0 | 0. 335 3 |
| 7 | World Economy and Politics Institute | 2nd prize | Economic reform | 0. 371 7 | 0. 446 2 | 0. 230 5 | 0. 353 3 | 0. 247 4 | 0 | 0. 314 4 | 0. 328 6 |
| 8 | World Economy and Politics Institute | 3rd prize | Investment | 0. 349 6 | 0. 400 2 | 0. 538 4 | 0. 559 5 | 1 | 0 | 0. 011 3 | 0. 318 6 |
| 9 | Institute of Industrial Economics | 3rd prize | Economic reform | 0. 194 7 | 0. 232 3 | 0. 379 6 | 0. 437 4 | 0. 401 0 | 0. 666 7 | 0. 208 1 | 0. 316 5 |
| 10 | Central Department | 1st prize | Politics | 0. 053 1 | 0. 048 5 | 0. 376 3 | 0. 319 1 | 0. 022 4 | 0. 666 7 | 1 | 0. 309 4 |
| 11 | Institute of Linguistics | 2nd prize | Chinese language | 0. 216 8 | 0. 192 9 | 0. 786 5 | 0. 585 5 | 0. 117 8 | 0. 666 7 | 0. 001 6 | 0. 274 4 |
| 12 | Institute of Population and Labor Economics | 3rd prize | Sociology and statistics | 0. 190 3 | 0. 189 1 | 0. 641 9 | 0. 446 2 | 0. 311 4 | 0. 666 7 | 0. 002 5 | 0. 268 8 |
| 13 | Institute of Contemporary China Studies | 3rd prize | Industrial economy | 0. 128 3 | 0. 212 4 | 0. 218 1 | 0. 314 5 | 0. 005 9 | 0. 666 7 | 0. 179 6 | 0. 265 0 |
| 14 | World History Institute | 3rd prize | World literature | 0. 088 5 | 0. 239 1 | 0. 280 6 | 0. 310 3 | 0. 004 1 | 0. 666 7 | 0. 108 9 | 0. 264 8 |
| 15 | Finance Institute | 3rd prize | Finance | 0. 203 5 | 0. 200 6 | 0. 125 0 | 0. 172 8 | 0. 150 2 | 0. 666 7 | 0. 047 6 | 0. 244 2 |

Continued

| Number | Institutes | Awards | Disciplines | Absolute citation | Relative citation | Absolute download | Relative downloading | Journal index | Digest index | Social echo | Synthetical score |
|---|---|---|---|---|---|---|---|---|---|---|---|
| 16 | Rural Development Institute | 3rd prize | Agricultural economy | 0. 283 2 | 0. 298 8 | 0. 298 8 | 0. 535 8 | 0. 223 8 | 0 | 0. 162 8 | 0. 238 9 |
| 17 | Institute of Philosophy | 3rd prize | Archeology | 0. 048 7 | 0. 121 3 | 0. 105 5 | 0. 230 2 | 0. 565 4 | 0. 666 7 | 0. 046 9 | 0. 216 4 |
| 18 | Institute of Sociology | 3rd prize | Economic theory and history of economic thought; philosophy | 0. 084 1 | 0. 098 7 | 0. 339 2 | 0. 296 6 | 0. 565 4 | 0. 666 7 | 0. 013 3 | 0. 214 5 |
| 19 | Central Department | 3rd prize | Chinese language | 0. 137 2 | 0. 122 1 | 0. 493 5 | 0. 363 1 | 0. 117 8 | 0. 666 7 | 0. 004 9 | 0. 214 0 |
| 20 | Institute of Ethnology and Anthropology | 3rd prize | Administration and state administration | 0. 137 2 | 0. 128 7 | 0. 148 4 | 0. 163 2 | 0. 075 0 | 0. 666 7 | 0. 118 7 | 0. 211 3 |
| 21 | History Institute | 3rd prize | Chinese ancient history | 0. 066 4 | 0. 169 0 | 0. 151 7 | 0. 259 7 | 0. 121 7 | 0. 500 0 | 0 | 0. 187 4 |
| 22 | Institute of Foreign Literature | 3rd prize | World literature | 0. 066 4 | 0. 137 7 | 0. 136 1 | 0. 101 6 | 0. 039 4 | 0. 666 7 | 0. 007 5 | 0. 186 9 |
| 23 | Institute of Latin American Studies | 3rd prize | Economic reform | 0. 070 8 | 0. 085 0 | 0. 051 4 | 0. 089 2 | 0. 049 5 | 0. 666 7 | 0. 143 3 | 0. 180 2 |
| 24 | Institute of History Institute | 2nd prize | History of China | 0. 048 7 | 0. 108 5 | 0 | 0. 009 6 | 0. 035 9 | 0. 666 7 | 0. 039 5 | 0. 167 3 |
| 25 | Rural Development Institute | 3rd prize | Economic reform | 0. 013 3 | 0. 015 8 | 0. 195 3 | 0. 226 0 | 0. 259 3 | 0. 666 7 | 0. 047 4 | 0. 150 4 |
| 26 | Institute of Ethnology and Anthropology | 2nd prize | Chinese politics and international politics | 0. 097 3 | 0. 147 7 | 0. 499 3 | 0. 476 4 | 0. 565 4 | 0 | 0. 005 7 | 0. 148 0 |
| 27 | Institute of Literature | 3rd prize | Chinese literature | 0. 057 5 | 0. 181 0 | 0. 314 5 | 0. 314 1 | 0. 018 0 | 0 | 0. 143 3 | 0. 138 1 |
| 28 | Institute of Modern History | 3rd prize | Chinese modern history | 0. 022 1 | 0. 049 8 | 0. 234 4 | 0. 341 3 | 0. 121 7 | 0. 333 3 | 0. 042 5 | 0. 114 5 |

Continued

| Number | Institutes | Awards | Disciplines | Absolute citation | Relative citation | Absolute download | Relative downloading | Journal index | Digest index | Social echo | Synthetical score |
|---|---|---|---|---|---|---|---|---|---|---|---|
| 29 | Institute of Western Asia and Africa Studies | 3rd prize | Chinese politics and international politics | 0.004 4 | 0.006 1 | 0.016 9 | 0.023 4 | 0.005 6 | 0.666 7 | 0.001 2 | 0.109 1 |
| 30 | World Economy and Politics Institute | 3rd prize | Finance | 0.031 0 | 0.034 5 | 0.151 0 | 0.150 4 | 0.347 4 | 0 | 0.080 5 | 0.061 4 |
| 31 | History Institute | 3rd prize | Travel; archeology | 0.031 0 | 0.046 6 | 0.112 0 | 0.119 2 | 0.121 7 | 0 | 0.096 9 | 0.055 0 |
| 32 | Institute of Finance and Trade Economics | 3rd prize | Information economy and post economy | 0.013 3 | 0.024 9 | 0.059 2 | 0.155 8 | 0.259 3 | 0 | 0.003 2 | 0.036 7 |
| 33 | Institute of Foreign Literature | 3rd prize | World literature | 0.017 7 | 0.036 7 | 0.054 7 | 0.035 2 | 0.039 4 | 0 | 0.061 6 | 0.033 5 |
| 34 | Institute of Foreign Literature | 3rd prize | World literature | 0.013 3 | 0.027 5 | 0.085 3 | 0.060 2 | 0.039 4 | 0 | 0.049 8 | 0.029 3 |
| 35 | Institute of Media Research | 3rd prize | Chinese literature | 0.008 8 | 0.025 6 | 0.076 2 | 0.103 2 | 0.100 5 | 0 | 0.004 1 | 0.026 3 |
| 36 | Institute of Ethnology and Anthropology | 3rd prize | Religion | 0.013 3 | 0.044 7 | 0.043 0 | 0.047 1 | 0 | 0 | 0.000 1 | 0.025 9 |
| 37 | Institute of Western Asia and Africa studies | 3rd prize | Chinese politics and international politics | 0.017 7 | 0.026 9 | 0.050 1 | 0.043 9 | 0.005 6 | 0 | 0.001 8 | 0.018 6 |
| 38 | Institute of Russian, Eastern European, Central Asia studies | 3rd prize | Chinese politics and international politics | 0.013 3 | 0.020 1 | 0.022 8 | 0.017 6 | 0.080 5 | 0 | 0.002 1 | 0.016 7 |
| 39 | Institute of Western Asia and Africa Studies | 3rd prize | Chinese politics and international politics | 0.013 3 | 0.020 1 | 0.004 6 | 0 | 0.005 6 | 0 | 0.035 5 | 0.016 4 |
| 40 | Institute of Archaeology | 3rd prize | Archeology | 0.004 4 | 0.013 2 | 0.028 6 | 0.080 1 | 0.066 4 | 0 | 0.006 0 | 0.015 9 |
| 41 | Institute of Archaeology | 3rd prize | Travel; archeology | 0.008 8 | 0.013 3 | 0.052 7 | 0.055 8 | 0.020 2 | 0 | 0.021 1 | 0.015 7 |
| 42 | Economy Institute | 3rd prize | Chinese modern history | 0.004 4 | 0.010 0 | 0.023 4 | 0.044 0 | 0.017 9 | 0 | 0.011 5 | 0.010 7 |
| 43 | Institute of Archaeology | 3rd prize | Archeology | 0 | 0 | 0.009 8 | 0.047 1 | 0.066 4 | 0 | 0.004 7 | 0.006 9 |

Notes: The evaluation vesult above used only for research and the titles of the papers are represented by number here.

(3) After sorting the papers by author's institutes, the ranking of papers in each institute is consistent with the awards in general, but inconformity exists in several institutes. For example, paper rankings in Central Department, Economy Institute, and Financial Institute are consistent with the paper awards, but not for World Economy and Politics Institute, Institute of Ethnology and Anthropology, and History Institute in which a second prize paper ranks behind the third prize paper respectively. Moreover, there are of great difference in the level and number of paper awards between the institutes. It is mainly because of the distribution in total number of papers in each institute, and the differences between the disciplines. These award-winning papers were selected by experts from all the papers recommended by the institutes subordinate to Chinese Academy of Social Sciences. All institutions have an equal chance to get an award. As the award-winning papers are classified according to the classification in CNKI, the papers in a discipline may be written by authors in different institutes, and the papers written by authors in an institute may be in different disciplines.

Thus, the synthetically evaluation ranking of the papers is in accordance with the paper awards in general. Inevitably, the inconsistencies exited in the meantime as a result of three main reasons: First of all, the differences between the disciplines are not eliminated completely; Further research on calculation methods of DACRP and DADRP is needed. Second, the score of social echo is calculated according to the retrieval results of search engine which lacks stability, miss and noise will affect the accuracy of the search results. Finally, the papers' awards arrangement may be influenced by some artificial factors or others, such as the academic status of the authors or total number of papers in each institute. Generally speaking, the evaluation index system constructed in this paper is rational and operational, and can provide a reference for paper prize awarding activities, but still needs to be verified and improved continuously through practice.

## 5 Conclusions

In this paper, the evaluation index system is constructed and verified, and the core paper zone, DACRP and DADRP are mainly discussed. We have obtained a series of meaningful research results and findings.

This research still leaves much to be desired. A main defect is that the size of the sample selected is not large enough. Larger sample would help to improve the evaluation index system to be more reasonable and credible, and can get some new scientific findings. Secondly, similar to Relative Citation Index and Downloaded Index, there are differences among disciplines on journal index, digest index and social echo because of the characteristics of disciplines. Relative Journal Impact Factor (RJIF), which takes the average journal impact factor in each discipline, can be adopted to determine the value of journal index instead of Absolute Journal Impact Factor (AJIF). And it still needs further study to reveal the dynamics of the different disciplinary communities in humanities and social sciences and explore more valuable indexes.

In addition, the evaluation system should be more open and flexible. In the modern international and network era, the criteria for evaluating researches in humanities and social sciences increasingly use some experiences from natural sciences for reference, such as standardization and citation index. But some criteria in humanities and social sciences, such as historical origins and the social value, still deserve our attention.

## References


Dalkey N, Helmer O. 1963. An experimental application of the Delphi method to the use of experts. *Management Science*, 9 (3): 458-467.

Finkenstaedt T. 1990. Measuring research performance in the humanities. *Scientometrics*, 19 (5/6): 409-417.

Forman E H, Gass S I. 2001. The analytical hierarchy process—an exposition. *Operations Research*, 49 (4): 469-486.

Glänzel W, Thijs B, Schubert A, et al. 2009. Subfield-specific normalized relative indicators and a new generation of relational charts: methodological foundations illustrated on the assessment of institutional research performance. *Scientometrics*, 78 (1): 165-188.

Hanney S, Frame I, Grant J, et al. 2005. Using categorizations of citations when assessing the outcomes from health research. *Scientometrics*, 65 (3): 357-379.

Jiang X, Yin G, Mo Z. 2009. *A Guide to the Humanities & Social Science Core Journals in China* (2008 *Edition*). Beijing: Social Sciences Academic Press.

Landeta J. 2006. Current validity of the Delphi method in social sciences. *Technological Forecasting and Social Change*, 73 (5): 467-482.

Moed H F, Luwel M, Nederhof A J. 2002. Towards research performance in the humanities. *Library Trends*, 50 (3): 498-520.

Norris M, Oppenheim C. 2003. Citation counts and the research assessment exercise V archaeology and the 2001 RAE. *Journal of Documentation*, 59 (6): 709-730.

Saaty T L. 1977. A scaling method for priorities in hierarchical structures. *Journal of Mathematical Psychology*, 15 (3): 234-281.

Saaty T L. 2008. Relative measurement and its generalization in decision making：why pairwise comparisons are central in mathematics for the measurement of intangible factors—the Analytic Hierarchy/Network Process. *Review of the Royal Spanish Academy of Sciences*，*Series A*，*Mathematics*，102（2）：251-318.

Skulmoski G J，Hartman F T，Krahn J. 2007. The Delphi method for graduate research. *The Journal of Information Technology Education*，6：1-21.

Vidal L A，Marle F，Bocquet J C. 2011. Using a Delphi process and the Analytic Hierarchy Process（AHP）to evaluate the complexity of projects. *Expert Systems with Applications*，38（5）：5388-5405.

Zhang L，Zhao H，Li Q，et al. 2010. Establishment of paper assessment system based on academic disciplinary benchmarks. *Scientomerics*，84：421-429.

# 2-5 Emerging Topic Detection Based on LDA Combined

Ma Jianxia[①], Fan Yunman[①]

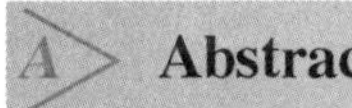

## Abstract

According to the study of the features of the emerging topic, we proposed a set of the emerging topic feature indices. We employ novelty index (NI), and published volume index (PVI) put forward by Tu Yining and Seng Jia lang, and propose a new index, cited volume index (CVI) to characterize the emerging topic. Then we proposed a met hod to identify the features of the emerging topic based on the LDA (Latent Dirichlet Allocation) model. The first step is to extract the topical words of the documents using the LDA model, the next is to build the mapping from topics to documents using the document-topic matrix, and then to visualize the life span of an emerging topic especially with novelty index (NI), published volume index (PVI), cited volume index and the detection point to characterize the emerging topics. According to the method a toolkit is developed to carry out emerging topic detection. With this method and tool, we carried out an experiment on the corpus covering "machine learning" downloaded from the Web of Science to prove after adding the time dimension into the indicators, we detect emerging topics from the corpus, and we depict the features of the development of topics in the period of the born, potential emerging, emerging in the topic life cycle. We verify utilizing LDA to extract topics can avoid the semantic ambiguity of frequency of words. We find combined cited volume index and with novelty index and published volume index, and we can detect the emerging topic earlier. And we analyze the effectiveness and validity of the indices and method we supposed.

① Lanzhou Branch of the National Science Library, CAS/Scientific Information Center for Resources and Environment, CAS (China).

**Keywords**: LDA (Latent Dirichlet Allocation); topic model; emerging topic feature; novelty index; published volume index; cited volume index

## 1 Introduction

Research frontier (research front) is the major key issues which represent trend of the discipline, and it's the fields that can have significant impact and play a leading role in the discipline. Research fronts include emerging topics which can be the solution of scientific problem in the future, and also include some hot topics which absorb much attention to carry out many exploration and have gotten a great breakthrough.

Tracking the evolution of a discipline and detecting theemerging topic is important for researchers and scholars to stay one step ahead in their research. Research topic goes through a life cycle of birth, growth, decay, and death, which is reflective of its popularity over time. Research frontier and emerging trend detection is dedicated in finding the newly developing topics and the topics which have been in early stage of steadily developing as soon as possible. However, before a new research topic can be identified, sometimes years can pass. Human experts who are tasked with identifying emerging trend need to rely on automated systems as the amount of information available in digital form increase.

An emerging trend detection application takesa collection of textual data as input and identifies topic areas that are either novel or are growing in importance within corpus. For detecting emerging topics, there are two important steps. The first is how to mine massive amounts of documents to identify topics of the documents, the second is how to determine which topic is emerging, i. e. , is worthwhile to continue to study.

This articlestudied the features of the emerging topic. We employed novelty index (NI), and published volume index (PVI) put forward by Tu Yining and Seng Jialang, and propose a new index, cited volume index (CVI) to characterize the topics in their growth stages to detect the emerging topics. Then we proposed a method to identify the features of the emerging topic. According to topic extraction, we use LDA (Latent Dirichlet Allocation) topic model rather than traditional keywords extraction according to frequency. Then we use curves of NI, PVI, and CVI, and form the detection period for topics according to the developing year. And we form the detection table for topics to detect the emerging topics. After that we make an experiment on documents searched from Web of Science related to machine learning to analyze the emerging topic and evaluate our method.

The remainder of this paper is organized as follows. Section 2 discusses therelated research of the study. Section 3 proposes indices of the emerging topic based on LDA combined with citation. Section 4 describes the method how to identify the features of the emerging topic based on the LDA model and introduce the tool we develop for emerging topic detection. Section 5 presents the experimental design, execution and experimental results. Section 6 discusses contributions of the study and presents the future research plan.

# 2 Related Research

## 2.1 Research Fronts and Emerging Trends

The concept of research frontin bibliometrics was proposed by Price in 1965 to describe the dynamic nature of the research field (Price, 1965). For more than 40 years, definition and comprehension that scholars made to concept and connotation of the research front can be roughly divided into three classes:

(1) a group of highly cited literatures (Price 1965; Small, 1973);

(2) a group of citing literatures (Persson, 1994; Morris et al., 2003);

(3) emerging and hot topics.

Kleinberg (2003) proposed a bursty detecting algorithm considering the changing density of word frequency to recognize the words with high concentration and high density, i. e., the emergent words. Morinaga and Yamanishi (2004) improved Kleinberg's approach and proposed a new topic analysis framework. Chen (2006) used Kleinberg's bursty detecting algorithm to achieve research front words. However, CiteSpace Ⅱ could detect a topic while it was still emerging but not before it emerged. Scholars, who define the research front as emerging and hot topics, mostly use methods of the word analysis such as frequency analysis and co-word analysis.

## 2.2 Emerging Trend Detection

Kontostathis et al. (2003) pointed out "An emerging trend is a topic area that is growing in interest and utility over time". Based on Kontostathis' definition, Hoang (2006) proposed a model for emerging trend detection, that is taking a scientific corpus as the input, after the topic extraction, representation, and identification procedure, in topic verification steps two functions for interest and utility were used to evaluate if the output is a set of emerging trends. Six steps are suggested by Glänzel

(2012) for the analyses of emerging topics in scientometric, i. e. , structural analysis of the discipline, dynamic analysis of the discipline, identification of emerging topics, delineation of the topic (optional), network analysis of the topic (optional), bibliometric study of the topic. Zhang and Gao (2011) pointed out, emerging trend detection is divided into 3 phases, the presentation of topics, identification and extraction of features of topics, and verification and judgement of topics. The efficiency of detection of emerging trends depends on these steps.

According to the features of emerging trends, Upham and Small (2010) analyzed the features of emerging trends. By associating each topic with its weight of mentioning, citation information, influence, author reputations and weight of sources Hoang (2006) computed the interest and utility more reasonably. Yin (2008) proposed the integrated index of emerging trends, considering national policy, support of fund, patent, and website of research organizations, but it depends on human beings and there is no practical application and verification in computer-assisted integrated solution by now. Tu and Seng (2011) proposed two indices of NI and PVI to detect emerging topics, and carried out practical detection in sensor network, semantic web and support vector machines. In this study we employ indices of Tu and Seng (2011) and made improvement by introducing cited volume index.

## 2.3 Identification of Topics

Normally, identification of topics is the basis of emerging trend detection. Bun andIshizuka (2001) use TF × PDF (Term Frequency × Proportional Document Frequency) in their emerging topic tracking system. Liu et al. (2010) extracted topics with TF × IDF in their emerging trend detection. This method extracts words according to word frequency and syntax structure rather than the semantics of the topics. Hofmann (1999) proposed PLSI ( Probabilistic Latent Semantic Indexing) as the first topic model. Blei et al. (2003) proposed Latent Dirichlet Allocation (LDA), which is a generative model based on statistical foundations and which also tends to produce clearly interpretable output in a more robust manner. Several related, augmented variants of LDA have been developed in recent years—including the author-topic model, for example, dynamic topic model (David and John, 2006), continuous time dynamic topic model (Wang et al. , 2008) , topic over time model (Wang and Mccallum, 2006), correlated topic model (Blei and Lafferty, 2007), bigram topic model (Wallach, 2006), author-topic model (Rosen-Zvi et al. , 2004), topic-author model (Wang, 2011). These and other methods built on the foundations of LDA have come to be known as "topic models".

One key strength of LDA over common document clustering methods based on simple mixtures fit by K-means is that LDA explicitly represents each document as being generated from a document-specific mixture of multiple topics (a multinomial over words), rather than a single topic. This flexibility tends to lead to more clear and interpretable topics. The models have generated significant interest because they create fine-grained, immediately interpretable topics that are robust against synonymy and polysemy. In addition, generating topics and document topic assignments requires virtually no human supervision (Mccallum et al. , 2006).

He and Li (2012) extracted the topics of scientific documents with topic models, and after computing the intensity and impact of topics, they carried out research trend analysis. Hall et al. (2008) map the changing history of ideas using topic models. Stanford Topic Modeling Toolbox can extract topics of documents, and according to the map of documents and topics, time slices, the theme of intensity change trend is formed.

Shibata et al. (2011) analyzed the topology of the citation networks which they collected. They investigated the factors which determine the capability of academic articles to be cited in the future using a topological analysis of citation networks. Their works suggested that the cited times is the main factor in explaining the near future cited times, and betweenness centrality is correlated with the distant future cited times. Dietz et al. (2007), He et al. (2009) think citation has effect on topic evolution. Nallapati et al. (2008) proposed a model to combine cited article with citing article as a couple, and analyzed the couples with topic model. Mccallum et al. (2006) proposed a set of bibliometric indexes of topic model combining topic model with co-citation. Although these studies took citation papers into the procedure of topic identification, they didn't introduce citation number into the prediction of topic evolution. All these studies show citations can help to understand topic evolution.

# 3 Indices of Emerging Topic Based on LDA

## 3.1 The Procedure of Extracting Topics from Document with LDA

LDA posits that each document is generated as a mixture of topics where the continuous-valued mixture proportions are distributed as a latent Dirichlet random variable, and every topic is composed by a group of words with different probability. Wang and McCallum (2005) proposed TNG (Topicical N-grams) as an extended model of LDA. Mann et al. (2006) testified rather than viewing a topic as a list of single words, TNG can present its user with multi-word phrases, which, especially in research domains, are much more interpretable. So in this study, we use TNG to extract topics from documents.

Table 1 shows the procedure of extracting topics from documents with LDA.

**Table 1 Procedure of extracting topics from documents with LDA**

| Steps | Procedure of extracting topics from document with LDA |
|---|---|
| Step 1 | Get the length of the document, N-Poission ($\beta$) |
| Step 2 | Get the distribution of the document on the topic, $\theta_m$-Dir ($\alpha$); |
| Step 3 | for $n=1$ to $N$ |
| Step 4 | (a) extract a topic $Z_{mn}$-Multinomial ($\theta_m$) |
| Step 5 | (b) get the words and multi-word phrases $w_{mn}$-Multinomial ($\varphi_{z\,mn}$) |

$\alpha$ and $\beta$ are anticipated parameters of LDA, $\theta$ and $\varphi$ are estimated parameters. In this study, we use Gibb sampling algorithm (Griffiths, 2002) to get the estimated parameters (Table 2).

$$\theta_j^i = \frac{n^{(d_i)} + \alpha}{n_{-i,j}^{(d_i)} + K_\alpha}, \quad \varphi_j^{wj} = \frac{n_{-ij}^{(w_i)} + \beta}{n_{-i,j}^{(.)} + W_\beta} \tag{1}$$

**Table 2 The symbols in the article**

| Symbols | Meaning of the symbols |
|---|---|
| $N$ | Length of the document $d$ |
| $\alpha$ | Anticipated parameters of the distribution of documents to topics |
| $\beta$ | Anticipated parameters of the distribution of topics to words |
| $\theta_{dz}$ | Probability of document $d$ isgenerated by topic $z$ |
| $\varphi_{z_{dn}}$ | Probability of word $w_{dn}$ represents topic $z$ in document $d$ |
| $W$ | Size of the dictionary |
| $K$ | Number of topics |
| $n_{-i,j}^{(d_i)}$ | Number of words represents topic $j$ in document $d_i$, not including current words |
| $n_{-i,\cdot}^{(d_i)}$ | Sum of words in document $d_i$, not including current words |
| $D_z^t$ | Number of supporting documents for topic $z$ in the period of $t$ |
| $C_z^t$ | Sum of cited numbers of supporting document for topic $z$ in the period of $t$ |
| $Sum_d(k)$ | Accumulated value of supporting documents of topic $z$ in the $k$-th year |
| $Sum_c(k)$ | Accumulated value of cited numbers of supporting documents of topic $Z$ in the $k$-th year |

## 3.2 Supporting Document of a Topic

According to the extraction results of topics in many documents, if the probability of a document $d$ generated by topic $z$ is more than or equal to 10%, then the document $d$ is the supporting document of topic $z$. One document can be the supporting document of several topics. Supporting document is the bridge for turning the computing of topics to the computing of documents.

Table 3 shows some parameters, $D_z^t$, $C_z^t$, $Sum_d(k)$ and $Sum_c(k)$ of supporting documents in topic 52 of our corpus from 1999 to 2005 ( $\theta_{dz} \geqslant 0.1$ )

Topic 52 is the topic we detected in the experiment, it represents "topic model" .

**Table 3 Some parameters of topic 52**

| Symbols | 1999 | 2000 | 2001 | 2002 | 2003 | 2004 | 2005 |
|---|---|---|---|---|---|---|---|
| $D_z^t$ | 0 | 2 | 2 | 13 | 12 | 1 | 5 |
| $C_z^t$ | 0 | 54 | 463 | 82 | 1 835 | 40 | 15 |
| $Sum_d(k)$ | — | 2 | 4 | 17 | 29 | 30 | 35 |
| $Sum_c(k)$ | — | 54 | 517 | 599 | 2 434 | 2 474 | 2 489 |

Tu and Seng (2012) supposed the novelty and volume of papers published on a topic are important indices for determining it has potential to become a hot new emerging topic. The novelty index (NI) and published volume index (PVI) are developed and utilized to identify the detection point (DP) of a topic. The DP in a topic's life cycle will move as later papers continue to be published. The moving DPs form a detection period which represents if the topic is emerging and valuable.

In our opinion, citation status also reflects the evolution of the topic during the topic life cycle. To some extent, it stands for the community attention. We employed novelty index (NI), published volume index (PVI) according to Tu Yining's study, and proposed a new index, cited volume index. After utilizing topic models to extract topics of documents, we investigate these bibliometric indexes of selected topics during the growth stage of their life cycle. An emerging topic is defined as a research topic that is important and in the growth stage but still not a hot topic. According to some criteria we defined for the emerging topic, we characterize the evolution of the topics and find the emerging topic.

### 3.2.1 Novelty index

Whether a topic has potential is partly based on its novelty. Generally, as the number of times a topic is accessed increases, its potential value increases. A valuable topic attracts the attention of many researchers. Novel topics have considerable content that has not yet discussed. Novel topics have not yet completely developed and there is large space in the real world for applications.

Arrange the number of supporting documents of the topic by the year, from the latest year, check if the number of supporting documents of the topic is 0. If it's 0, then the year after the year with zero supporting documents is the first year (FY) of the topic, potential development year (PDY) is defined as the period from the first year to the current year when a topic becomes a research topic that does not include any year with zero supporting documents in the following years.

The novelty index of topic $z$ in year $k$ is (Tu and Seng, 2012):

$$NI_k = \frac{1}{k - \mathrm{FY} + 1} \tag{2}$$

Assuming that a research topic is new when it is first published, then, NI should be normalized to 1 ( = 100%). In its second year, the NI should be 1/2 ( = 50%). The value of the NI should be normalized to 0-1.

According to Table 3, We can get FY = 2000, $\mathrm{NI}_{2000} = 1/(2000-2000+1) = 1$, $\mathrm{NI}_{2001} = 1/(2001-2000+1) = 0.5$, $\mathrm{NI}_{2002} = 0.33$, $\mathrm{NI}_{2003} = 0.25$, $\mathrm{NI}_{2004} = 0.20$, $\mathrm{NI}_{2005} = 0.17$. The NI is a measurement of novelty. Normally after a topic emerges, because more and more articles are published on this topic, the novelty index of the topic will decrease as time goes on.

### 3.2.2 Published volume index

$$\mathrm{PVI}_k^z = \frac{\mathrm{Sum}_d(k)}{\mathrm{Sum}_d(z)} \tag{3}$$

$\mathrm{Sum}_d(z)$ is the accumulated number of articles on topic $z$ published from FY to current year. $\mathrm{Sum}_d(k)$ is the accumulated number of articles on topic $z$ published from FY to the $k$ year. The PVI is defined as the accumulative relative frequency of the $k$ th development year normalized to 0-1.

We not only consider the volume of published papers but also a topic's hotness over time. The accumulative relative frequency curve refects the variation in published volume. This method is better than the frequency of a topic at detecting topic status based on comparing the growth situation without waiting until it really becomes mature. If the focus is solely on the volume of published papers, one cannot determine whether the topic has potential research value. Notably, as the volume of papers increases, topic impact decreases. The conventional frequency curve method for determining the volume of published papers lacks the ability to determine whether a topic is hot or emerging.

$\mathrm{Sum}_d(k)$ in Table 3 shows the PVI on topic 52 from 2000 to 2005. Then we can get the PVI every year:

$\mathrm{PVI}_{2000} = 2/35 = 0.06$, $\mathrm{PVI}_{2001} = 4/35 = 0.11$, $\mathrm{PVI}_{2002} = 0.49$, $\mathrm{PVI}_{2003} = 0.83$, $\mathrm{PVI}_{2004} = 0.86$, $\mathrm{PVI}_{2005} = 1$.

### 3.2.3 Cited times

Cited times (CT) of topic $z$ is the accumulated number of cited times of supporting documents of topic $z$ in time slice $t$.

### 3.2.4 Cited Volume index

We think the cited number of articles on the topic also represents the attention on the topic in the scientific community. It varies during the evolution of the topics.

$$\mathrm{CVI}_k^z = \frac{\mathrm{Sum}_c(k)}{\mathrm{Sum}_c(z)} \tag{4}$$

$\mathrm{Sum}_c(z)$ is the accumulated number of cited times for supporting documents of topic $z$ from FY to current year. $Sum_c(k)$ is the accumulated number of cited times for supporting documents of topic $z$ from FY to the $k$ year. According to $\mathrm{Sum}_c(k)$ of topics in Table 3, we can get CVI for the topic from 2000 to 2005:

$\mathrm{CVI}_{2000} = 0.02$, $\mathrm{CVI}_{2001} = 0.21$, $\mathrm{CVI}_{2002} = 0.24$, $\mathrm{CVI}_{2003} = 0.98$, $\mathrm{CVI}_{2004} = 0.99$, $\mathrm{CVI}_{2005} = 1$.

### 3.2.5 Detection point

As time goes on, NI of a topic is decreasing gradually. Furthermore, with the increase of PVI and CVI, the topic is getting more and more mature. At the same time, the attention of the topic should decrease gruadually. As shown in Figure 1, NI is a declining curve with time, while PVI and CVI are going-up curves which are at their minimum at PY, and reach their maximum 1 at the current year. The intersection of NI with PVI or CVI is called detection point (DP). At these points the novelty and hotness of the topic are at their maximum.

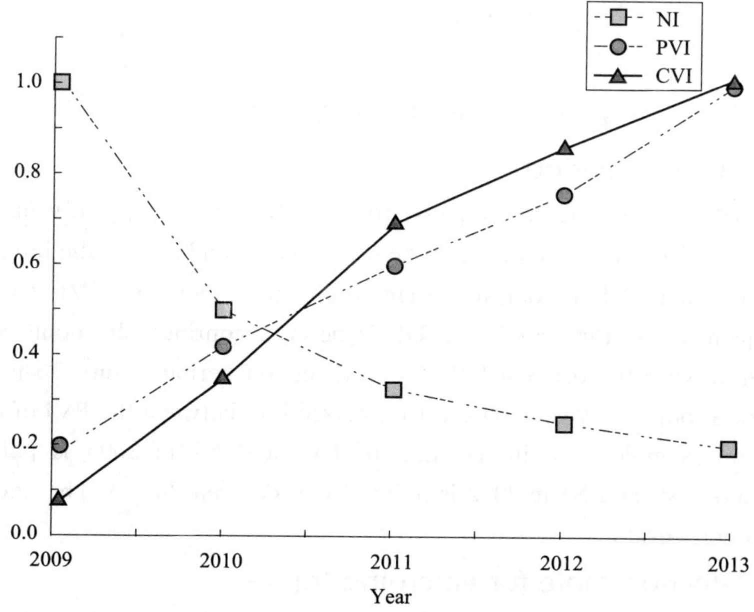


Figure 1 The curves of NI, PVI and CVI for topic "machine learning"

The NI, PVI and CVI are trade-off curves, that is, a new topic lacks the volume needed to be a hot topic, and when a hot topic exists for a period of time, it loses its novelty. Consequently, the maximal NI, PVI and CVI must be obtained from the intersection of curves, which is called the DP. We suggest that the DP can be used to determine whether a topic is hot and emerging, as it is the maximal value of 3 indices for novelty and hotness.

The X-axis of DP is the year of detection point (YDP). The Y-axis equals to the NI, PVI or CVI, which is called VDP. VDP of published volume index is $VDP_p$, VDP of cited volume index is $VDP_c$. The algorithm of YDP and VDP is shown below (Figure 2).

算法 3-1. 计算 YDP
Input A group of Ni, PVI, CVI of a topic
Output YDP. VDP
1. For $i$ = Current Year to FY
2. IF $PVI_i = NI_i$
3. YDP = $i$
   VDP = PVI
   Return;
4. ELSE IF $PVI_i > NI_i$ and $PVI_i < NI_{i-1}$
5. $YDP = \frac{2i-1}{2}$
   $VDP = \frac{PVI + PVI_{i-1} + NI + NI_{i-1}}{4}$
   Return;
6. End-IF
7. Next

Figure 2 Algorithm of YDP and VDP

### 3.2.6 Detection Period

For a topic $z$, as time goes on, NI forms a steadily curve, the value is $1/(k-FY+1)$. While PVI and CVI will form new curves $PVI'$ and $CVI'$, the intersection of $PVI'$ ($CVI'$) and NI is delayed than the intersection of PVI (CVI) and NI. Consequently, as time goes on, DP is delayed continuously along NI, which forms a period with the original DP, i. e., detection period. Figure 3 shows the NI and PVI for a topic every year. There are intersections between the PVI of every year with NI. For example, the intersection of PVI and NI in 2003 is point A. The intersection of PVI and NI in 2012 is point C. C is delayed than A. The line $\overline{AB}+\overline{BC}$ is the detection period.

### 3.2.7 Detection table for emerging topics

Computing VDP on a topic yearly during the detection period, and then filling all these data in a table, generates the detection table. By using the median VDP for each research topic generated by LDA of the collection, one can construct a baseline. When we consider published volume index, the baseline is $VDP_p$, when we consider cited volume index, the baseline is $VDP_c$. In this study, we combine the

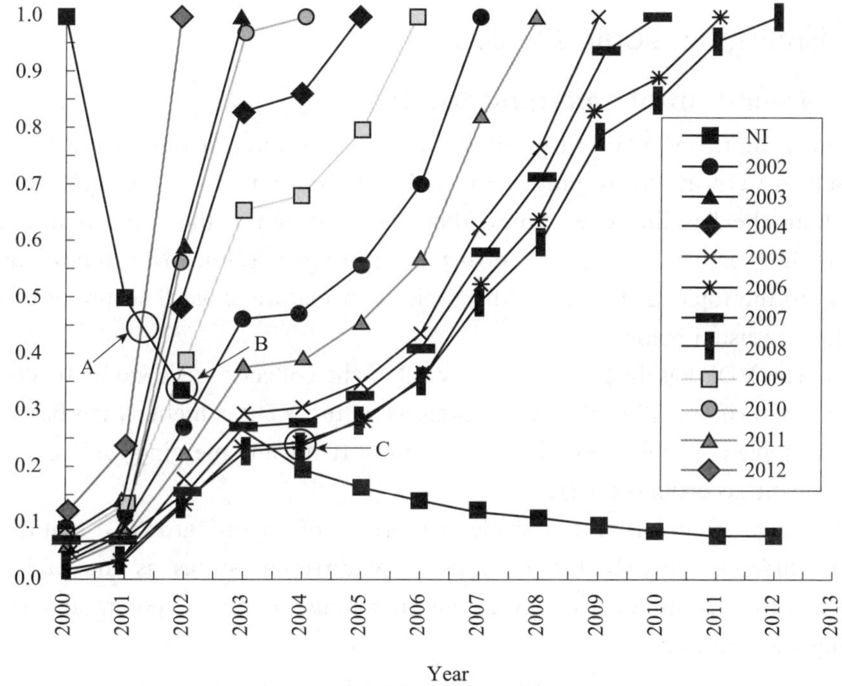


Figure 3 Examples of detection period

published volume with cited volume, so the baseline VDP is the arithmetic mean of linear superposition of $\text{VDP}_p$ and $\text{VDP}_c$. This is shown in formula 5.

$$\overline{\text{VDP}} = \lambda \times \text{VDP}_p + (1-\lambda)\text{VDP}_c, \quad \lambda \in [0,1] \tag{6}$$

In the experiment, we find the detection method combined cited volume with published volume is better than the method only considering published volume index.

Therefore, when the life cycle stage of a topic is unknown, the emerging topic detection table can be used to compare the topic with the baseline. When the VDP of a topic is higher than the baseline, it is worth keeping; when the VDP of a topic is lower than the baseline, the topic is worthless. Comparatively, if the VDP of a topic from the start to end of its life cycle is never higher than the baseline, it cannot be an important topic (Table 4) .

**Table 4 Examples of Detection Table**

| Year | 3rd year | 4th year | 5th year | 6th year | 7th year | 8th year |
|---|---|---|---|---|---|---|
| Baseline VDP | 0. 628 187 | 0. 515 402 | 0. 447 796 | 0. 402 806 | 0. 372 456 | 0. 350 284 |
| VDP of topic 1 | 0. 632 46 | 0. 516 419 | 0. 445 659 | 0. 397 516 | 0. 369 681 | 0. 346 77 |
| YDP of topic 1 | 2 000. 5 | 2 001. 5 | 2 001. 5 | 2 001. 5 | 2 001. 5 | 2 002. 5 |

## 3.3 Emerging Topic Detection

### 3.3.1 Definition of emerging topics

Comparing the VDP of a topic with the VDP baseline, if the VDP is lower than the baseline VDP at the beginning of its life cycle, and then the VDP is getting higher than the baseline over time, then, at that time, the topic is a emerging topic; if the VDP of a topic is getting lower than baseline over time, then the attention to the topic is decrease, the topic is in mature stage. The premises of the definition are listed below:

(1) The VDP baseline is the mean VDP of the collections we study. It represents the mean maturity of all topics in the datasets. If the VDP is higher than baseline, it means the topic is in the period of mature stage. If it's an emerging topic depends on the judgement according to (2).

(2) Each topic has a life cycle comprised of birth, growth, maturity and death. In different period, the features show different status as the table below (Table 5). In this study, we pay attention mainly to birth, potential emerging, emerging and mature.

**Table 5 Features of NI, PVI, CVI, VDP and baseline VDP on different phases of a topic**

| Item | NI ∈ [0, 1] | | PVI ∈ [0, 1] | | CVI ∈ [0, 1] | | Relationship of VDP and baseline VDP |
|---|---|---|---|---|---|---|---|
| | Value | Trend | Value | Trend | Value | Trend | |
| Birth | 1 | — | 0 | - | 0 | - | - |
| Potential emerging | ≈1 | ↓ | ≈0 | ↑ | ≈0 | ↑ | VDP is below baseline VDP |
| Emerging | Medium, (0, 1) | ↓ | Near to Medium, (0, 1) | ↑ | Near to Medium, (0, 1) | ↑ | VDP has intersected with baseline, and keep upward. Before the DP intersection, the topic is a latent emerging topic. After the DP intersection, the topic is a emerging topic |
| Mature | From near medium to 0 | ↓ | Near to 1 | steadily raise | Near to 1, (0,1) | steadily raise | After VDP is up to maximum, it starts down |

### 3. 3. 2 Judgement of emerging topics

An emerging topics is an important research topic which has the potential of developing, but it is still not a hot topic. For the researchers, to find potential emerging topics and emerging topics is very valuable. We pay much attention to the topics with low-up VDP shape, i. e. , the VDP is lower than the baseline VDP at the beginning of its life cycle, and then the VDP is getting higher than the baseline over time. As Table 5 shows, for those topics with low-up VDP shape, before the DP, the topic is a potential emerging topic, while after the DP, the topic is an emerging topic.

# 4 Method to Detect Emerging Topics

Based on the indices as mentioned in section 3. 3, we put forward a method to detect emerging topics based on LDA andemerging topic indices. There are 3 steps as shown in Figure 4. The first step is the data pre-processing. We download data from Web of Science, excluding the words frequency>900 or frequency<5. The second step is topic extraction. In this step, TNG topic model is chosen to get better semantic presentation. And the determination of the suitable topic number is also important. The third step is to detect the emerging topic by analyzing the indices through emerging topic detection table and the curves we form according to the table.

## 4. 1 The TNG Topic Model and the Toolkit

TNG topic model is an expansion of LDA, it presents topics with words and phrases so as to getting better semantic presentation. So, in this study, we use TNG topic model with the parameters $\alpha = 50$ , $\beta = 0.01$ , $\gamma = 0.01$ , $\delta = 0.03$ , $\delta_1 = 0.2$ , $\delta_2 = 1000$ .

## 4. 2 Choosing the Suitable Number of Topics for LDA

Because LDA is a kind of bag-of-words model, the conditional probability of word generating topics is conditional independence. According to perplexity proposed by David et al. (2003), when the complexity is the minimum, the number of topics (K) for LDA means the best representative of semantics. We estimate the number of topics in the corpus is about 200. We choose $K$ varying from 25 to 200 with the step of 25. Each $K$ is iterated 1, 000 times. The perplexity of each $K$ is shown in Figure 5. When $K$ is 25, 50, 75, the perplexity decreased rapidly. When $K$ is 75 the decrease is getting slower. In our study, we choose the number of topics as 75.

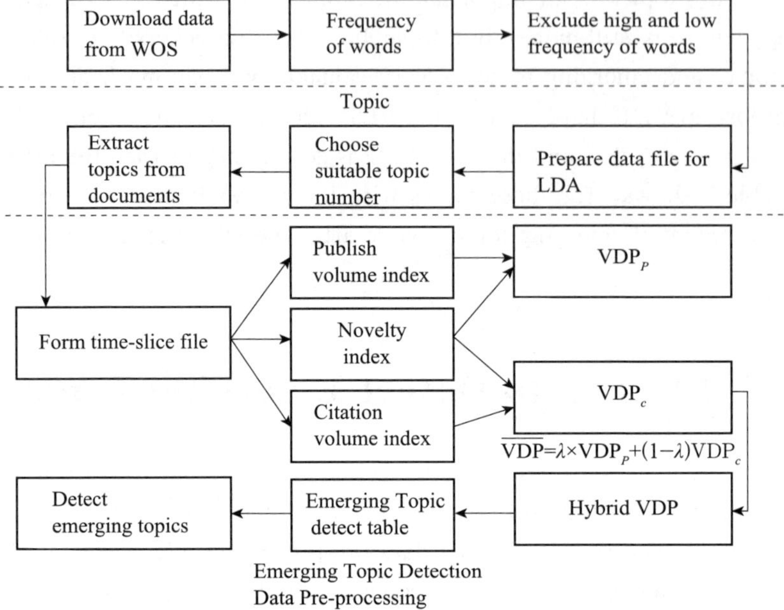


Figure 4 The method of emerging topic detection

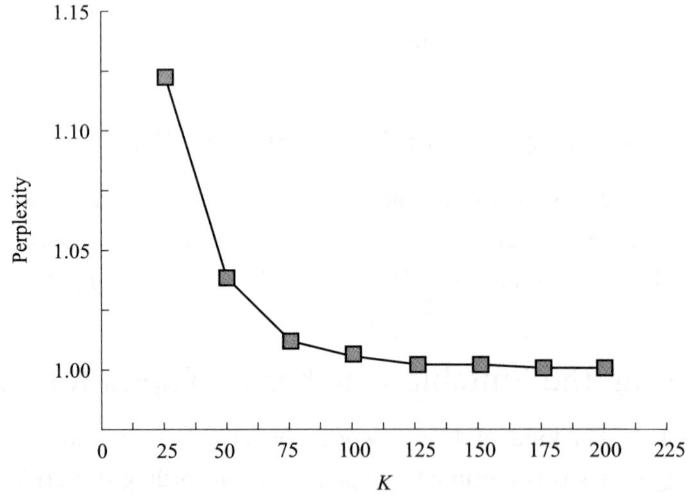


Figure 5 The curve for the number of the topics and the model perplexity

## 4.3 Procedures of Emerging Topic Detection

Utilizing MALLET with input of the collection of documents pre-processed, we get the document-topic matrix generated by topic model. On the basis of the reference files and the matrix we form the topic properties time slice file as Table 6.

**Table 6　The format of topic properties' time slice file**

| Number of topics | Number of documents | Published year | Cited times | Probability of topic generating documents |
|---|---|---|---|---|
| Topic 0 | 9 023 | 2000 | 2 | 0. 382 352 941 176 47 |
| Topic 0 | 5 952 | 2006 | 0 | 0. 370 370 370 370 37 |
| Topic 0 | 5 709 | 2006 | 12 | 0. 369 747 899 159 663 |

## 4. 4　Toolkits Developed for Emerging Topic Detection

We choose MALLET (Machine Learning for Language Toolkit), an open source machine learning toolkit, to extract topics. Based on MALLET, we develop a tool to emerging topic detection (Figure 6).



Figure 6　Function framework of the emerging detection tool

Figure 7 to Figure 9 show the mainfunction snapshot of the tool.

Figure 7　Topic model for the document

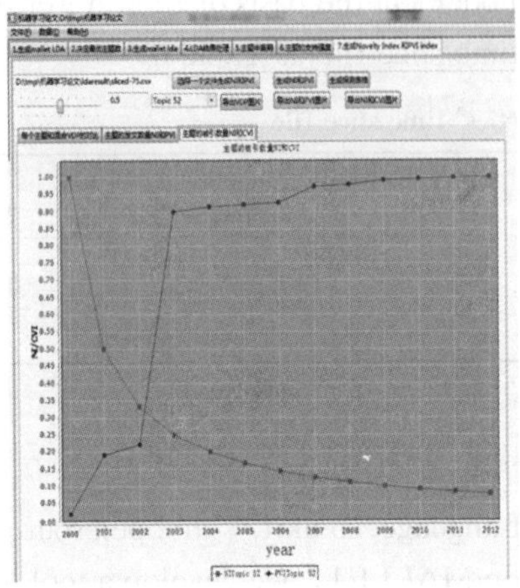


Figure 8 The curves for NI and CVI

Figure 9 Display of the VDP

# 5 Research Experiment

To verify the accuracy and effectiveness of the proposed indices and methods we proposed for emerging topic detection, this study uses research papers on machine learning searched and downloaded from Web of Sciences.

## 5.1 Research Data Preparation

The aim of the experiment is to verify the effectiveness and accuracy of the indices and methods we proposed for emerging topic detection, we choose "machine learning" as the research field. We choose 64 journals in WOS related to "machine learning" and 6 journals recommended by Wikipedia, the time period is from year 2000 to 2012, and we get 10, 156 articles as the test corpus.

We download the bibliographies from WOS, merge the title, abstract, author keywords, keywords plus as a document $d$ as input of LDA for an article, and get the article number, published year, cited times as $d$'s reference file as the next step for time slice.

## 5.2 Results of the Experiment

### 5.2.1 The emerging topics we detected in machine learning

According to the topics we extracted with LDA combined with verification indices, NI, CVI and PVI we choose, we get the topic properties time slice table including FY, number of supporting document, published volume, cited volume per year of each topic and form the novelty curve, published volume curve, cited volume curve, consequently, we get 15 emerging topics we detected in machine learning. i. e. , (1) neural network; (2) decision process; (3) fisher criterion application in data visualization; (4) support vector machine; (5) self organizing map; (6) regression model; (7) spatial data analysis; (8) function stability; (9) concept lattice in spatial and temporal reasoning; (10) topic model; (11) algorithm optimization; (12) high dimensional data analysis; (13) Bayesian inference; (14) voice recognition; (15) mobile agent.

### 5. 2. 2 Analysis on the number of topics detected by the topic model

In order to evaluate the effectiveness of indices and methods we propose, we compare the result in 3 groups, group 1 with indices NI and PVI, group 2 with indices NI and CVI, group 3 with indices NI, PVI and CVI with $\lambda = 0.5$ (Table 7).

**Table 7 Comparing table of emerging topics' detected results with different indices**

| Indices | Number of emerging topics detected | Emerging topics detected |
| --- | --- | --- |
| NI and PVI | 37 | Topic 0, Topic 2, Topic 5, Topic 7, Topic 10, Topic 11, Topic 13, Topic 19, Topic 22, Topic 23, Topic 24, Topic 25, Topic 27, Topic 30, Topic 32, Topic 33, Topic 34, Topic 35, Topic 38, Topic 39, Topic 40, Topic 46, Topic 49, Topic 50, Topic 52, Topic 53, Topic 56, Topic 57, Topic 58, Topic 60, Topic 62, Topic 63, Topic 67, Topic 68, Topic 69, Topic 72, Topic 73 |
| NI and CVI | 12 | Topic 5, Topic 11, Topic12 , Topic 21, Topic 23, Topic 26, Topic 28, Topic 32, Topic 52, Topic 58, Topic 60, Topic 67 |
| NI, PVI and CVI | 15 | Topic 5, Topic 11, Topic 12, Topic 21, Topic 23, Topic 26, Topic 28, Topic 32, Topic 47, Topic 52, Topic 57, Topic 58, Topic 60, Topic 67, Topic 72 |

As shown in Table 7, we find:

(1) The number of emerging topics detected is NI and PVI group>NI, and PVI and CVI group> NI and CVI group.

(2) Although NI and PVI group gets 37 topics, we find if only consider NI and PVI, as Figure 10, Topic 0, Topic 7 and Topic 10 only have one point above the baseline VDP, and these topics are not emerging topics. For CVI and NI group, the number of emerging topics detected is the least, and also there are some topics have only one point above the baseline VDP. NI, PVI and CVI group detected more topics than NI and CVI group, and there is only one topic has only one point above VDP (Figure 11, Figure 12).

| 第n年 | 3rd Year | 4rd Year | 5rd Year | 6rd Year | 7rd Year | 8rd Year | 9rd Yea | 10rd Ye | 11rd Ye | 12rd Ye | 13rd Ye |
|---|---|---|---|---|---|---|---|---|---|---|---|
| Topic 0 | 0.526515152 | 0.333333333 | 0.424404762 | 0.344812925 | 0.28125 | 0.2541667 | 0.2482759 | 0.221121 | 0.212184 | 0.206811 | 0.2 |
| Topic 2 | 0.558333333 | 0.526515152 | 0.5 | 0.477564103 | 0.325 | 0.3479487 | 0.3049242 | 0.25 | 0.2375 | 0.229847 | 0.2275 |
| Topic 5 | 0.540333333 | 0.379820937 | 0.364348371 | 0.350456621 | 0.37920389 | 0.3373792 | 0.3082227 | 0.280417 | 0.280124 | 0.27956 | 0.277848 |
| Topic 7 | 0.520833333 | 0.375 | 0.335451977 | 0.361458333 | 0.30123874 | 0.263984 | 0.2538043 | 0.218191 | 0.214862 | 0.20569 | 0.202465 |
| Topic 10 | 0.526515152 | 0.362745098 | 0.325 | 0.367324561 | 0.30169753 | 0.2546695 | 0.2482388 | 0.222835 | 0.217275 | 0.210031 | 0.206102 |
| Topic 11 | 0.543439716 | 0.421171171 | 0.364273927 | 0.335349462 | 0.32916667 | 0.2948481 | 0.2893027 | 0.259806 | 0.24837 | 0.233356 | 0.227632 |
| Topic 13 | 0.554347826 | 0.448059361 | 0.379901961 | 0.345052083 | 0.33532378 | 0.3049242 | 0.2752414 | 0.2625 | 0.248775 | 0.246702 | 0.233356 |
| Topic 19 | 0.491666667 | 0.365740741 | 0.341145833 | 0.323198198 | 0.35171569 | 0.3025498 | 0.2679264 | 0.243031 | 0.23645 | 0.2305 | 0.225095 |
| Topic 22 | 0.556547619 | 0.503787879 | 0.479166667 | 0.373587571 | 0.33333333 | 0.3020674 | 0.2844697 | 0.281994 | 0.276175 | 0.272917 | |
| Topic 23 | 0.526515152 | 0.348333333 | 0.342948718 | 0.329022989 | 0.37669256 | 0.3458333 | 0.2983757 | 0.272504 | 0.267456 | 0.252976 | 0.249282 |
| Topic 24 | 0.503787879 | 0.411458333 | 0.379385965 | 0.356060606 | 0.31317204 | 0.3194673 | 0.275641 | 0.254704 | 0.25 | 0.245076 | 0.243037 |
| Topic 25 | 0.510658915 | 0.432471264 | 0.408333333 | 0.318502825 | 0.2884887 | 0.2727178 | 0.2294202 | 0.217449 | 0.216035 | 0.21043 | 0.231975 |
| Topic 27 | 0.56127451 | 0.395833333 | 0.358333333 | 0.333333333 | 0.32598039 | 0.2909583 | 0.272122 | 0.255208 | 0.253289 | 0.245427 | 0.24229 |
| Topic 30 | 0.5 | 0.458333333 | 0.433192564 | 0.422619048 | 0.33333333 | 0.2809685 | 0.254529 | 0.220259 | 0.218432 | 0.214959 | 0.190931 |
| Topic 32 | 0.514784946 | 0.364071038 | 0.354487179 | 0.344047619 | 0.32275641 | 0.2896832 | 0.2650043 | 0.243706 | 0.242068 | 0.240875 | 0.25 |
| Topic 33 | 0.512681159 | 0.323087432 | 0.313596491 | 0.425833333 | 0.31867284 | 0.2766745 | 0.2560696 | 0.240641 | 0.237939 | 0.227134 | 0.235677 |
| Topic 34 | 0.484068627 | 0.324792961 | 0.310235507 | 0.409858218 | 0.30268265 | 0.2666139 | 0.2398399 | 0.23196 | 0.22874 | 0.221264 | 0.218069 |
| Topic 35 | 0.518939394 | 0.361318408 | 0.345 | 0.328921569 | 0.308171 | 0.2733844 | 0.253592 | 0.24053 | 0.25 | 0.241591 | 0.23974 |
| Topic 38 | 0.5 | 0.476190476 | 0.442708333 | 0.358333333 | 0.32765152 | 0.2908333 | 0.2549242 | 0.239619 | 0.235451 | 0.227885 | |
| Topic 39 | 0.552083333 | 0.4375 | 0.353070175 | 0.345833333 | 0.29289216 | 0.264881 | 0.25 | 0.253409 | 0.246121 | 0.233094 | 0.228172 |
| Topic 40 | 0.477564103 | 0.337962963 | 0.325 | 0.319444444 | 0.32921287 | 0.2757035 | 0.2527852 | 0.240303 | 0.239732 | 0.238051 | 0.235877 |
| Topic 46 | 0.528846154 | 0.339912281 | 0.429731658 | 0.370665549 | 0.31333333 | 0.2761835 | 0.2532051 | 0.235887 | 0.250453 | 0.244207 | 0.242696 |
| Topic 49 | 0.544871795 | 0.473484848 | 0.390625 | 0.323464912 | 0.26383333 | 0.2202128 | 0.2001623 | 0.215173 | 0.203177 | 0.195 | |
| Topic 50 | 0.553571429 | 0.462878788 | 0.390151515 | 0.385648523 | 0.37302922 | 0.3692529 | | | | | |
| Topic 52 | 0.517156863 | 0.389367816 | 0.383333333 | 0.358333333 | 0.32765152 | 0.328373 | 0.295144 | 0.259695 | 0.252315 | 0.234461 | 0.229563 |
| Topic 53 | 0.563333333 | 0.537037037 | 0.49931694 | 0.362681159 | 0.31357527 | 0.2758333 | 0.2270418 | 0.225689 | 0.223934 | 0.222653 | |
| Topic 56 | 0.494047619 | 0.341666667 | 0.322619048 | 0.390277778 | 0.30297619 | 0.2641129 | 0.2476852 | 0.235606 | 0.226937 | 0.222297 | 0.217339 |
| Topic 57 | 0.476190476 | 0.337643678 | 0.329301075 | 0.321969697 | 0.3545712 | 0.3033425 | 0.2789603 | 0.258399 | 0.256608 | 0.254634 | 0.253603 |
| Topic 58 | 0.5 | 0.481060606 | 0.427849523 | 0.348958333 | 0.31348039 | 0.2719395 | 0.25 | 0.245594 | 0.235017 | 0.225997 | |
| Topic 60 | 0.545454545 | 0.40719697 | 0.367424242 | 0.347052846 | 0.30759804 | 0.2799797 | 0.2561049 | 0.251545 | 0.238135 | 0.217819 | 0.2115 |
| Topic 62 | 0.525406504 | 0.363095238 | 0.32982866 | 0.354418367 | 0.2939376 | 0.261147 | 0.2399598 | 0.228398 | 0.225652 | 0.217676 | 0.213451 |
| Topic 63 | 0.558333333 | 0.41025641 | 0.389367816 | 0.362745098 | 0.36391844 | 0.3225575 | 0.2847468 | 0.251503 | 0.246324 | 0.239815 | 0.232065 |
| Topic 67 | 0.536458333 | 0.350225225 | 0.333333333 | 0.416241497 | 0.36304615 | 0.3179513 | 0.2867903 | 0.273237 | 0.268665 | 0.265293 | 0.258121 |
| Topic 68 | 0.525641026 | 0.370098039 | 0.341397849 | 0.324530516 | 0.33455882 | 0.3010753 | 0.276592 | 0.260417 | 0.25459 | 0.250453 | 0.247688 |
| Topic 69 | 0.517857143 | 0.356060606 | 0.330974843 | 0.392045455 | 0.3304924 | 0.2922297 | 0.2653186 | 0.241422 | 0.246476 | 0.231958 | 0.223739 |
| Topic 72 | 0.56284153 | 0.532521303 | 0.500563063 | 0.389295676 | 0.3469551 | 0.3109808 | 0.2789231 | 0.278095 | 0.274362 | 0.272572 | |
| Topic 73 | 0.492816092 | 0.333333333 | 0.316885965 | 0.304263566 | 0.31427305 | 0.2714947 | 0.2490942 | 0.23974 | 0.2375 | 0.23366 | 0.231529 |

Figure 10 Emerging Topic Detection Table for NI and PVI Group
(the black ones mean the point is above the baseline VDP. Topic 0, topic 7 and topic 10 only have one point above the baseline VDP)

| 第n年 | 3rd Ye | 4rd Ye | 5rd Ye | 6rd Ye | 7rd Ye | 8rd Ye | 9rd Ye | 10rd Ye | 11rd Ye | 12rd Ye | 13rd Ye |
|---|---|---|---|---|---|---|---|---|---|---|---|
| Topic 5 | 0.603118 | 0.5065 | 0.457844 | 0.438466 | 0.384811 | 0.370993 | 0.355522 | 0.349281 | 0.340583 | 0.3386 | 0.337993 |
| Topic 11 | 0.580089 | 0.539191 | 0.504202 | 0.447024 | 0.406 | 0.390055 | 0.374954 | 0.369033 | 0.366042 | 0.363559 | 0.363319 |
| Topic 21 | 0.594386 | 0.570806 | 0.506522 | 0.467248 | 0.437956 | 0.42433 | 0.412118 | 0.405605 | 0.40101 | 0.399342 | 0.398842 |
| Topic 23 | 0.592593 | 0.558594 | 0.535959 | 0.533784 | 0.523734 | 0.521875 | 0 | 0 | 0 | 0 | 0 |
| Topic 26 | 0.593991 | 0.591484 | 0.580596 | 0.527641 | 0.510867 | 0.48238 | 0.447227 | 0.436487 | 0.434265 | 0.433015 | 0.432953 |
| Topic 28 | 0.531272 | 0.438379 | 0.415312 | 0.407486 | 0.386988 | 0.38411 | 0.379386 | 0.374366 | 0.372404 | 0.37118 | 0.371006 |
| Topic 32 | 0.602679 | 0.483929 | 0.483274 | 0.430929 | 0.384245 | 0.37369 | 0.364121 | 0.361038 | 0.358075 | 0.357112 | 0.356921 |
| Topic 48 | 0.569719 | 0.517655 | 0.45948 | 0.434107 | 0.41891 | 0.399767 | 0.384662 | 0.382974 | 0.381593 | | |
| Topic 52 | 0.613314 | 0.457358 | 0.452321 | 0.450474 | 0.448528 | 0.434032 | 0.432506 | 0.428657 | 0.427711 | 0.426563 | 0.426251 |
| Topic 58 | 0.513413 | 0.476522 | 0.435683 | 0.429567 | 0.419348 | 0.409198 | 0.40615 | 0.403335 | 0.402626 | 0.402133 | |
| Topic 60 | 0.585597 | 0.445233 | 0.422861 | 0.401625 | 0.393812 | 0.386814 | 0.380068 | 0.378919 | 0.378067 | 0.376266 | 0.375898 |
| Topic 65 | 0.601198 | 0.466025 | 0.431855 | 0.413204 | 0.390837 | 0.377421 | 0.368647 | 0.362114 | 0.359048 | 0.35732 | 0.356878 |

Figure 11 Emerging Topic Detection Table for NI and CVI Group
(the black ones mean the point is above the baseline VDP.
Topic 5 only has one point above the baseline VDP)

| 第n年 | 3rd Ye | 4rd Ye | 5rd Ye | 6rd Ye | 7rd Ye | 8rd Ye | 9rd Ye | 10rd Ye | 11rd Ye | 12rd Ye | 13rd Ye |
|---|---|---|---|---|---|---|---|---|---|---|---|
| Topic 5 | 0.571726 | 0.443161 | 0.411096 | 0.394461 | 0.382052 | 0.354186 | 0.331873 | 0.319349 | 0.313354 | 0.30909 | 0.30792 |
| Topic 11 | 0.561764 | 0.480181 | 0.434238 | 0.391187 | 0.367583 | 0.342452 | 0.322128 | 0.31442 | 0.307206 | 0.298458 | 0.295475 |
| Topic 12 | 0.59607 | 0.418565 | 0.389041 | 0.376785 | 0.359501 | 0.351313 | | | | | |
| Topic 21 | 0.576157 | 0.499544 | 0.436327 | 0.396695 | 0.371202 | 0.347484 | 0.335894 | 0.319832 | 0.313359 | 0.307668 | 0.304279 |
| Topic 23 | 0.559554 | 0.453464 | 0.439454 | 0.431403 | 0.450168 | 0.433854 | 0.149188 | 0.136297 | 0.133727 | 0.126488 | 0.124641 |
| Topic 26 | 0.585686 | 0.524909 | 0.502103 | 0.443968 | 0.418628 | 0.38797 | 0.356163 | 0.336531 | 0.331542 | 0.327141 | 0.323178 |
| Topic 28 | 0.551053 | 0.426927 | 0.394209 | 0.36935 | 0.349242 | 0.331077 | 0.318504 | 0.31107 | 0.308285 | 0.302797 | 0.299841 |
| Topic 32 | 0.558732 | 0.424 | 0.418881 | 0.387489 | 0.353501 | 0.331637 | 0.314562 | 0.302372 | 0.300071 | 0.298993 | 0.303461 |
| Topic 47 | 0.585 | 0.498106 | 0.395423 | 0.376515 | 0.339197 | 0.133211 | 0.150694 | 0.142109 | 0.138774 | | |
| Topic 52 | 0.565235 | 0.423363 | 0.417827 | 0.404404 | 0.38809 | 0.381202 | 0.363845 | 0.344176 | 0.340013 | 0.330482 | 0.327907 |
| Topic 57 | 0.591941 | 0.479893 | 0.400893 | 0.393592 | 0.367606 | 0.308379 | 0.283926 | 0.269594 | 0.267666 | 0.266224 | 0.265564 |
| Topic 58 | 0.506707 | 0.478791 | 0.431764 | 0.389263 | 0.366414 | 0.340569 | 0.328075 | 0.324464 | 0.318821 | 0.314065 | |
| Topic 60 | 0.565526 | 0.426215 | 0.395143 | 0.374339 | 0.350705 | 0.333397 | 0.317586 | 0.315232 | 0.308101 | 0.297042 | 0.293699 |
| Topic 67 | 0.589658 | 0.476249 | 0.452851 | 0.437287 | 0.403993 | 0.305901 | 0.255816 | 0.248686 | 0.24487 | 0.241057 | 0.237211 |
| Topic 72 | 0.556506 | 0.456387 | 0.427274 | 0.380295 | 0.341111 | 0.322161 | 0.303796 | 0.302929 | 0.137181 | 0.136286 | |

Figure 12 Emerging topic detection table for NI, CVI and PVI group (Topic 47 has one point above the baseline VDP)

### 5.2.3 Analysis on detection of emerging time

In order to evaluate the effectiveness of indices and methods we propose, we compare the detection tables of topic 52 with different indices, NI and PVI, NI and CVI, NI, PVI and CV as shown in Figure 13. We sign the intersection of VDP and baseline VDP of different indices as Dp, Dc, Dpc. We find Dc is earlier than Dpc, Dp is later than Dpc. Figure 14 shows the curves of NI, PVI and CVI for topic 52. The increase of citation is slower than the increase of publication, PVI is concave type, CVI is convex type, while NI is fixed. So when we compute the $VDP_p$ and $VDP_c$ of some year, we can get the intersection of CVI and NI locates left of the intersection of PVI and NI, Year ($VDP_c$) < Year ($VDP_p$). But for NI and CVI, as discussed before, the sum of emerging topics we detected is the least. Synthesizing each kind of situation, in this study, we choose NI, CVI and PVI as the indices of emerging topic detection. As the results show, NI, CVI and PVI, we can detect the emerging topic earlier than NI and PVI. This will be helpful for the researchers to track the newly development, grasp the newly research direction earlier.

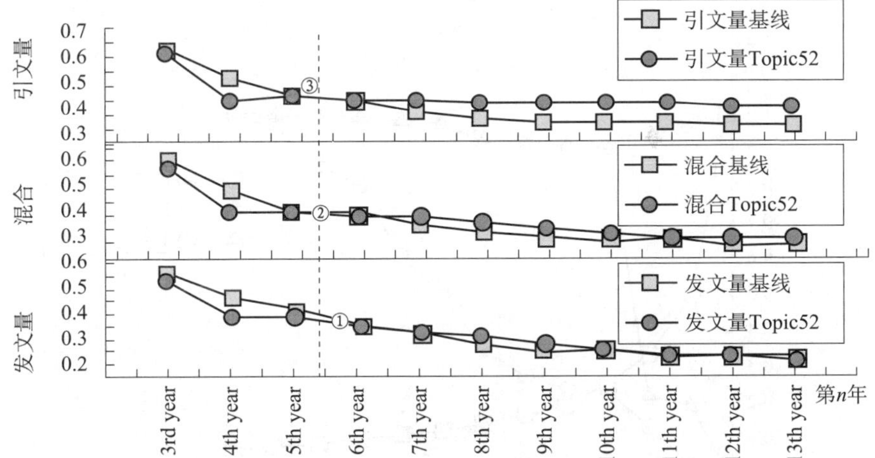


Figure 13 Comparative curves on DP of NI and PVI, NI and CVI, NI, PVI and CVI

### 5.2.4 Analysis on expressivity of topics extracted by TNG topic model

In order to analyze the efficiency of the indices and methods for detecting emerging topics we haveproposed, we choose topic 52 to analyze further.

As shown in Table 8 and Table 9, TNG topic model for *n*-gram words (multi-word phrases) has much more expressiveness, and to some extent, it will avoid the vagueness of the topic model for 1-gram words (single-word). In Table 9, there are multi-word phrases such as topic models, topic model, data streams, evolving data streams, Latent Dirichlet Allocation, semantic indexing, topic level, and so on. While in Table 8, there are single word lists including application, examples, term, latent, modelling, etc. It's hard to determine the meaning of Table 8, while we can understand that is a topic related to topic model, especially related to the application of semantic analysis on Latent Dirichlet Allocation.

**Table 8 Topic with 1-gram words** (single word)

| Number of topics | Words for the topic |
| --- | --- |
| Topic 52 | application examples topic term latent assumption specific modeling present concept negative widely LDA positive effectively terms direct difficulty water address |

**Table 9 The topic with *n*-gram words** (multi-word phrases)

| Number of topics | Words for the topic |
| --- | --- |
| Topic 52 | topicmodels; data streams; latent Dirichlet Allocation; concept drift; topic modeling; topic model; evolving data streams; dialogue management; semantic analysis; latent class models; text corpora; parameter poisson; model evaluation; combination weighting; semantic indexing; probabilistic latent semantic analysis; technology innovation capability; multiple attribute; drifting data streams; topic level |

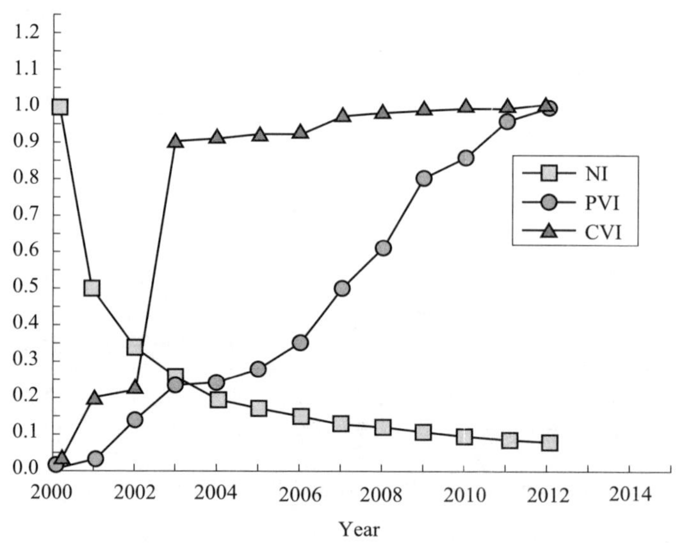


Figure 14 The curves of NI, PVI and CVI for topic 52 in 2012

TNG topic model for $n$-gram words (multi-word phrases) has the ability of disambiguity. For example, in Table 8 we can see LDA, but LDA has several meanings, such as linear discriminant analysis, local density approximation, lauryl dimethylaminoacetic acid, if it's a 1-gram word topic model, people will not be confused.

Topic 52 is topic model, especially Latent Dirichlet Allocation's application in semantic analysis of data. Topic model includes LDA, PLSI, and other extended models of LDA. LDA is put forward by David M. Blei in 2003, and has widespread use in image analysis, text classification, information retrieval and so on. It is still a hot topic and research frontier now. As the result shows, with the indices and methods we proposed we can detect "topic model" from the document of "machine learning", and we can get the time of becoming an emerging topic, 2003.

## 6 Conclusions

Based on the methods proposed by Tu and Seng (2012), which represent the development status of the topic with the intersection of the curve of publishing volume index and the curve of novelty index, we combined LDA topic model with bibliometrics, and optimized the indices with cited volume index. After an experiment on a corpus related with "machine learning", we detect the emerging topics in this field with the optimized indices and methods we pointed out. We carry out analysis of the effectiveness and validity of the indices and methods. At last, we forecast the development trend of the detected emerging topics with a method of curve fitting.

This study made some steps forward:

(1) We extract topics from documents with TNG topic model, rather than extract keywords based on word frequency. The experiment shows in this way, the topics we extracted have better capability in semantic expressions and better capability in disambiguation.

(2) On the basis of novelty index and published volume index, we introduce cited volume index. We compare the effectiveness of emerging topic detection with different indices NI and PVI, NI and CVI, and NI, PVI and CVI. The experiment shows after introducing CVI we can detect the emerging topic earlier.

There are some research plans of our research:

(1) Although through introducing cited volume index to indices of emerging topics, we get some optimization, we are thinking the possibility and effectiveness of introducing author and affiliation to the indices.

(2) For emerging topic detection, it's important to forecast the future trend of a topic. We are trying to choose time series analysis or some method of curve fitting to anticipate the trend.

(3) The combination of NI, PVI and CVI depends on simple linear superposition. How

to choose the value of coefficients of the indices to detect emerging topics is also an open question.

## Acknowledgement

This study is supported by NSFC project (Grant number 71373260) and the Light of West Project of CAS (Grant number Y200201001).

## References

贺亮，李芳 . 2012. 基于话题模型的科技文献话题发现和趋势分析 . 中文信息学报（2）：109-115.

刘红霞，乔晓东，张运良 . 2010. 新兴趋势监测指标体系探索，情报杂志，6：93-97.

王萍 . 2011. 基于概率主题模型的文献知识挖掘 . 情报学报，30（6）：583-590.

殷蜀梅 . 2008. 判断新兴研究趋势的技术框架研究 . 情报科学，123（5）：76-80.

张云秋，高歌 . 2011. 基于文献的新兴趋势探测方法的问题及对策研究 . 情报理论与实践，34（1）：47-50.

Blei D M，Ng A Y，Jordan M I. 2003. Latent Dirichlet allocation. *Journal of Machine Learning Research*，3：993-1022.

Blei D M，Lafferty J D. 2007. A correlated topic model of science. *Annals of Applied Statistics*，1（1）：17-35.

Bun K K，Ishizuka M. 2001. Emerging topic tracking system. *Third International Workshop on Advanced Issues of E-commerce and Web-based Information Systems* （WECWIS'o）. San Juan，California：2-11.

Chen C M. 2006. CiteSpace Ⅱ：detecting and visualizing emerging trends and transient patterns in scientific literature. *Journal of the American Society for Informati on Science and Technology*，57（3）：359-377.

David M B，John D L. 2006. Dynamic topic models. *Proceedings of the 23rd International Conference on Machine Learning*. Pittsburgh，Pennsylvania：ACM.

Dietz L，Bickel S，Scheffer T. 2007. Unsupervised prediction of citation influences. *Proceedings of the 18th ACM Conference on Information and Knowledge Management*. Hong Kong，China. New York：ACM：957-966.

Glänzel W. 2012. Bibliometric methods for detecting and analysing emerging research topics. *El Profesional de la Información*，21（2）：194-201.

Griffiths T. 2002. Gibbs sampling in the generative model of latent Dirichlet allocation. Standford University.

Hall D，Jurafsky D，Manning C D. 2008. Studying the history of ideas using topic models. *Proceedings of the Conference on Empirical Methods in Natural Language Processing*. Honolulu，Hawaii：363-371.

He Q, Chen B, Pei J, et al. 2009. Detecting topic evolution in scientific literature: how can citations help? *Machine Learning and knowledge Discover in Database*, 7524: 597-612

Hoang L M. 2006. Emerging trends detection from scientific online documents. Japan Advanced Institute of Science and Technology.

Hofmann T. 1999. Probabilistic latent semantic indexing. *SIGIR'qq: Proceedings of the 22nd Annual International ACM SIGIR Conference of Research and Development in Information Retieval*. Berkerley, California. New York: ACM.

Kleinberg J. 2003. Bursty and hierarchical structure in streams. *Data Mining and Knowledge Discovery*, 7 (4): 373-397.

Kontostathis A, Galitsky L M, Pottenger W M, et al. 2003. A survey of emerging trend detection in textual data mining. In: Berry M W. A *Comprehensive Survey of Text Mining* (pp: 185-219). Berlin: Springer-Verlag.

Mccallum A, Mann G S, Mimno D. 2006. Bibliometric impact measures leveraging topic analysis. *In: Proceedings of the 6th ACM/IEEE-CS Joint Conference of Digital Libraries, JCDL'06*. Chapel Hill, NC, New York: ACM: 65-74.

Morinaga S, Yamanishi K. 2014. Tracking dynamics of topic trends using a finite mixture model. *In: The 10th ACM SIGKDD International Conference on Knowledge Discovery and Data Mining*. Seattle, USA, 811-816.

Morris S , Yen G, Wu Z , et al. 2003. Timeline visualization of research fronts. *Journal of American Society for Information Science and Technology*, 54 (5): 4131.

Nallapati R M, Ahmed A, Xing E P, et al. 2008. Joint latent topic models for text and citations. *In*: Proceeding of the 14th ACM SIGKDD International Conference on knowledge Discovery and Data Mining. New York, USA, 542-550.

Persson O. 1994. The intellectual base and research fronts of JASIS 1986-1990. *Journal of the American Society for Information Science*, 45 (1): 31-38.

Price D. 1965. Networks of scientific papers. *Science*, 149 (3683): 5121.

Rosen-Zvi M, Griffiths T, Steyvers M, et al. 2004. The author-topic model for authors and documents. In: Chickering D M and Halpern J Y (eds), *UAI'u4, Proceedings of the 20th Conference on Uncertainty in Artificial Intelligence.* Banff, Canada. Corvallis, OR: AVAI Press: 487-494.

Shibata N, Kajikawa Y, Takeda, Y, et al. 2011. Detecting emerging research fronts in regenerative medicine by the citation network analysis of scientific publications. *Technological Forecasting and Social Change*, 78 (2): 274-282.

Small H. 1973. Co-citation in the scientific literature: a new measure of the relationship between two documents. *Journal of the American Society for Information Science*, 24 (4): 265-269.

Tu Y N, Seng J L. 2012. Indices of novelty for emerging topic detection. *Information Processing & Management*, 48 (2): 303-325.

Upham S P, Small H. 2010. Emerging research fronts in science and technology: patterns of new knowledge development. *Scientometrics*, (83): 15-38.

Wallach H M. 2006. Topic modeling: beyond bag-of-words. *In: Proceedings of the 23rd International Conference on Machine Learning*. ACM: 977-984.

Wang C，Blei D，Heckerman D. 2008. Continuous time dynamic topic models. *In*：*Proceedings of the 23rd Conference on Uncertainty in Artificial Intelligence*. Helsink，Finland：579-586.

Wang X，McCallum A. 2005. A note on topical *n*-grams. Technical Report. UM-CS-2005-071，University of Massachusetts，Amherst，December.

Wang X，Mccallum A. 2006. Topics over time：a non-Markov continuous-time model of topical trends. *In*：*Proceedings of the 12th ACM SIGKDD International Conference on Knowledge Discovery and Data Mining*. Philadelphia，PA. New York：ACM：424-433.

# 2-6 肿瘤学学科结果的计量学分析

邵红芳①

## 1 引　言

近半个世纪以来，恶性肿瘤的发病率一直呈上升趋势，它与心脑血管疾病一并成为人类致死的两类主要疾病。恶性肿瘤对人类生命的威胁已经很大而且还在继续加大，人类与之抗争的迫切性也随之不断增加，进而催生出了医学学科领域的一个独立的重要学科——肿瘤学，并已进一步形成若干分支。肿瘤学的发展愈发呈现出壮观的知识图景，我们有必要从计量学视角分析肿瘤学的学科发展现状，探析学科内在的结构和学科地位，分析肿瘤学的发展态势，揭示肿瘤学学科知识结构演进的规律性。计量学研究肿瘤学学科结构对科学、客观、合理评价肿瘤学研究工作和成果，促进学科健康发展有着重要的参考价值。

## 2 数据与方法

### 2.1 数据的选择

本文所使用的数据，来源于美国 ISI 的《科学引文索引》数据库（SCI），数据最后的更新时间为 2012 年 12 月 25 日。根据 2011 年 ISI 数据库中的 JCR（*Journal of Citation Report*）报告可知，肿瘤学共有 196 种期刊。选取影响因子最高的 40 种肿瘤学期刊（表 1）作为肿瘤学的重要期刊，检索并下载这 40 种期刊 2008 年 1 月至 2012 年 12 月期间的文献记录作为本文期刊共被引分析的原始数据。检索到 88 686 篇期刊文献，共有 2 231 675 条引文记录。

**表 1　肿瘤学领域顶级期刊名录**

| 序号 | 缩写 | 期刊名 | 2011 年影响因子 | 发表文章数（2008～2012 年） |
|---|---|---|---|---|
| 1 | CCJC | CA-CANCER J CLIN | 101.78 | 186 |
| 2 | NRC | NAT REV CANCER | 37.545 | 990 |

---

① 邵红芳，山西医科大学，Shftanya@126.com。

续表

| 序号 | 缩写 | 期刊名 | 2011 年影响因子 | 发表文章数（2008~2012 年） |
|---|---|---|---|---|
| 3 | CC | CANCEL CELL | 26. 566 | 788 |
| 4 | LO | LANCET ONCOL | 22. 589 | 1 767 |
| 5 | JCO | J CLIN ONCOL | 18. 372 | 13 946 |
| 6 | JNCI | J NATL CANCER I | 13. 757 | 1 793 |
| 7 | NRCO | NAT REV CLIN ONCOL | 11. 963 | 487 |
| 8 | CMR | CANCER METAST REV | 10. 573 | 282 |
| 9 | LEUK | LEUKEMIA | 9. 561 | 1 762 |
| 10 | BRC | BBA-REV CANCER | 9. 38 | 226 |
| 11 | NCPO | NAT CLIN PRACT ONCOL | 8 | 174 |
| 12 | CR | CANCER RES | 7. 856 | 7 365 |
| 13 | SC | STEM CELLS | 7. 781 | 1 399 |
| 14 | CCR | CLIN CANCER RES | 7. 742 | 4 295 |
| 15 | JMGB | J MAMMARY GLAND BIOL | 6. 741 | 185 |
| 16 | ACR | ADV CANCER RES | 6. 733 | 109 |
| 17 | SCB | SEMIN CANCER BIOL | 6. 475 | 276 |
| 18 | AO | ANN ONCOL | 6. 425 | 12 791 |
| 19 | ONCOG | ONCOGENE | 6. 373 | 2 712 |
| 20 | JP | J PATHOL | 6. 318 | 1 431 |
| 21 | CTR | CANCER TREAT REV | 6. 054 | 625 |
| 22 | NEOPL | NEOPLASIA | 5. 946 | 576 |
| 23 | NO | NEURO-ONCOLOGY | 5. 723 | 7 103 |
| 24 | CARCIN | CARCINOGENESIS | 5. 702 | 1 440 |
| 25 | RO | RADIOTHER ONCOL | 5. 58 | 1 447 |
| 26 | EJC | EUR J CANCER | 5. 536 | 5 237 |
| 27 | IJC | INT J CANCER | 5. 444 | 3 667 |
| 28 | BCR | BREAST CANCER RES | 5. 245 | 1 311 |
| 29 | MCT | MOL CANCER THER | 5. 226 | 1 610 |
| 30 | JESHC | J ENVIRON SCI HEAL C | 5. 16 | 60 |
| 31 | MO | MOL ONCOL | 5. 082 | 263 |
| 32 | PCMR | PIGMCELL MELANOMA R | 5. 059 | 1 271 |
| 33 | BJC | BRIT J CANCER | 5. 042 | 3 041 |
| 34 | CPR | CANCER PREV RES | 4. 908 | 835 |
| 35 | ONCOT | ONCOTARGET | 4. 784 | 379 |
| 36 | CANCERACS | CANCER-AM CANCER SOC | 4. 771 | 2 370 |
| 37 | BCRT | BREAST CANCER RES TR | 4. 431 | 2 780 |
| 38 | CROH | CRIT REV ONCOL HEMAT | 4. 411 | 759 |
| 39 | JNCCN | J NATL COMPR CANC NE | 4. 409 | 460 |
| 40 | ERC | ENDOCR-RELAT CANCER | 4. 364 | 488 |

## 2. 2 研究方法

### 2. 2. 1 期刊共被引分析

期刊共被引分析是指以期刊为基本单元而建立的共被引关系，把引用次数较多的期刊按被引证关系联结起来，揭示了期刊所表征的学科之间的相互关系和结构特征（Goldberg et al.，2007；Chen et al，2012；Cobo et al.，2011）。

#### 2.2.2 多元统计分析方法

多元统计分析最主要的特征之一是“降低维度技术”（通常是二维），其中主要包括相关分析、多维尺度分析和主成分分析。这些方法与科学计量学中的引文分析相结合可用来分析某学科的学科结构和地位以及不同学科领域的分布状况（刘则渊等，2008）。

此处采用的多元统计分析主要有以下几种：一是相关分析。相关分析是研究变量之间密切程度的一种统计方法，本研究中，利用距离相关分析计算变量之间的相似性测度（McCain，1990）。二是主成分分析（principal component analysis，PCA）。利用PCA对期刊共被引频次相关矩阵进行数学变换，并通过斜交转换来简化因子结构。在期刊共被引分析中，位于某导出主成分因子下的所有期刊共同决定或解释了该因子的命名。三是多维尺度分析（multi-dimensional scaling，MDS）。通过某种非线性变换，把高维数据转换成低维数据，并利用平面距离来反映期刊之间的相似程度。具体操作步骤是通过SPSS中的一个分析模块——共被引分析中最常使用的克拉斯卡尔（Kruskal）的非度量（non-metric）多维测度来进行分析的。同时绘制MDS期刊共被引图谱，MDS的图形可以直观和形象地看出类间的关系（李志辉和罗平，2000；刘林青，2005；马庆国，2002）

本研究中利用Bibexcel实现期刊共被引的原始矩阵，通过距离相关分析，利用SAS软件将原始共被引矩阵转化为相关矩阵并实现多维尺度分析。

## 3 结果分析

学术期刊的内容可以反映某学科的发展。因为如果某个科学研究领域成为研究热点，就会有越来越多的人追踪研究这个领域的科学问题，而随着研究的逐渐广泛和深入，以该领域为主题的学术期刊便会随之产生。本文借鉴期刊共被引分析方法，以表1中所列出的肿瘤学研究领域顶级期刊为分析对象，来确定肿瘤学的主流分支学科与学科结构，并进一步分析其学科地位和作用。

### 3.1 计量学分析

#### 3.1.1 期刊共被引分析

期刊共被引分析的前提是构建期刊共被引矩阵。在进行数据分析之前，首先进行了数据的标准化，即将同一期刊的不同名称进行统一，如期刊 *Clinical Cancer Research* 在SCI数据库存中就有 *Clin Cancer Res*、*CLIN CANCER RES*、*Clin Cancer Research* 等多种表达方式，本研究将其统一为CCR。数据标准化后，通过Bibexcel软件的操作，进行共被引分析，形成期刊共被引频次矩阵，即原始矩阵（部分数据见表2）。此矩阵为对称矩阵，对角线的值为缺省值，非对角线上的值为期刊共被引频次。在此矩阵中，共被引频次的范围是 0～55 336。

**表 2 期刊共被引频次矩阵（部分）**

| 期刊名称 | 临床肿瘤学杂志 | 血液 | 癌症研究 | 新英格兰医学杂志 | 临床癌症研究 | 国际放射肿瘤期刊 | 肿瘤年刊 | 癌症 |
|---|---|---|---|---|---|---|---|---|
| 临床肿瘤学杂志 | 0 | 3 723 | 355 | 55 336 | 39 895 | 16 871 | 329 | 33 287 |
| 血液 | 3 723 | 0 | 18 525 | 16 956 | 12 073 | 905 | 51 | 7 936 |
| 癌症研究 | 355 | 18 525 | 0 | 1 651 | 39 108 | 2 860 | 55 | 888 |
| 新英格兰医学杂志 | 55 336 | 16 956 | 1 651 | 0 | 1 883 | 63 | 7 920 | 9 563 |
| 临床癌症研究 | 39 895 | 12 073 | 39 108 | 1 883 | 0 | 2 582 | 6 678 | 738 |
| 国际放射肿瘤期刊 | 16 871 | 905 | 2 860 | 63 | 2 582 | 0 | 2 038 | 5 657 |
| 肿瘤年刊 | 329 | 51 | 55 | 7 920 | 6 678 | 2 038 | 0 | 5 568 |
| 癌症 | 33 287 | 7 936 | 888 | 9 563 | 738 | 5 657 | 5 568 | 0 |

### 3.1.2 相关性分析

将原始矩阵，即期刊共被引频次矩阵转化为期刊相似性矩阵。通过 SAS 统计分析软件的交互式矩阵程序设计语言模块（IML）功能将原始矩阵进行皮尔逊相关矩阵转化（部分数据见表 3），转化为相关系数矩阵，这个步骤即对原始矩阵进行了标准化，进而消除了因期刊被引频次高低差别所带来的相似性影响。表 3 可以看作是期刊相似性矩阵，在这里的相似性是由相关系数 $r$ 值来衡量的，即两个期刊的正相关性越强（$r$ 的绝对值越大），表明它们的科学研究领域或研究视角就越相似，亦表明它们所表征的学科知识背景就越相近。一般情况下，从统计学意义上来说，$r>0.7$ 为两者相关性较高。

由于篇幅所限，我们在表 3 中只列出期刊共被引相关矩阵的部分，原相关矩阵是 38×38。我们可以清晰地看到，期刊《新英格兰医学杂志》（*The New England Journal of Medicine*，NEJM）与《临床肿瘤学杂志》（*Journal of Clinical Oncology*，JCO）有较强的相关性（$r\approx0.99$），其值几乎接近于 1，而与（《癌症研究》*Cancer Research*，CR）的相关性却较弱（$r=0.35$）。这就说明，《新英格兰医学杂志》作为世界四大权威综合医学杂志之一，与临床肿瘤学的相关性较大，多刊载与肿瘤临床研究相关的文献，而《癌症研究》作为肿瘤学方面的老牌杂志，主要偏向肿瘤基础研究，所以与其存在着较大的差异。

我们对原始 38×38 矩阵分析并归类，与期刊《临床肿瘤学杂志》（JCO）相关的期刊（$r$ 接近于 0.7）主要有《血液》（*Blood*）、《癌症研究》（CR）、《新英格兰医学杂志》（NEJM）、《临床癌症研究》（CCR）、《肿瘤学纪事》（AO）、《癌症》（*Cancer*）、《美国国家癌症研究所杂志》（JNCI）和《柳叶刀》（*Lancet*），说明这些期刊所表征的学科是肿瘤学的一类子学科群，而 JCO 是临床肿瘤学的代表性期刊之一，由此，从相关性分析，我们可以初步判定这一子学科群是临床肿瘤学；与期刊《血液》相关系数 $r>0.7$ 的明显只有《白血病》（$r=0.822$），显然与《血液》关系最密切的当属《白血病》；CR 是肿瘤学研究领域内偏重于肿瘤基础研究的代表性期刊之一，与 CR 主要相关的期刊（$r$ 接近于 0.7）除了 JCO 和 CCR 这两种期刊外（说明临床肿瘤学实践与肿瘤基础研究关系直接，基础研究总是服务并应用于临床实践的），还有《美国国家科学院院刊》（PNASU）、《细胞》（*Cell*）、《生物化学杂志》（JBC）、《国际癌症杂志》（IJC）、《原癌基因》（*Oncogene*）、《癌症自然评论》（NRC）、《科学》（*Science*）和《自然》（*Nature*），这组期刊群中大都是以肿瘤基础研究为主的期刊，收录的大部分是肿瘤分子生物学、细胞生物学等方面的文献（其中 PNASU、*Nature* 和 *Science* 三个综合性世界级权威性杂志除外）；与期刊《新英格兰医学杂志》相关系数 $r>0.7$ 的只有 JCO，说明 NEJM 具有偏重临床医学研究的性质；与《临床肿瘤研究》相关系数 $r>0.7$ 的只有 JCO 和 CR，进一步证明 CCR 这种期刊以“临床肿瘤学”为主的办刊宗旨以及根植于肿瘤基础研究的特性；与《中华放射肿瘤学》（IJRO）相关系数 $r>0.7$ 的只有《肿瘤放射治疗》，这也不用多作说明，这两个期刊同属肿瘤放射学范围，属于肿瘤治疗学范畴的同类期刊；与《肿瘤学纪事》相关系数 $r>0.7$ 的只有 JCO，说明了 AO 的临床肿瘤学的本质；与《癌症》相关系数 $r>0.7$ 的也只有 JCO，更证明了肿瘤基础研究与肿瘤临床实践的密不可分性。

**表 3　期刊共被引相关矩阵（部分）**

| 期刊名称 | 临床肿瘤学杂志 | 血液 | 癌症研究 | 新英格兰医学杂志 | 临床癌症研究 | 国际放射肿瘤期刊 | 肿瘤年刊 | 癌症 |
|---|---|---|---|---|---|---|---|---|
| 临床肿瘤学杂志 | 1 | 0.820 | 0.762 11 | 0.999 981 9 | 0.890 7 | 0.552 11 | 0.766 | 0.775 57 |
| 血液 | 0.820 37 | 1 | 0.853 9 | 0.373 973 | 0.313 2 | 0.036 103 | 0.13 | 0.225 39 |
| 癌症研究 | 0.762 11 | 0.85 | 1 | 0.353 627 6 | 0.990 06 | 0.111 338 | 0.153 | 0.235 25 |
| 新英格兰医学杂志 | 0.999 98 | 0.373 5 | 0.353 63 | 1 | 0.316 75 | 0.160 62 | 0.187 | 0.222 81 |
| 临床癌症研究 | 0.890 7 | 0.313 2 | 0.990 06 | 0.316 751 | 1 | 0.099 52 | 0.186 | 0.201 63 |
| 国际放射肿瘤期刊 | 0.552 1 | 0.036 1 | 0.111 3 | 0.160 615 | 0.099 52 | 1 | 0.087 | 0.238 7 |
| 肿瘤年刊 | 0.765 75 | 0.129 8 | 0.152 85 | 0.186 900 3 | 0.185 6 | 0.087 099 | 1 | 0.169 1 |
| 癌症 | 0.775 57 | 0.225 | 0.235 25 | 0.222 81 | 0.201 63 | 0.238 70 | 0.169 | 1 |

### 3.1.3　多维尺度分析

在生成原始矩阵，即期刊共被引频次矩阵的基础上，我们借助多维尺度排列分析绘制出期刊共被引多维尺度分析（MDS）图谱（图 1）。其中 Stress 值等于 0.1772，RSQ 值等于 0.7851，反映了一定的适合度。Stress 值是用来衡量原始矩阵和转化矩阵的适合度的术语，一般认为，Stress 值越低，RSQ 值越高，适合度越高，在此类共被引研究中，Stress 值只要低于 0.2 都可被接受（李志辉和罗平，2000）。

图谱中清晰地将肿瘤学期刊分成六大部分，即六大期刊群。

期刊群［1］包含8种期刊，其中，《生物化学杂志》（JBC）重点收录分子生物学、细胞生物学、蛋白质等学科或专业领域的研究成果；《国际癌症杂志》（IJC）主要关心使用新的分子探针诊断、检测和治疗各种癌症，以及分子遗传学、病毒学等方面的研究成果；《原癌基因》主要发表包括癌基因的结构和功能的所有方面的细胞分子层面的肿瘤原创性成果；《癌症自然评论》是《自然》旗下的一种肿瘤方面研究的权威杂志，其2011年影响因子为37.55。主要发表如细胞永生和端粒维持、细胞死亡及细胞凋亡、血管生成-肿瘤的生长、基因组不稳定性等方面的研究成果，着重于癌症的分子基础方面的研究。根据这些期刊所代表的内容与性质，我们认为这一肿瘤学主流分支学科群为肿瘤分子生物学。

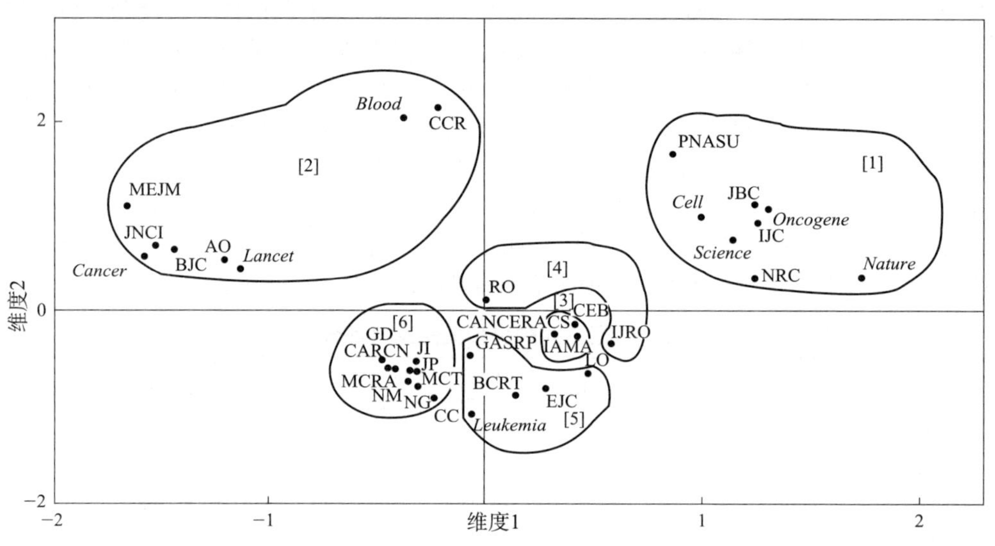


Stress = 0.1772，RSQ = 0.7851

［1］肿瘤分子生物学　［2］临床肿瘤学　［3］肿瘤流行病学

［4］肿瘤放射治疗学　［5］肿瘤诊断学　［6］肿瘤免疫学

图1　肿瘤学期刊共被引多维尺度分析图谱

而在另外的几种期刊中，《细胞》是与《科学》、《自然》等齐名的世界权威杂志，是近10年来在分子生物学和遗传学研究领域中最具影响力和最热门的期刊。《细胞》在这一期刊群中出现，说明了肿瘤学是以“分子生物学”为其母本学科的。这一期刊群中另一期刊——《美国科学院院刊》（PNAS）在SCI综合科学类中排名第三位，收录涵盖生物、物理和社会科学等多学科的研究成果。这两种著名的期刊与《科学》和《自然》同时出现在肿瘤学这一母本学科群，充分证明了现代肿瘤学是一直以来以分子生物学为母本学科的特性，同时具有与生物、物理、化学甚至社会科学等多学科交叉的跨学科研究的性质。

这一结果也可从前面的相关分析（表3）中得到验证，在表3中与CR相关性较高的期刊大都集中在多维尺度分析图中的第一象限。同时第一象限里的这几种期刊的研究的关注点大都在肿瘤分子生物学，说明肿瘤分子生物学是肿瘤学主要的主流分支学科，从而也推衍出分子生物学是肿瘤学的母本学科。

期刊群［2］包含 8 种期刊，它们分别是《临床肿瘤研究》（CCR）、《血液》（Blood）、《癌症》（Cancer）、《新英格兰医学杂志》（NEJM）、《英国癌症杂志》（BJC）、《美国国家癌症研究所杂志》（JNCI）、《肿瘤学纪事》（AO）、《柳叶刀》（*Lancet*）。根据这些期刊的性质，可以认为它们代表着临床肿瘤学。而这一结果可以从前面的相关分析中得到验证——在表 3 中，与 JCO 相关性较大的期刊就是图 1 中期刊群［1］中的期刊，它们大都集中在第二象限。

期刊群［3］包含 3 种期刊。它们分别是《癌症流行病学生物标记与预防》（CEBP）、《美国医学会杂志》（JAMA）和《美国癌症协会杂志》（CANCER - AM CANCER SOC）。这一学科群为肿瘤流行病学。

期刊群［4］只包含了 2 种期刊《肿瘤放射治疗》（RO）和《中华放射肿瘤学》（IJRO）。我们注意到，《肿瘤放射治疗》这个期刊所代表的研究领域占据了图谱接近于中心的地位，位于期刊群［1］和期刊群［2］之间，这说明肿瘤放射治疗学研究在肿瘤基础研究与肿瘤临床实践之间起着桥梁、纽带作用，从某种角度可以认为肿瘤放射治疗学是将肿瘤基础研究与肿瘤临床医学实践连接起来的纽带。尽管如此，肿瘤放射治疗学仍与肿瘤分子生物学所在的期刊群［1］的关系更为密切。这说明肿瘤放射治疗学始终是基础研究与应用研究（临床实践研究）连接的纽带，这一期刊群为肿瘤放射治疗学。

期刊群［5］包含 5 种期刊，分别是《欧洲肿瘤》（EJC）、《乳腺癌研究与治疗》（BCRT）、《胃肠病学》（GASTRO）、《白血病》（*Eeukemia*）和《柳叶刀肿瘤学》（LO）。这一期刊群为肿瘤诊断学。

期刊群［6］包含了 10 种期刊，它们分别是《免疫学杂志》（JI）、《分子细胞生物学杂志》（MCB）、《癌变》（CARCIN）、《基因与发育学杂志》（GD）、《临床调查》（JCI）、《癌症分子疗法》（MCT）、《自然遗传学》（NG）、《癌细胞》（CC）、《美国病理学杂志》（AJP）和《自然医学专刊》（NM）。这一期刊群为肿瘤免疫学。

综上，从多维尺度分析来看，我们认为这六大期刊群所表征的子学科群代表着肿瘤学的主流分支学科。

### 3. 1. 4 主成分分析

对原始矩阵，即期刊共被引频次矩阵作了相关分析和多维尺度分析后，再利用主成分分析作为聚类的依据以完善多维尺度分析图并进一步验证相关分析的结果。也就是通过创立更少数量的导出期刊主成分来解释原始期刊数据间的关系（表 4）（刘则渊，2008）。

对肿瘤学 top 40 种顶级期刊共被引频次矩阵进行主成分分析，根据特征根值（>1）共提取出 8 个导出主成分，累计贡献率达到 72. 16%，即 8 个导出主成分代表了原始指标的 72. 16%的信息，它们分别代表了肿瘤学学科群中不同的分支学科（表 4）。

表 4　期刊共被引主成分提取表（2008～2012 年）

| 导出主成分因子 | 标签（负载） | | | | | | | |
|---|---|---|---|---|---|---|---|---|
| 主成分 1：肿瘤分子生物学 | CR | *Nature* | PNASU | JBC | *Oncogene* | *Cell* | *Science* * | NRC |
| | 0.386 | 0.25 | 0.23 | 0.213 | 0.209 | 0.208 | 0.195 | 0.16 |
| 主成分 2：临床肿瘤学 | JCO | NEJM | *Cancer* | JNCI | AO | *Lancet* | BJC | |
| | 0.28 | 0.232 | 0.23 | 0.226 | 0.202 | 0.19 | 0.15 | |
| 主成分 3：肿瘤流行病学 | CEB | *Blood* | IJC | JAMA | | | | |
| | 0.1 | 0.37 | 0.35 | 0.11 | | | | |
| 主成分 4：肿瘤放射治疗学 | RO | IJRO | | | | | | |
| | 0.651 | 0.621 | | | | | | |
| 主成分 5：肿瘤诊断学 | *Leukemia* | EJC | GASTRO | BCRT | | | | |
| | 0.82 | 0.181 | 0.155 | 0.13 | | | | |
| 主成分 6：肿瘤免疫学 | JI | GD | NG | MCB | JCI | MCT | | |
| | 0.397 | 0.339 | 0.263 | 0.251 | 0.29 | 0.237 | | |
| 主成分 7：肿瘤病理学 | AJP | NM | MCB | *Science* * | *Oncogene* | JBC | | |
| | 0.329 | 0.318 | 0.271 | 0.25 | 0.253 | 0.223 | | |
| 主成分 8：实验肿瘤学 | MCT | *Cancer* | CANCER-ACS | CC | MCT | MCB | JBC | NRC |
| | 0.68 | 0.311 | 0.295 | 0.2 | 0.237 | 0.23 | 0.228 | 0.201 |

* 表示在 0.05 水平上显著。

表 4 中列出了每个主成分中起主要作用的期刊及其主成分期刊系数，系数越大，说明它提供给这组主成分的信息越大，它在这组主成分中所起到的作用就越大。

我们可以清晰地看到，主成分 1 包括了 8 种期刊，与前面的多维尺度分析图中结果作对比分析，主成分 1 基本对应了多维尺度分析图中的期刊群［1］所包含的期刊。根据这些主要期刊所反映的学科信息，我们称主成分 1 为肿瘤分子生物学。

主成分 2 包含了 7 种期刊，与前面的多维尺度分析图中结果作对比分析，主成分 2 基本对应了多维尺度分析图中的期刊群［2］所包含的期刊，JCO 在主成分 2 中的系数最大，说明它的期刊性质代表了这一学科群的主要信息。我们称主成分 2 为临床肿瘤学。

主成分 3 包含了 4 种期刊，*Blood* 最大，为 0.37。我们称主成分 3 所代表的学科群为肿瘤流行病学。与多维尺度分析图中的期刊群［3］基本对应。

主成分 4 仅包含了 2 种期刊——RO 和 IJRO，与多维尺度分析图中的期刊群［4］对应。在整个主成分期刊系数值系列中，明显比其他期刊高很多，分别为 0.651 和 0.621。这两种期刊是肿瘤放射治疗研究领域的主要代表期刊，我们称这个学科群为肿瘤放射治疗学。

主成分 5 包含了 4 种期刊，与多维尺度分析图中的期刊群［5］基本对应。我们称这一学科群为肿瘤诊断学。

主成分 6 包含了 6 种期刊，这一主成分期刊系数中 JI 的值最高，达 0.397。我们称这一学科群为肿瘤免疫学。与多维尺度分析图中的期刊群［6］基本对应。

主成分 7 包含了 6 种期刊，这一组主成分期刊系数中 AJP 的值最高，达 0.329。我们称这一学科群为肿瘤病理学。在多维尺度分析图中并不能找到基本与这一主成分对应的学科群，应包含在期刊群［6］中。

主成分8包含了8种期刊，其中系数值最高的《癌症分子疗法》（MCT）达0.68，它偏重于从方法学视角寻求实验癌症治疗的分子靶标，开展肿瘤预防的目标及新模式、新方法、新技术等方面的研究。我们称这一分支学科群为实验肿瘤学。

## 3.2 肿瘤学学科结构

从对肿瘤学 top 40 种期刊的期刊共被引频次矩阵的多维尺度分析的结果来看，生成的多维尺度分析图谱中存在清晰可辨的六大期刊群组。在图1中，我们发现：《新英格兰医学杂志》《柳叶刀》《血液》三种重要的医学综合类期刊都距原点较近，也就是都处于整个期刊网络图的核心位置，说明肿瘤学具有交叉学科、跨学科研究的学科发展特征。在肿瘤学学科衍生与发展过程中，生命科学的各门学科，如分子生物学、细胞生物学、遗传学和免疫学等与临床肿瘤研究紧密结合，产生了肿瘤分子生物学、肿瘤免疫学、肿瘤诊断学、肿瘤放射治疗学等生物医学学科；随着分子生物学、免疫学、肿瘤学及细胞因子等研究技术的进展，产生了更多的新兴学科和交叉学科，如肿瘤细胞生物学、肿瘤遗传学、实验肿瘤学、肿瘤影像学、肿瘤核医学等，使基于经验的传统临床肿瘤学转变成为以现代生物学知识和试验方法为基础的肿瘤各分支学科、新兴学科和交叉学科，肿瘤学也在其自身衍生发展中构成了一个复杂的肿瘤学学科群系统。在这六大期刊群中，期刊群［1］、期刊群［2］和期刊群［6］各自占的比例较大，说明肿瘤分子生物学、临床肿瘤学和肿瘤免疫学是最主要的肿瘤学主流分支学科。而其他3个期刊群相比之下所占成分较弱，为仅次于前3种分支学科的肿瘤学主要分支学科。

再利用主成分分析作为聚类的依据以完善多维尺度分析图并进一步验证相关分析的结果，根据特征根值（>1）共导出8个主成分，8个主成分占到原始指标72.16%的信息。仅主成分1——肿瘤分子生物学就占到原始指标的20.6%的信息，累积贡献率达到20.6%，说明在肿瘤学衍生与发展过程中，肿瘤分子生物学是最主要的主流分支学科，决定着肿瘤学发展演进的方向，是肿瘤学的基础学科或母本学科。临床肿瘤学对原始指标总量的累计贡献率达到9.9%，说明了它是肿瘤学又一主流分支学科。在我国，肿瘤学是设置在临床医学这个一级学科目录下的一个二级学科，肿瘤是一种全身性的疾病，人类与恶性肿瘤的斗争开始于有人类的那一天起，这就决定了肿瘤学学科发展与临床医学的密切相关性，可以说临床医学一直以来就是肿瘤学发展的源头和根本。其他6个主成分，从主成分3至主成分8共占到约40%的原始指标信息，其中主成分7和主成分8对原始指标各自仅起到4%的累计贡献率，这也充分验证了在多维尺度分析图中，出现六大子学科群，而主成分7和主成分8对应的应包含在多维尺度分析图的期刊群［6］中，说明这两个学科群——肿瘤病理学和实验肿瘤学是肿瘤学的分支学科，但其作用比起前面的6种学科要小得多。

综合分析可以基本得出：在肿瘤学整个学科结构中，临床医学和分子生物学是肿瘤学发展的母本理论学科；肿瘤分子生物学、临床肿瘤学、肿瘤流行病学、肿瘤放射治疗学、肿瘤诊断学、肿瘤免疫学是其主流分支学科；肿瘤遗传学、肿瘤病理学、肿瘤病因学、实验肿瘤学和表观遗传学是其主要的新兴学科和交叉学科。

# 4 结　语

## 4.1 发展对策

通过计量学分析肿瘤学学科结构，我们认为，未来要推进肿瘤学的有序发展，应从以下两方面加强研究。

**1. 强化基础，核心集结**

目前，肿瘤学研究在很大意义上还只是粗浅的，对其基本理论问题缺乏应有的关注。今后应对肿瘤学的相关学科所涉及的具有一般性、普遍性的问题进行集纳式的总结和概括，加强临床医学、分子生物学等基础学科的研究。

学科发展的过程是不断地会聚集结的，故其学科结构关系不应是固定化的，而应是柔性化、灵活化、变动性的，只有这样，学科集结才能突破任何设定的边界和任何固定的模式而有效展开。学科集结之后，参与的每个学科都会影响其他学科，又受惠于其他学科。

肿瘤学学科的核心集结是要将肿瘤学基础研究成果迅速有效地转化为可在临床实际应用的理论、技术、方法和药物，在实验室到病房（Bench to Bedside）之间架起一条快速通道，其实质是理论与实际相结合，是基础与临床的整合（Weber and Hamm，2008）。

**2. 促进交叉，均衡发展**

作为一个开放的科学知识体系，随着研究工作的逐步深入和精细化，肿瘤学在广延和纵深的各个方向上拓展、延伸，出现明显的学科交叉与渗透趋势。主要体现在其与免疫学、流行病学、放射学等学科的交叠、合流之势。同时，肿瘤学的学科交叉性，决定了其理论生长中的综合性。以学科交叉为基础，走综合创新之路，将是肿瘤学学科发展的契机和优势所在（Hood，2007）。肿瘤学在今后的发展中，应当广泛吸纳相关学科的研究，借鉴和利用各种学科的理论和方法，既要促进不同学科间的内外交融，又要促成各分支学科之间的均衡发展。

## 4.2 研究的不足与展望

由于本研究所依据的数据以肿瘤学 SCI 期刊论文为主，我们假定期刊文献能够反映学科发展的实际水平和学者的主要思想，但其信息传递的可靠性和有效性问题需要进一步研究。此外，对国内外肿瘤学论文的计量对比分析也是下一步研究的重点。

总之，我们认为肿瘤学的研究、分析、整理和利用，通过定量与定性、规范与实证方法的综合，才能真正实现在研究方法上的经验与理念、历史与逻辑的统一。

## 参考文献

李志辉，罗平 . 2000. SPSS for Windows 统计分析教程 . 2 版. 北京：电子工业出版社 .

刘林青 . 2005. 作品共被引分析与科学地图的绘制 . 科学学研究，23（2）：155-159.

刘泽渊，陈悦，侯海燕，等 . 2008. 科学知识图谱：方法与应用 . 北京：人民出版社 .

马庆国 . 2002. 管理统计 . 北京：科学出版社 .

Chen C M，Hu Z G，Liu S B，et al. 2012. Emerging trends in regenerative medicine：a scientometric analysis in CiteSpace. *Expert Opinion on Biological Therapy*，12：593-608.

Cobo M J，Lopez-Herrera A G，Herrera-Viedma E，et al. 2011. Science mapping software tools：review，analysis，and cooperative study among tools. *Journal of the American Society for Information Science and Technology*，62：1382-1402.

Goldberg A D，Allis C D，Bernstein E. 2007. Epigenetics：a landscape takes shape. *Cell*，128：635-638.

Hood L. 2007. *Systems Biology and Medicine*：*From Reactive to Predictive*，*Personalised*，*Preventive and Participatory Medicine*. The UK Focus for Biomedical Engineering Annual Lecture. London：The Royal Society of Medicine.

McCain K W. 1990. Mapping authors in intellectual space：a technical overview. *Journal of the American Society for Information Science*，1（6）：33-36.

Weber M，Hamm C. 2008. Novel biomarkers—the long march from bench to bedside. *European Heart Journal*，29（9）：1079-1081.

# 三、科学计量指标研究

# 3-1 科研影响力的全局 $h$ 指数*

徐 芳[①] 杨立英[②] 刘文斌[③]

**摘 要**：随着政府对科技投入的增加，科研评价已然成为科技战略管理的重要工具。在实际操作中，从时间和经济条件考虑，大样本的评价主要依靠定量指标，如 $h$ 指数。虽然该指标学术意义重大，但本研究指出 $h$ 指数本质问题之一在于 $h$ 核选取标准的局部性使得学术期刊进行分类或排名时缺乏可比性，导致基于 $h$ 指数的评价方法有理论上的缺陷。针对该问题，本研究提出构造全局性的评价指标（Gh 指数和 $Gh_{adj}$ 指标），并采用英国 RAE 数据进行实例研究，可为科技管理层提供科研评价的新思路，更好地服务于科技管理。

**关键词**：科研评价；RAE；$h$ 指数；全局指数 Gh

## 1 引 言

在知识经济时代，科技发展水平和科技创新能力已经成为提高国家竞争力、促进经济发展和改善社会福祉的核心要素。鉴于科技与经济社会各方面息息相关，各国政府纷纷加大了对科技的投入。同时，科技发展的全球化，以及政府绩效管理的压力促使各界人士对科研评价越来越重视。科研评价不仅仅是公众理解科技的重要渠道，它已经成为科技战略管理的重要工具。一般来说，大到国家层面或学科层面（朱明和杨晓江，2013），小到单篇论文（王孝宁等，2004）或研究员个人，科研评价由于其价值导向不同，侧重点也不尽相同，如有注重科技资源使用效率的评价，或是针对科研产出效果的评价。前者关心如何用最少的投入获取最多的产出，而后者则关注的是科研产出的学术价值、科研活动的影响力，这是科研评价中最根本而又最艰巨的难点。目前在高校或科研机构，科研评价不仅仅关注科研产出本身，

---

* 本研究得到 2012 年度国家自然科学基金青年科学项目的资助（项目批准号：71201159）。

① 徐芳（1984— ），女，汉族，英国肯特大学商学院管理学博士，中国科学院科技政策与管理科学研究所助理研究员。研究方向：科技评价和科技管理。电话：010-59358415。邮箱：xufang@casipm.ac.cn。

② 杨立英（1973— ），女，汉族，中国科学院文献情报中心研究员。研究方向：科学计量、科研管理和科技政策。

③ 刘文斌，英国肯特大学商学院。

而且更关注科研产出的影响力。从评价方法看，目前常用的方法主要有两大类：一是同行评议（或定性方法）。该方法主要通过同领域专家的学术知识和专业经验进行评价。二是文献计量学方法（或定量方法）。大样本的科技评价更青睐于定量方法，研究较多的属于各种文献计量学指标的研究如 SJR 指标（Gonzalez-Pereira et al.，2010）、Eigenfactor 指标（Bergstrom，2007）等。流行较广的有期刊影响因子（journal impact factor，JIF）（Garfield，1972）和 $h$ 指数。该指标提出后在各领域的研究应用层出不穷，如中国 SCI 期刊的 $h$ 指数与影响因子比较（陈红光和雷二庆，2008），$h$ 指数对图书馆学情报学领域期刊的解读（许新军，2009），以及医学类期刊 $h$ 指数与影响因子、总被引频次的相关性研究（孙慧和汤先忻，2009）等。

虽然 $h$ 指数的适用性颇为广泛，但如果从其构造原理仔细推敲，不难发现 $h$ 指数用于影响力评价有理论上的缺陷。与任一被评单元（如研究员个人、学术期刊或科研机构）的其他论文相比，学术界称单篇论文引用数不低于 $h$ 的 $h$ 篇论文为 $h$ 核论文。由此得知，不同被评单元的 $h$ 核是独特的，构造标准是局部性的。同一篇论文可能入选被评单元 A 的 $h$ 核（即论文引用数达到 $h^A$核的构造标准），但未必入选被评单元 B 的 $h$ 核（即论文引用数未能达到期刊 $h^B$核的构造标准）。以期刊 A 和 B 举例说明：期刊 A 和 B 的前 8 篇论文引用数分别为：65、55、40、24、10、8、5、5 和 15、12、12、10、7、6、6、3。两者 $h$ 指数同为 6。期刊 B 的第五、第六篇论文引用数分别为 7 和 6，可以入选 $h^B$核，但不能入选 $h^A$核。这一例子再次表明 $h$ 核的构造采用的是基于期刊论文数和引用数的独特标准、局部性标准。显然在此基础上根据 $h$ 指数进行期刊的全局优劣排名，可能会有失评价的公正。值得指出的是，这一问题不仅仅是 $h$ 指数本身的问题，很多 $h$ 指数的扩展，如 $h$（2）指标、$g$ 指标等都有此弊端。所以该问题具有更大范围的普适性，研究颇有意义。

## 2 Gh 指数的提出与测度介绍

如上所述，我们认为基于局部标准的 $h$ 指数直接用于局部或全局评价（以下统称为全局评价）时，不同被评单元之间缺乏可比性，导致评价结果有理论上的缺陷。但这并不代表 $h$ 指数就无法用于全局评价。改进方案之一是构造全局标准，保证不同被评单元进行全局评价时标准的统一性。在此理念下，我们提出全局 $h$ 指数的概念。以期刊为例，该指标定义如下：

假设所有被评期刊的论文按引用数降序排列，如果该全局论文集中的 $N$ 篇论文中，$h$ 篇论文每篇至少有 $h$ 的引用数，其余 $N-h$ 篇论文每篇引用数小于 $h$，那么该全局论文集的 $h$ 指数为 $h$（记作 Gh 指数），相应的 $h$ 核称为全局 $h$ 核（记作 Gh 核）。期刊 $i$ 在 Gh 核中的论文数即为该期刊的全局 $h$ 指数（记作 $\mathrm{Gh}^i$ 指标）。

$\mathrm{Gh}^i$ 指标用数学形式表示如下：

$$\mathrm{Gh}^i = \sum_{m=1}^{n} \mathrm{sing}\Big(c(p^m) - \mathrm{Gh}\Big), \mathrm{sing}(x) = \begin{cases} 1, x \geqslant 0 \\ 0, x < 0 \end{cases}$$

$$\text{Gh}^i\text{-core} = \{\bigcup_{m=1}^{n} \{p^m\} : c(p^m) \geqslant \text{Gh}\} \tag{1}$$

其中，出 $c$（$p^m$）是第 $m$ 篇论文的引用数，$n$ 是期刊 $i$ 的论文总数。一般来说，GH 指数是针对全局论文集而言的，而 Gh 指数是针对某一期刊而言的。

Gh 指数借鉴的是 $h$ 指数的构造理念，我们将其归为 $h$ 指数的拓展之一。我们首先通过性质1，分析 Gh 指数和 $h$ 指数之间的关系，帮助读者从理论上更好地理解该指标。对于任何两个期刊 $i$ 和 $j$，如果 $h^i > h^j$ 并不一定有 $\text{Gh}^i \geqslant \text{Gh}^j$，但是我们有如下的相容性结果：

**性质** 对于任何两个被评单元 $i$ 和 $j$，如果 $\text{Gh}^j < h^j$，$h^i < h^j$，那么 $\text{Gh}^i \leqslant h^i$。

我们通过反证法来证明该性质。

**证明：**

$$\text{如果 Gh}^i > h^i$$

$$\Rightarrow h^i - \text{core} \cup \{p^{h^i+1}\} \subset \text{GH} - \text{core}$$

根据 $h$ 指数的定义，

$$\forall p^i \in h^i - \text{core} \Rightarrow c(p^i) \geqslant h^i$$

$$\forall p^j \in h^j - \text{core} \Rightarrow c(p^j) \geqslant h^j$$

$$\text{a)} \because h^i = \min_{p^i \in h^i - \text{core}} c(p^i), h^i < h^j$$

$$\therefore min_{p^j \in h^j - \text{core}} c(p^j) \geqslant h^j > h^i = \min_{p^i \in h^i - \text{core}} c(p^i)$$

$$\because (1) \& (2)$$

$$\therefore h^j - \text{core} \in \text{Gh} - \text{core}$$

$$\therefore \text{Gh}^j \geqslant h^j$$

与原假设相矛盾。

$$\text{a)} \because h^i < \min_{p^i \in h^i - \text{core}} c(p^i), h^i < h^j$$

$$\therefore c(p^{h^i+1}) \leqslant h^i < h^j \leqslant \min_{p^i \in h^i - \text{core}} c(p^i)$$

$$\because (1) \& (3)$$

$$\therefore h^j - \text{core} \in \text{Gh} - \text{core}$$

与原假设相矛盾，倒推之则假设得证。

证明完毕。

上述性质不仅表明了 Gh 指数的单调性，同时也论证了 Gh 指数和 $h$ 指数之间的相容性。

以下我们研究 Gh 指数在实例中的应用效果如何。实例研究以笔者之前收集的英国 RAE2008 管理和商业方向的 50 个科研机构（具体机构名单见附表 1）为样本（徐芳等，2007)。RAE 是英国每隔 5 年左右进行的国家高等教育机构科研评估项目。最近一次的 RAE，即 RAE2008（或 REF）要求各评估单位参与论文提交的研究人员每人提供最优秀的 4 篇论文。专家组通过对研究机构学术论文的评审，公布各机构在研究产出（即学术论文）的科研质量评估报告。这些报告给出了科研机构研究质量从“国际领先水平（4 星）”至“无科研贡献（0 星）”等 5 档的百分比分布。以下是计算该机构的 GPA 得分，用以排名。

$$\text{GPA}=\frac{(4\text{ 星论文数}\times 4+3\text{ 星论文数}\times 3+2\text{ 星论文数}\times 2+1\text{ 星论文数}\times 1+0\text{ 星论文数}\times 0)}{100}$$

科研质量评估报告及研究人员提交的论文资料可以从 RAE2008 官方网站（http：//www. rae. ac. uk/）获得。论文引用数据收集于 Google Scholar，2010 年年底。根据 Gh 指数的构造原理，我们计算各机构的 Gh 指数以及其他常用的文献计量学指数，如表 1 所示。

**表 1　RAE50 个机构 GPA 和 *h*、*g*、Gh 指数**

| 代码 | GPA | *h* 指数 | 排名（*h*） | *g* 指数 | 排名（*g*） | Gh 指数 | 排名（Gh） | 代码 | GPA | *h* 指数 | 排名（*h*） | *g* 指数 | 排名（*g*） | Gh 指数 | 排名（Gh） |
|---|---|---|---|---|---|---|---|---|---|---|---|---|---|---|---|
| 135 | 3.191 | 82 | 1 | 144 | 1 | 42 | 1 | 139 | 2.678 | 24 | 25 | 55 | 17 | 1 | 21 |
| 204 | 2.567 | 66 | 2 | 105 | 2 | 19 | 2 | 50 | 2.000 | 23 | 27 | 39 | 27 | 1 | 21 |
| 163 | 2.744 | 62 | 3 | 99 | 3 | 10 | 3 | 185 | 2.314 | 22 | 28 | 38 | 28 | 1 | 21 |
| 108 | 2.611 | 53 | 4 | 78 | 9 | 4 | 12 | 81 | 2.214 | 22 | 28 | 33 | 33 | 0 | 34 |
| 155 | 2.644 | 52 | 5 | 93 | 4 | 7 | 7 | 73 | 2.123 | 22 | 28 | 36 | 31 | 0 | 34 |
| 156 | 2.838 | 49 | 6 | 86 | 6 | 9 | 5 | 167 | 2.355 | 21 | 31 | 44 | 23 | 3 | 14 |
| 114 | 2.828 | 47 | 7 | 79 | 7 | 10 | 3 | 53 | 2.069 | 20 | 32 | 33 | 33 | 0 | 34 |
| 115 | 2.538 | 47 | 7 | 75 | 10 | 5 | 11 | 164 | 2.585 | 19 | 33 | 38 | 28 | 1 | 21 |
| 169 | 2.684 | 46 | 9 | 67 | 13 | 1 | 21 | 68 | 2.314 | 19 | 33 | 33 | 33 | 1 | 21 |
| 124 | 2.758 | 45 | 10 | 91 | 5 | 7 | 7 | 66 | 2.308 | 19 | 33 | 28 | 38 | 0 | 34 |
| 109 | 2.769 | 44 | 11 | 79 | 7 | 9 | 5 | 83 | 2.167 | 19 | 33 | 30 | 37 | 0 | 34 |
| 2 | 2.778 | 42 | 12 | 74 | 11 | 6 | 9 | 69 | 1.844 | 18 | 37 | 25 | 40 | 0 | 34 |
| 116 | 2.672 | 42 | 12 | 66 | 14 | 3 | 14 | 202 | 1.698 | 18 | 37 | 31 | 36 | 0 | 34 |
| 152 | 2.664 | 40 | 14 | 68 | 12 | 2 | 18 | 146 | 2.306 | 17 | 39 | 27 | 39 | 0 | 34 |
| 113 | 2.123 | 38 | 15 | 63 | 16 | 3 | 14 | 72 | 2.125 | 17 | 39 | 34 | 32 | 1 | 21 |
| 134 | 2.803 | 37 | 16 | 66 | 14 | 6 | 9 | 75 | 1.846 | 16 | 41 | 22 | 41 | 0 | 34 |
| 111 | 2.497 | 32 | 17 | 49 | 19 | 1 | 21 | 177 | 2.022 | 15 | 42 | 20 | 46 | 0 | 34 |
| 160 | 2.479 | 31 | 18 | 47 | 21 | 3 | 14 | 64 | 1.926 | 15 | 42 | 22 | 41 | 0 | 34 |
| 127 | 2.426 | 30 | 19 | 55 | 17 | 4 | 12 | 79 | 1.854 | 14 | 44 | 22 | 41 | 0 | 34 |
| 122 | 2.465 | 28 | 20 | 49 | 19 | 1 | 21 | 100 | 1.225 | 12 | 45 | 21 | 45 | 0 | 34 |
| 51 | 2.346 | 27 | 21 | 47 | 21 | 2 | 18 | 112 | 2.27 | 11 | 46 | 22 | 41 | 1 | 21 |
| 125 | 2.604 | 26 | 22 | 42 | 25 | 2 | 18 | 76 | 1.889 | 11 | 46 | 17 | 48 | 0 | 34 |
| 168 | 2.309 | 26 | 22 | 43 | 24 | 1 | 21 | 56 | 1.900 | 10 | 48 | 18 | 47 | 0 | 34 |
| 71 | 2.194 | 25 | 24 | 38 | 28 | 0 | 34 | 95 | 1.546 | 9 | 49 | 12 | 50 | 0 | 34 |
| 119 | 2.726 | 24 | 25 | 40 | 26 | 1 | 21 | 89 | 1.527 | 7 | 50 | 16 | 49 | 1 | 21 |

我们发现，*h* 指数的机构排名和 Gh 指数的机构排名结果有一定的一致性，特别是排名比较靠前或靠后的机构。但是我们同时观察到，两指标的排名结果也有很大的不一致性，部分机构在 *h* 指数排名较靠后但在 Gh 指数排名却较靠前。例如，机构 112 号的 *h* 指数排名是 46 位，但在 Gh 指数排名中位列第 21 位。可能原因在于，相对来说，全局论文集的 Gh 核的构造标准低于该机构的 *h* 核的构造标准，导致该现象的产生。研究同时发现，Gh 指数数值较小且区分能力还有待提高，因此无法保证评价或排名的精确性，

评价结果的参考意义也受影响。在下文中我们将集中讨论如何解决该问题。

## 3 $Gh_{adj}$指标的提出与测度介绍

上文提到，由于全局论文集的 Gh 指数过小，各机构的 Gh 指数数值过小，最终限制了该指标的排名区分能力。接下来我们从指标定义入手分析 Gh 指数数值过小的原因。Gh 指数是 $h$ 指数的变形，它是论文数递增和引用数递减过程中相遇后产生的。如果引用数递减幅度过大，随着论文数的逐渐增加，对引用数的需求也越来越高（因为构造 Gh 指数的需求）。实际情况是往往又满足不了，最终会导致 Gh 指数过小。所以研究问题就聚焦于如何“放慢”论文数的增长速度。参考笔者之前提出的 $X$ 指标和 $L$ 指标（徐芳等，2007）的做法，我们建议运用如下的方程“缓减”论文数的增长速度，这样使得构造 Gh 指数时对引用数的需求也降低，从而提高 Gh 指数的数值。对在这种调整方法下产生的 Gh 指数我们称之为 $Gh_{adj}$指标，其数学表达式如下所示：

$$
\begin{gathered}
Gh_{adj} = n^* \\
\text{iff c}\,(n^*) \geqslant (n^*)^a \text{ and c}\,(n^*+1) < (n^*+1)^a \\
\text{where } 0 < a < 1
\end{gathered} \tag{2}
$$

各期刊的 $Gh_{adj}$指数计算方法参照定义中 Gh 指数的计算方法。

上述的参数 α 用以调整论文数增长的速度。以论文数 1 至 5 为例，如果 α 设为 0.5，那么调整后的论文数为 1、1.4、1.7、2、2.2，增长速度明显减慢。问题是如何合理地设置 α 的数值？任何的人为设置都会影响指标的数值。我们的逻辑是保证全局比较下的“好论文”（即 Gh 核论文）的数量与局部比较下“好论文”（即 $h$ 核论文）的总量一致，所以我们建议调整 $\alpha$ 数值使全局论文集的 Gh 指数等于被评单元 $h$ 指数的总和。50 个机构的 $Gh_{adj}$指数计算结果如表 2 所示。

我们发现经过 $\alpha$ 调整后产生的 $Gh_{adj}$指数满足我们的需求，数值较 Gh 指数有明显的增长，同时排名区分能力更是大幅提升，特别是对排名在后的机构。

## 4 $L_j$科研影响力新分类方法的提出

通过观察 $Gh_{adj}$指数和 $h$ 指数，我们发现，这两个指数存在以下几种关系。

一是 $Gh_{adj}$指数大于 $h$ 指数。这个不难理解：$\alpha$ 参数的设置确保全局比较时的 $h$ 指数（GH 指数）与所有单元 $h$ 指数之和相等。那么，由于来源于其他单元的低引论文的“衬托”作用，某些优秀或高质量的机构会有更多的论文被选入 Gh 核中，从而导致该单元的 $Gh_{adj}$指数大于自身的 $h$ 指数。

二是 $Gh_{adj}$指数小于 $h$ 指数。这个关系则针对低质量或较差的机构而言。如果优秀机构的 $Gh_{adj}$指数大于 $h$ 指数，那么必然会有一部分低质量或差的机构的 $Gh_{adj}$指数小于 $h$ 指数。在某些极端例子中，某些机构的 $Gh_{adj}$指数也可能为 0。

表 2 RAE50 个机构文献计量学指标

| 代码 | GPA | $h$ 指数 | $g$ 指数 | Gh 指数 | $Gh_{adj}$ 指数 | 代码 | GPA | $h$ 指数 | $g$ 指数 | Gh 指数 | $Gh_{adj}$ 指数 |
|---|---|---|---|---|---|---|---|---|---|---|---|
| 135 | 3.191 | 82 | 144 | 42 | 152 | 139 | 2.678 | 24 | 55 | 1 | 17 |
| 204 | 2.567 | 66 | 105 | 19 | 165 | 50 | 2.000 | 23 | 39 | 1 | 13 |
| 163 | 2.744 | 62 | 99 | 10 | 116 | 185 | 2.314 | 22 | 38 | 1 | 11 |
| 108 | 2.611 | 53 | 78 | 4 | 68 | 81 | 2.214 | 22 | 33 | 0 | 15 |
| 155 | 2.644 | 52 | 93 | 7 | 75 | 73 | 2.123 | 22 | 36 | 0 | 10 |
| 156 | 2.838 | 49 | 86 | 9 | 61 | 167 | 2.355 | 21 | 44 | 3 | 24 |
| 114 | 2.828 | 47 | 79 | 10 | 62 | 53 | 2.069 | 20 | 33 | 0 | 8 |
| 115 | 2.538 | 47 | 75 | 5 | 62 | 164 | 2.585 | 19 | 38 | 1 | 14 |
| 169 | 2.684 | 46 | 67 | 1 | 56 | 68 | 2.314 | 19 | 33 | 1 | 9 |
| 124 | 2.758 | 45 | 91 | 7 | 56 | 66 | 2.308 | 19 | 28 | 0 | 6 |
| 109 | 2.769 | 44 | 79 | 9 | 53 | 83 | 2.167 | 19 | 30 | 0 | 5 |
| 2 | 2.778 | 42 | 74 | 6 | 48 | 69 | 1.844 | 18 | 25 | 0 | 3 |
| 116 | 2.672 | 42 | 66 | 3 | 46 | 202 | 1.698 | 18 | 31 | 0 | 7 |
| 152 | 2.664 | 40 | 68 | 2 | 41 | 146 | 2.306 | 17 | 27 | 0 | 6 |
| 113 | 2.123 | 38 | 63 | 3 | 34 | 72 | 2.125 | 17 | 34 | 1 | 7 |
| 134 | 2.803 | 37 | 66 | 6 | 35 | 75 | 1.846 | 16 | 22 | 0 | 3 |
| 111 | 2.497 | 32 | 49 | 1 | 22 | 177 | 2.022 | 15 | 20 | 0 | 0 |
| 160 | 2.479 | 31 | 47 | 3 | 23 | 64 | 1.926 | 15 | 22 | 0 | 3 |
| 127 | 2.426 | 30 | 55 | 4 | 24 | 79 | 1.854 | 14 | 22 | 0 | 3 |
| 122 | 2.465 | 28 | 49 | 1 | 19 | 100 | 1.225 | 12 | 21 | 0 | 3 |
| 51 | 2.346 | 27 | 47 | 2 | 22 | 112 | 2.27 | 11 | 22 | 1 | 3 |
| 125 | 2.604 | 26 | 42 | 2 | 18 | 76 | 1.889 | 11 | 17 | 0 | 2 |
| 168 | 2.309 | 26 | 43 | 1 | 14 | 56 | 1.900 | 10 | 18 | 0 | 3 |
| 71 | 2.194 | 25 | 38 | 0 | 11 | 95 | 1.546 | 9 | 12 | 0 | 0 |
| 119 | 2.726 | 24 | 40 | 1 | 14 | 89 | 1.527 | 7 | 16 | 1 | 1 |

还有一种情况是 $Gh_{adj}$ 指数与 $h$ 指数相等，即局部比较和全局比较时都不会影响该机构论文入选 $h$ 核的数量。在实际情况中，该例子较少。

通过上述分析，我们发现通过观察 $Gh_{adj}$ 指数和 $h$ 指数数值，有助于我们进一步辨别科研影响力，由此我们设想可否根据这两指数，对科研影响力进行等级划分。具体操作如下：

首先我们将 $Gh_{adj}$ 指数大于 $h$ 指数的机构划分为 4 星。然后将余下的样本的论文重新整合为全局论文集，根据定义再次计算 $Gh_{adj}$ 指数，此时再筛选出 $Gh_{adj}$ 指数大于 $h$ 指数的机构，将其划分为 3 星。按此方法重复操作，直至我们找到 2 星和 1 星机构。我们将该等级划分方法称为 $L_j$ 科研影响力等级划分法。该方法的最大优点在于利用 $Gh_{adj}$ 指数和 $h$ 指数实现了对科研影响力的自然分类，摈弃了以前常用的人为设置分类区间的方法。分类结果相关性分析见表 3。

表 3　RAE50 个机构文献计量学指标相关性表

| | | GPA | $h$ 指数 | $g$ 指数 | $L_j$ 分类 |
|---|---|---|---|---|---|
| GPA | Pearson 相关性 | 1 | | | |
| | 显著性（双侧） | | | | |
| | $N$ | 50 | | | |
| $h$ 指数 | Pearson 相关性 | 0.777** | 1 | | |
| | 显著性（双侧） | 0.000 | | | |
| | $N$ | 50 | 50 | | |
| $g$ 指数 | Pearson 相关性 | 0.804** | 0.982** | 1 | |
| | 显著性（双侧） | 0.000 | 0.000 | | |
| | $N$ | 50 | 50 | 50 | |
| | 显著性（双侧） | 0.000 | 0.000 | 0.000 | |
| | $N$ | 50 | 50 | 50 | |
| $L_j$ 分类 | Pearson 相关性 | 0.803** | 0.836** | 0.835** | 1 |
| | 显著性（双侧） | 0.000 | 0.000 | 0.000 | |
| | $N$ | 50 | 50 | 50 | 50 |

**表示在 0.01 水平上显著。

上述结果显示 $L_j$ 分类结果与 GPA 具有更高的相关性，说明该指标与专家综合评审结果具有内在一致性。

## 5 结　论

科技投入的大幅度增加带来了对科研评价问题的再度探讨。去年，党中央、国务院发布的《关于深化科技体制改革，加快国家创新体系建设的意见》一文中，明确提出深化科技评价。党的十八大报告中，也提出要完善科技创新评价标准。因此，科技评价不仅是文献计量学领域的热点问题，也是科研人员和管理层关注的焦点。在实际中 $h$ 指数在评价中的应用不胜枚举。尽管如此，我们通过分析 $h$ 指数的构造，认为 $h$ 指数用于评价的本质问题之一在于 $h$ 核构造标准的局部性使得不同被评单元之间缺乏可比性，导致评价有理论上的缺陷。本文从该问题出发，提出解决该问题的可行方案之一是保证参评单元进行评价时使用统一的全局标准而非局部标准。具体地，我们构造了全局论文集、全局意义的 Gh 指数，以及反映研究质量的 Gh 指数。

但我们注意到，本研究也存在一定的局限性，如研究发现是否同样适用于其他被评单元如期刊或研究人员个人，以及学科的特殊性是否会影响整个研究，导致得出不同的研究结果，这在今后的研究中值得进一步探索。

## 参考文献

陈红光，雷二庆 . 2008. 中国 SCI 期刊的 $h$ 指数与影响因子比较 . 中国科技期刊研究，19（3）：402-404.

孙慧，汤先忻 . 2009. 医学类期刊 *h* 指数与影响因子、总被引频次的相关性研究 . 中国科技期刊研究，20（3）：469-471.

王孝宁，何苗，何钦成，等 . 2004. 基于文献计量学研究方法的科技论文定量评价 . 科学学与科学技术管理，25（4）：22-29.

徐芳，李晓轩，刘文斌 . 2007. 等同论文数（EPN）：学术论文质量评估的新指标 . 科研管理，32（7）：50-156.

许新军 . 2009. *h* 指数对期刊的解读——以图书馆学情报学期刊为例 . 图书情报工作，53（4）：140-143.

朱明，杨晓江 . 2013. 大学学科水平评价之辨析 . 科学学与科学技术管理，34（3）：80-88.

Bergstrom C E. 2007. Eigenfactor，measuring the value and prestige of scholarly journals. *College & Research Libraries News*，68（5）：314-316.

Garfield E. 1972. Citation analysis as a tool in journal evaluation. *Science*，178（60）：471-479.

Gonzalez-Pereira B，Guerrero-Bote V P，Moya-Anegón F. 2010. A new approach to the metric of journals' scientific prestige：the SJR indicator. *Journal of Informetrics*，4（3）：379-391.

**附表 1　50 个 RAE 机构名单**

| 代码 | 机构 | 代码 | 机构 |
|---|---|---|---|
| 135 | London Business School | 68 | De Montfort University |
| 156 | University of Oxford | 185 | University of Ulster |
| 114 | University of Cambridge | 168 | University of Glasgow |
| 134 | King's College London | 66 | Manchester Metropolitan University |
| 2 | Cranfield University | 146 | School of Oriental and African Studies |
| 109 | University of Bath | 112 | University of Bristol |
| 124 | University of Leeds | 81 | University of the West of England，Bristol |
| 163 | University of Warwick | 71 | Nottingham Trent University |
| 119 | University of Exeter | 83 | University of Westminster |
| 169 | University of Strathclyde | 72 | Oxford Brookes University |
| 139 | Queen Mary，University of London | 113 | Brunel University |
| 116 | University of Durham | 73 | University of Plymouth |
| 152 | Loughborough University | 53 | University of Central Lancashire |
| 155 | University of Nottingham | 177 | Aberystwyth University |
| 108 | Aston University | 50 | Bournemouth University |
| 125 | University of Leicester | 64 | Leeds Metropolitan University |
| 164 | University of York | 56 | Coventry University |
| 204 | University of Manchester | 76 | London South Bank University |
| 115 | City University，London | 79 | University of Teesside |
| 111 | University of Bradford | 75 | Sheffield Hallam University |
| 160 | University of Southampton | 69 | University of Northumbria at Newcastle |
| 122 | University of Kent | 202 | London Metropolitan University |
| 127 | Birkbeck College | 95 | University of Abertay Dundee |
| 167 | University of Edinburgh | 89 | University of Wales Institute，Cardiff |
| 51 | University of Brighton | 100 | Queen Margaret University Edinburgh |

# 3-2 Construction of Journal Retraction Impact Factor

Fan Shaoping[①], Zhang Zhiqiang[②]

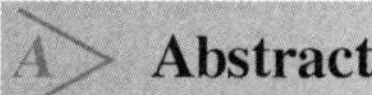

## Abstract

Academic misconduct has been increasing all the time in past several decades, especially in the last decade, which has attracted more and more attention in academic communities. In order to reduce the impact, academic journals are gradually strengthening the processing to retract misconduct articles on time. In this paper, a journal retraction impact factor (RIF) has been put forward to judge the negative impact of academic journals, based on the existed quantitative research results and the idea of journal impact factor (JIF). Then, we calculated 18 journals' retraction impact factor, respectively, by the data of retracted articles from Web of Science in 2011. Finally, the relationship between journal retraction impact factor and journal impact factor has been discussed. We hope to attract both academic communities and public attention to the academic misconduct.

**Keywords**: academic misconduct; journal impact factor; journal retraction impact factor

## 1 Introduction

Academic misconduct is increasing rapidly all over the world in these years, which has attracted more and more attention in both academic communities and the public. It has become a hot area in academia and journals. Because of the current assessment of researchers and institutions are based on the number of their output, such as published

① Lanzhou Branch of National Science Library/Scientific Information Center for Resources and Environment, Chinese Academy of Sciences, Lanzhou (China).

② University of Chinese Academy of Sciences, Beijing (China).

articles and patents. Some researchers overlooked the ethics of science, to gain more benefits by misconduct behavior. Facing with this truth, academic journals are taking actions by retracting articles and publishing retraction notice to fight back.

Academic databases such as Web of Science and PubMed have started to mark the retracted articles on their titles in the database. This makes readers easier to access to the information of the retracted articles (retracted time, reason, etc.) and provides help on the quantitative research of retracted articles. Based on the retracted information provided by the databases, some conclusions have been drawn. In "Top Journals' Top Retraction Rates" by Liu (2006), we have known that articles published in the high impact factor journals are more likely to be retracted (Table 1).

The article retractions caused by academic misconduct are greatly increased. These misconduct articles are harmful to the researchers and journal's readers, destroying the whole research atmosphere, and finally cutting down the quality of science research. Especially in medical domain, academic misconduct may endanger the patients' health. However, no matter what method we used in the evaluation of journals, like impact factor or some other evaluation indicators in JCR, for example, IF (JCR), IF (Scopus), $h$-Index, SJR indieator (SCImago Journal Rankings), SNIP (Source Normalized Impact per Paper)(Wang, 2011), eigenfactor (Bergstrom et al., 2008), are all based on the positive impact of journals, which means the higher the indicator score is, the better the journal is. But researchers overlooked the negative impact on journals, like academic misconduct. As pointed by Shi (2012), the academic journals should be dismissed if they contain academic misconduct articles or have never been cited by others. Academic journals are the platform to show results of science to the public, if there are no valid approaches to restrain the behavior of academic misconduct, these journals must be pushed out of journal rank.

**Table 1 Retractions and impact factors of the major journals**

| Journal | Retraction (notices) | Impact factor (2003) |
|---|---|---|
| *Science* | 38 | 29.162 |
| *Nature* | 32 | 30.979 |
| PNAS | 32 | 10.272 |
| *Cell* | 13 | 26.626 |
| J Immunol | 13 | 6.702 |
| J BiolChem | 12 | 6.482 |
| EMBO J | 11 | 10.456 |
| J Clin Invest | 11 | 14.307 |
| N Engl J Med | 10 | 34.833 |
| *Lancet* | 9 | 18.316 |

Notes: The data is based on a search in PubMed on May 6, 2006 and impact factors released for 2003.

In this paper, we tried to construct a retraction impact factor (RIF) to evaluate journals based on the data from Web of Science. Then we showed the relationship between retraction impact factor and journal impact factor. The main objective of our study is to help journal readers to forecast the general situation of article retraction by retraction impact factor, and alert academic communities to pay more attention to articles which citing the retracted articles. A secondary objective is to push publishers, editors, reviewers and authors to work more carefully, precisely and responsibly.

We searched in Web of Science with title ( = retracted article), time span ( = all the years), database ( = SCI-expanded and SSCI and CPCI-S and CCR-expanded and IC), the first retracted article we got, is *Cello Scrotum*, published on *British Medical Journal* in 1974, and retracted in 2009.

The United States is one of the few countries that have national evaluating system for evaluating scientific fraud. A lot of scientific misconduct cases in research sponsored by the National Institutes of Health were widely exposed. This led the U. S. Congress to create the Office of Scientific Integrity in 1989 (later renamed to Office of Research Integrity). The ORI published regulations for disposing scientific misconduct in research supported by the Department of Health and Human Services. The ORI has also developed a set of case law and fine-tuned its regulations. It has received 30 to 40 new cases each year, which is a sharp reminder that dishonesty in scientific research is a substantial ongoing issue (Sox and Rennie, 2006).

In order to increase the transparency of retraction processing, science writers Ivan Oransky (executive editor of Reuters Health) and Adam Marcus (managing editor of *Anesthesiology News*) launched a blog named Retraction Watch in August 2010, focusing on the reports of scientific articles' retractions. They observed that retractions are always not announced, and the reasons for retractions are not published. One result of this fact is researchers and the public who are unaware of the retractions may be mislead by these invalid articles. The blog said that retractions provide a window into the self-correcting nature of science (Wikipedia, 2013).

Most of recent quantitative studies on retracted articles are limited to medical field, and based on PubMed (Fang et al. , 2012; Samp et al. , 2012; Steen, 2012; Foo, 2011). This creates a wrong impression that academic misconduct only happened in medical field. To get a whole picture of retractions about the full spectrum of scholarly disciplines, Grieneisen and Zhang (2012) surveyed 42 largest bibliograghic databases for major academic fields and publisher websites to identify retracted articles. They found 4449 retracted articles from 1928 to 2011. Unlike Math, Physics, Engineering and Social Sciences, the percentages of retractions in Medicines, Life Science and Chemistry exceeded the percentages in Web of Science records. Fifteen prolific authors contributed more than half of all misconduct retractions. The number of retracted articles increased by a factor of 19. 06 each year from 2001 to 2010, and at the same time, the number of published articles decreased by a factor of 11. 36.

Trikalinos et al. (2008) performed an empirical evaluation. They found median time from publication to retraction is 28 months; it will be 79 months for articles authored senior researchers before retraction and 22 months by junior researchers. Retractions due to falsification may take an even longer time, especially by senior researchers. The senior rank includes professors, lab directors and experienced investigators. Junior rank includes students, scientists in training, technicians, supporting personnel, and young investigators.

Murat et al. ( 2007 ) introduced a four-parameter stochastic model for publication process. The model divides publication-retraction process into 2 parts: the acceptance strict level of a journal in accepting flawed manuscripts, and the post-publication examination level on corresponding academic communities. It needed four parameters to calculate the possibility of both IF-dependent changes in acceptance quality and examination level. This model can also be used to estimate the number of articles that should be retracted. For example, for *Nature* (1999-2004 average IF = 29. 5), the estimation is 45-67 articles should be retracted, however only 30 were actually retracted. There is still a big gap between the real and the estimation.

To determine whether journalsare different in retraction frequency and whether there is a relationship between retraction frequency and journal impact factor, Fang and Casadevall (2011) searched PubMed for retracted articles among 17 journals with impact factors ranging from 2 to 53. 484. They defined a "retraction index" for each journal by taken the number of retractions from 2001 to 2010, then, multiplied by 1 000, and finally divided by the number of published articles with abstracts. The journal retraction index versus the impact factor shows a surprisingly robust correlationship. Although this correlation did not imply causality, this preliminary investigation still suggests that the probability of retraction in a higher-impact journal is higher than that in a lower one.

# 2 Materials and Methods

## 2. 1 Construction of Retraction Impact Factor

Chen et al. (2013) investigated data in PubMed and defined retraction rate as the number of retraction notices issued each year divided by the total number of new publications added to PubMed in the same year. The retraction rate in 2001 was

0.000，05，which has doubled three times since then，in 2003，2006，and 2011，respectively. The retraction rate was up to 0.000，46 in 2011. He supposed that this number will continuously grow because there will be a delay from initial publication to recognize potential flaws.

Besides above，he loaded 1 721 retrieved records of retracted articles contained retraction time in Web of Science. The mean time between retracted article's publication and its retraction is 2.57 years，or 30 months. The median time to retraction is 2 years，or 24 months. However，a retracted article will continuously be cited after its retraction. The estimated duration time of citation since the first publication date is over 6 years (median = 5 years). All of the above have shown，it tends to take 2 years to retract an article and another 2 years to see a significant decrease of citations to the retracted article.

Journal impact factor was designed by Garfield，and has become an important indicator for evaluating journals. It is a relative indicator which can reduce the impact of high citing times and numerous published articles by famous journals to general ones. It will be balanced and revised by calculating the average citation rate at the citing peak time of each journal (Wikipedia，2013). Therefore the result is closer to the reality. However，IF is only an indicator for evaluating the journal's influence，it can't be used to evaluate the influence of single article in the journal.

Price，once claimed that，scientificarticle will reach its citing peak one or two years after published. The formula of IF is as follows：

$$\mathrm{IF} = \frac{A}{B} \tag{1}$$

where $A$ = the number ofcited times in the given year for articles published in desired journal during previous two years；$B$ = the total number of articles published by that journal previous two years.

Thomson Reuters published the new version of JCR since 2009，in which the author added 5-year impact factor ($\mathrm{IF}_5$) to reduce the issues that the short statistical period of IF couldn't measure the whole influence of journal rationally (Zhao，2010). The formula of $\mathrm{IF}_5$ is as follows：

$$\mathrm{IF}_5 = \frac{C}{D} \tag{2}$$

where $C$ = the number of cited times in the given year for articles published in desired journal during previous five years；$D$ = the total number of articles published by that journal during previous five years.

Based on the quantitative study on retracted articles by Chen Chaomei，we supposed the average of citation time for a retracted article from the original publication date is 6 years. At the same time, like the idea of journal impact factor, we constructed journal retraction impact factor, or RIF for further quantitative study on retracted articles. The formula of RIF is as follows:

$$\mathrm{RIF} = \frac{M}{N} \tag{3}$$

where $M =$ the number of cited times in the past six years for retracted articles which first published in six years ago, and these cited times are in indexed journals; $N =$ the total number of retracted articles which published in six years ago.

In this paper, we give a definition for "the life cycle of retracted article" . It refers to the time duration from the first published date to the last citation date of a retracted article. From the first day it published to the public, the retracted article has begun to have an impact in academic communities and the public. Due to the reasons that databases didn't update on time, publishers didn't publish retracted note, users didn't access to the retracted information, some researchers still cited the wrong data, wrong conclusions and wrong attitudes after an article has been retracted. Although, some scientists cited the retracted article as a negative case to demonstrate their own statement, this citation still took its negative effect. Therefore, the life cycle of retracted article was considered as 6 years in this paper.

Strictly speaking, the definition of retracted article in RIF is the article retracted 2 years later from its original publication date, and ended its citation in 6 years. However, there is limited data of retracted articles, and in order to avoid too much "zero" in the sheet, we analyzed the situation with retracted articles which first published in six years ago for the given year, and retracted before the given year, not exactly in the third year after published. Take the year 2013 for example:

$$\mathrm{RIF}_{2013} = \frac{M_{2013}}{N_{2013}} \tag{4}$$

where $M_{2013} =$ the number of cited times in 2008-2013 for retracted articles published in 2008; $N_{2013} =$ the total number of retracted articles published in 2008.

Obviously, the higher RIF indicates the more negative impact on academia. In that case, users should pay more attention to the journal. Meanwhile, users should also give more time to articles that cited retracted articles, for they may have more possiblity to cite wrong data, wrong conclusion or wrong attitude. In this way, the falsity or mistake maybe amplified. If these questionable articles were overlooked, a bad cycle of destructive impact on academia will continue, even expand. Therefore, both retracted articles and articles that cited them need our attention, in order to discover potential issues immediately, and decrease their serious impact in academic communities.

## 2.2 Data

Our case study focuses on retracted articles in Web of Science. At the time this article is written, the latest JCR (science edition) can be found is 2011, so we set the statistic year be 2011. The retrieval formula is as follows: title = retracted article, time span = 1900-01-01 to 2011-12-31, database = SCI-expanded and CCR-expanded and IC. Based on above criteria, we got 1980 records in total. There are two types of record style for these records, one is formal record (Figure 1), and the other is informal record (Figure 2, Figure 3). We chose 1 814 formal records that contain abstracts for analysis.

THE PRE-SYNAPTIC METABOTROPIC GLUTAMATE RECEPTOR 7 "mGluR7" IS A CRITICAL MODULATOR OF ETHANOL SENSITIVITY IN MICE (Retracted article. See vol. 223, pg. 488, 2012)

作者: Bahi, A (Bahi, A.)

来源出版物: NEUROSCIENCE 卷: 199 页: 13-23 DOI: 10.1016/j.neuroscience.2011.10.029 出版年: DEC 29 2011

被引频次: 5 (来自 Web of Science)

引用的参考文献: 66 [查看 Related Records] 引证关系图

摘要: Recent studies demonstrated that the metabotropic glutamate receptor subtype 7 "mGluR7" activation may reduce motivational aspects of ethanol dependence. We investigated the role of mGlu7 receptor in ethanol-related behaviors using the allosteric agonist AMN082 in mice. Results have shown that mGluR7 activation increased the sedative effect of ethanol as measured by the duration of loss of righting reflex (LORR) and reduced the severity of ethanol-induced withdrawal. Importantly, the protective effect of the drug on alcohol-induced withdrawal was found when the AMN082 was injected before, but not after, injection of ethanol suggesting that mGluR7 activation prevented development of dependence rather than producing an anti-convulsant effect. In addition, ethanol-induced locomotor stimulation was blocked by following mGluR7 activation. Furthermore, mice injected with AMN082 consumed less ethanol in a two-bottle free-choice paradigm and in a drinking in the dark (DID) model. Impairment in reward mechanisms in AMN082-injected mice was confirmed by the lack of ethanol-induced conditioned place preference (CPP). Follow-up control experiments have shown that plasma alcohol concentrations of AMN082 and vehicle-treated mice were similar. Taken together, these findings provide evidence for the crucial role of mGluR7 in ethanol-related behaviors, especially in voluntary alcohol drinking and alcohol reward. Thus, pharmacological targeting mGluR7 with AMN082-like compounds might be a potential means to tackle ethanol abuse and alcoholism in the future. (C) 2011 IBRO. Published by Elsevier Ltd. All rights reserved.

入藏号: WOS:000298206300002

文献类型: Article

语种: English

作者关键词: AMN082; CPP; drinking in the dark; ethanol; mGluR7; two-bottle choice

Figure 1 The formal record of retracted articles in Web of Science

WITHDRAWN: Dual targeting of glioma U251 cells and neovasculature with a nanoparticle encoding vasohibin and RGD peptides prevents tumor angiogenesis and inhibits tumor growth (Retracted Article)

作者: Chen, HJ (Chen, Hongjie)[1,2]; Yuan, BQ (Yuan, Bangqing)[3]; Wang, SS (Wang, Shousen)[1]; Zhen, ZC (Zhen, Zhaocong)[1]; Liu, Z (Liu, Zheng)[1]

来源出版物: INTERNATIONAL IMMUNOPHARMACOLOGY 卷: 11 期: 6 页: 778-778 DOI: 10.1016/j.intimp.2011.01.035 出版年: JUN 2011

被引频次: 1 (来自 Web of Science)

引用的参考文献: 1 [查看 Related Records] 引证关系图

入藏号: WOS:000291505900021

文献类型: Correction

语种: English

通讯作者地址: Wang, SS (通讯作者),Fuzhou Gen Hosp Nanjing Command, Dept Neurosurg, Fuzhou 350025, Fujian, Peoples R China.

地址:
[1] Fuzhou Gen Hosp Nanjing Command, Dept Neurosurg, Fuzhou 350025, Fujian, Peoples R China
[2] Fujian Med Univ, Fuzong Med Coll, Dept Neurosurg, Fuzhou 350025, Fujian, Peoples R China
[3] 476th Hosp PLA, Fuzhou 350025, Fujian, Peoples R China

Figure 2 The informal record of retracted articles that without retracted time in Web of Science

DOMINANT-NEGATIVE EFFECTS OF A NOVEL MUTATION IN THE FILAMIN MYOPATHY (Retracted article. See vol. 76, pg. 202, 2011)

作者: Kono, S (Kono, Satoshi); Miyajima, H (Miyajima, Hiroaki)

来源出版物: NEUROLOGY 卷: 75 期: 23 页: 2137-2138 出版年: DEC 7 2010

被引频次: 2 (来自 Web of Science)

引用的参考文献: 2 [查看 Related Records] 引证关系图

入藏号: WOS:000285044300028

文献类型: Letter

语种: English

出版商: LIPPINCOTT WILLIAMS & WILKINS, 530 WALNUT ST, PHILADELPHIA, PA 19106-3621 USA

Web of Science 类别: Clinical Neurology

研究方向: Neurosciences & Neurology

IDS 号: 690XV

ISSN: 0028-3878

Figure 3 The informal record of retracted articles that without abstracts in Web of Science

All the 1814 formal records were published in 820 journals, where 18 of 820 journals have at least 10 retracted articles, as shown in Table 2. We take these 18 journals as the research samples and calculate the RIF for each of them. The results are shown in Table 2.

$$RIF_{2011} = \frac{M_{2011}}{N_{2011}} \tag{5}$$

where $M_{2011}$ = the number of cited times in 2006-2011 for retracted articles published in 2006; $N_{2011}$ = the total number of retracted articles published in 2006.

# 3 Results and Discussion

## 3.1 The Possible Reasons for the Relationship Between $IF_5$ and RIF

By the data in Table 2, we can see journals with high $IF_5$, such as *Nature*, *Science*, *Cell* and *Proceedings of the National Academy of Sciences of the United States of America*, have higher RIF compared with others. On the other hand, there are 5 journals with RIF equal to 0, and at the same time, their $IF_5$ were less than 10, especially less than 5. This comparison shows us the retracted articles published in high $IF_5$ also have more impact than the lower one. The possible reasons for this relationship between $IF_5$ and RIF are explained as follows.

First, although it is well known that $IF_5$ is a flawed measure of scientific quality and importance, articles published in high impact journals still can offer a disproportionate benefit to the authors, for example, improved their job opportunities, grant success, peer recognition, honorific rewards, etc. (Szklo, 2008). In this regard, the disproportionate payoff associated with publishing in higher impact journals could encourage risk behavior by authors in study design, data presentation, data analysis, and interpretation that subsequently leads to the retraction of the work. Therefore, it is reasonable that the ratio of fraud and academic misconduct is higher in articles submitted and accepted to high impact journals.

**Table 2 Journals with retracted articles at least 10 until 2011 in Web of Science**

| Journal's full name | The total number of retracted articles until 2011 | The number of retracted articles published in 2006 | The number of cited times between 2006-2011 | RIF in 2011 | $IF_5$ in 2011 |
|---|---|---|---|---|---|
| *Acta Crystallographica Section E-Structure Reports Online* | 101 | 11 | 34 | 3.09 | 0.278 |
| *Proceedings of the National Academy of Sciences of the United States of America* | 44 | 4 | 36 | 9.00 | 10.472 |
| *Science* | 37 | 3 | 194 | 64.67 | 32.452 |
| *Journal of Biological Chemistry* | 32 | 3 | 49 | 16.33 | 5.117 |
| Nature | 32 | 3 | 331 | 110.33 | 36.235 |
| *Journal of Immunology* | 26 | 2 | 38 | 19.00 | 5.859 |
| *Blood* | 18 | 1 | 24 | 24.00 | 9.785 |
| *Canadian Journal of Anesthesia-Journal canadien d' anesthesie* | 17 | 0 | 0 | 0.00 | 2.100 |
| *Cell* | 16 | 2 | 47 | 23.50 | 34.774 |
| *Journal of Hazardous Materials* | 16 | 1 | 7 | 7.00 | 4.553 |
| *Biochemical and Biophysical Research Communications* | 14 | 2 | 9 | 4.50 | 2.523 |
| *European Journal of Anaesthesiology* | 11 | 0 | 0 | 0.00 | 1.821 |
| *Infection and Immunity* | 11 | 0 | 0 | 0.00 | 4.055 |
| *Molecular and Cellular Biology* | 11 | 2 | 42 | 21.00 | 5.766 |
| *Annals of Thoracic Surgery* | 10 | 1 | 14 | 14.00 | 3.503 |
| *Cancer Research* | 10 | 0 | 0 | 0.00 | 8.164 |
| *Journal of Cardiothoracic and Vascular Anesthesia* | 10 | 0 | 0 | 0.00 | 1.358 |
| *Journal of Virology* | 10 | 3 | 58 | 19.33 | 5.324 |

Second, the requirement for clear and definitive reports by high impact journals may encourage authors to manipulate their data to meet this expectation. Compared with the clean and beautiful results, disorder results and fail observations are closer to the reality. In such situation, desperate authors may be enticed to take short cuts, withhold data from the review process, over-interpret results, manipulate images, and engage in behavior ranging from questionable practices to outright fraud (Fanelli, 2009).

Third, publications in high impact journals have more visibility and may attract greater scrutiny accordingly. Readers and researchers interested in that research and result may have more thorough check and then discover the problems, eventually leading to retraction.

Whatever the reason is, the relationship between RIF and $IF_5$ shows us, RIF is another measurement just like $IF_5$, to represent scientific performance. On the other hand, this relationship also gives us some insight into journals and science.

## 3.2 The Advantage of RIF

$IF_5$ uses cited times to reflect the level of journals' reference and influence. RIF also relies on the cited times, but for retracted articles to evaluate the level of negative impact. RIF analyzes the cited situation of retracted articles, and this is feasible and effective in theory. Compared with the previous studies, RIF does not focus on the number of retracted articles in journals, but on the citing situation of retracted articles, which could deeply find the impact of retracted articles on science.

The advantages of the new indicator RIF are described as follows. First, the previous studies on retracted articles are always aimed at the number of them in certain time period, so the results are more absolute, without balancing the difference between famous journals and general or new born journals. Second, RIF will detect the degree of impact in academic communities by analyzing the citation of retracted articles. This is easier to find out the severity, thus causing more attention on retracted articles. What's more, there is plenty of data in academic databases, which makes the information collected feasibly. Finally, apart from detecting impact of retracted articles, the discovery of articles that citing retracted articles is also important. Scientists can find out the relationship within articles and draw the road map of the retracted articles. Our work is a base stone for the next step study on the impacts of retracted articles.

# 4 Conclusions

In this paper, based on the existed quantitative research results and the idea of journal impact factor, we put forward a new indicator of retraction impact factor (RIF) to judge journals. Then, we calculated 18 journals' retraction impact factor, respectively, with the data of retracted articles from Web of Science in 2011. The results indicate that the RIF could represent the impact of retracted articles in academic communities. Finally, we showed the relationship between retraction impact factor and journal impact factor.

The RIF used in the case study is only a single year indicator, in other words, it represents specific journal's RIF in the certain year, and we just analyzed the situation in 2011 in Web of Science. In the further study, we will try to calculate specific journal's RIF for some successive years, find out the trends of journals and retractions, and construct a dynamic model about RIF and $IF_5$. These will help users to use journals and articles more effectively, and evaluate them more comprehensively.

## References

Bergstrom C T, West J D, Wiseman M A. 2008. The eigenfactor metrics. *Journal of Neuroscience*, 28 (45): 11433-11434.

Chen C, Hu Z, Milbank J, et al. 2013. A visual analytic study of retracted articles in scientific literature. *Journal of American Society for Information Science and Technology*, 64 (2): 234-253.

Fanelli D. 2009. How many scientists fabricate and falsify research? A systematic review and meta-analysis of survey data. PLoS ONE, 4 (5): 1-11.

Fang F C, Casadevall A. 2011. Retracted science and the retraction index. *Infection and Immunity*, 79 (10): 3855-3859.

Fang F C, Steen R G, Casadevall A. 2012. Misconduct acounts for the mjority of rtracted sientific pblications. *Proceedings of the National Academy of Sciences*, 109 (42): 7028-17033.

Foo J Y A. 2011. A retrospective analysis of the trend of retracted publications in the field of biomedical and life sciences. *Science and Engineering Ethics*, 17 (3): 459-468.

Grieneisen M L, Zhang M. 2012. Comprehensive survey of retracted articles from the scholarly literature. *PLoS ONE*, 7 (10): 1-15.

Liu S V. 2006. Top journals' top retraction rates. *Scientific Ethics*, 1: 91-93.

Murat C, Ivan I, Raul R E, et al. 2007. How many scientific papers should be retracted? *European Molecular Biology Organization*, 8 (5): 422-423.

Samp J C, Schumock G T, Pickard A S. 2012. Retracted publications in the drug literature. *The Journal of Human Pharmacology and Drug Therapy*, 32 (7): 586-595.

Shi Q. 2012. The analysis of pros and cons of current academic journal evaluation model and a new evaluation index. *Editorial Friend*, (12): 32-34.

Sox H C, Rennie D. 2006. Research misconduct, retraction, and cleansing the medical literature: lessons from the Poehlman case. *Annals of Internal Medicine*, 144 (8): 609-613.

Steen R G. 2012. Retractions in the medical literature: how can patients be protected from Risk? *Journal of Medical Ethics*, 38 (4): 228-232.

Szklo M. 2008. Impact factor: good reasons for concern. *Epidemiology*, 19 (3): 369.

Trikalinos N A, Evangelou, E, Ioannidis J P A. 2008. Falsified papers in high-impact journals were slow to retract and indistinguishable from non-fraudulent papers. *Journal of Clinical Epidemiology*, 61: 464-470.

Wang Y. 2011. Comparison of IF (JCR), IF (Scopus), *h*-Index, SJR score, SNIP score with peer judgment for journal evaluation. *Library and Information Service*, 55 (16): 144-148.

Wikipedia. Impact factor. http: //en. wikipedia. org/wiki/Impact_factor [2013-06-13] .

Wikipedia. Retraction watch. http: //en. wikipedia. org/wiki/Retraction _ Watch [2013-05-25] .

Zhao X. 2010. An analysis of 5-year impact factor in JCR. *Journal of Library Science in China*, 36 (187): 120-126.

# 3-3 ID Index to Calculate the Influence of an Academic Paper Under the Context of Citation Network*

**Xia Hui, Tong Ying, Han Yi**[①]

## Abstract

The influence of an academic paper should be assessed under the context of citation network, in which a literature's references reveal the source of knowledge and its inheritance relationship, and a literature's citing papers reflect the flow and diffusion of knowledge. ID index and its derived RID index are designed on the basis of such inheritance and diffusion to evaluate the influence of an academic paper, which not only takes into account direct references and direct citing papers but also the contribution of indirect references and indirect citing papers. With the academic papers of Library and Information Science in Web of Science as a sample, the present research selects 12 sample vertexes in main path of its citation network to calculate their ID and RID. The correlation coefficient between the ID and traversal value verifies that the ID index has higher validity and effectiveness to measure the influence of an academic paper, and tells that $\alpha = 2$ and $\beta = 2$ are the optimistic adjusting factors.

**Keywords**: single academic paper; academic influence assessment; ID index; citation network; main path

## 1 Introduction

A scientific, objective and fair measure of the influence of academic literature is

* Research for this article has been funded by scientific and technological foundation of Southwest University (granted number SWU112001).

① The corresponding Author, No. 1 of Tiansheng Road, Beibei, Chongqing, 400715 (China), Hanyi72@swu. edu. cn.

of great significance for its objective assessment of scientific creation and academic works of researchers as well as its guidance to management work, such as selection of scientific research project, establishment of the assessing system of researcher performance, application of research funds, formulation of scientific plan and policy, and so on. However, it is always a puzzling problem to get a scientific, objective and fair evaluation method to measure the influence of academic literature, especially that of a single academic paper.

Measuring the influence of academic literature in the context of citation network is of great necessity since every literature is firmly grounded in corresponding citation network system. In citation network citing and cited relationship among literatures can be included, the characteristics of direct and indirect citation can be reflected, and the particular position of a specific literature in the structure of citation network can be located. An influential evaluation index designed on the basis of citation network structure would be undoubtedly an objective mapping for the holographic information of citation network. Literature in citation network would inherit the influence of cited literature by citing and simultaneously diffuse its own academic influence if cited. Inheritance, therefore, is the accumulation of influence; diffusion, meanwhile, the penetration of influence. There would be an increase in an academic paper's influence, academic strength and the power of back penetration if with more authorized knowledge source and more creative theories and knowledge. Thus, the influence of an academic paper is the integration of knowledge accumulation and penetration ability. Citation network is a faithful record of the process of influence accumulation and penetration of an academic paper. So how to quantify the influence of an academic paper based on this process is the major problem to be solved in this paper.

## 2 Related Researches on the Measurement of the Influence of an Academic Paper

Various methods are adopted by scholars to assess the influence of a single academic paper scientifically, objectively and fairly. Generally speaking, two kinds of methods are in current use: qualitative evaluation and quantitative evaluation.

### 2.1 Qualitative Evaluation

Commonly used method of qualitative evaluation is peer review which refers to the process of evaluating a scientific activity and its result by an evaluation committee composed of experts in a given scientific field (Geisler, 2000). Peer review can provide quality control for the distribution of scientific and technological resources in any level, from individual to institutions and to nation; and it is one of the methods to do after-evaluations for the scientific and technological activities and their performers. With definite goals and standards, objective and fair results can be achieved through peer review. But peer review usually costs much time and is liable to the influence of subjective and emotional factors of reviewers, which causes difficulty in ensuring an objective, fair and impartial evaluation.

In the biomedical field, F1000 (faculty of 1, 000) system was developed based on peer review (Bornmann and Leydsdorff, 2013). The system is the quantitative manifestation of peer review with the aim of an objective evaluation of important papers collected in the biomedical database, such as SCI and PubMed, on the basis of academic achievement rather than its source journals. The evaluation results have been catalogued into three grades: outstanding literature (score more than 9), the required ones (score between 6 to 9), and recommended ones (score between 3 to 6). An asserted involvement of more than ten thousand experts in the evaluation processes cannot cover the drawback of peer review itself. Moreover, evaluated papers are limited to the biomedical field and the number of the evaluated paper is relatively small (up till the present altogether about 130 000 articles). The application of F1000 is restricted as the result of its charging service system. Despite these drawbacks, the evaluation system is of great value to be popularized, especially so far as the OA (open access) periodical papers are concerned.

## 2.2 Quantitative Evaluation

With regard to the quantitative evaluation, scholars study the evaluation method mainly on two levels: using a single index or using integrated indices.

Easily obtainable, the journal impact factor becomes one of the common indices to evaluate the academic quality and influence of an academic paper, which makes it the simplest method to evaluate a single academic paper by using journal impact factor directly (Hoeffel, 1998; Garfield, 2006). But the inter-causal relationship between impact factors and paper quality renders this method a target of criticism by scholars. Then why is this method still used in practice? As Hoeffel (1998) and Garfield (2006) argued, the journal impact factor is not the perfect tool to evaluate paper quality, but no better ones are available at present. Owing to the great differences in impact factors among different disciplines, impact factor point average (IFPA) is proposed to solve the problem of the comparison of impact factors among different disciplines (Sombatsompop et al., 2005).

Except the journal impact factor, there are some other single indexes to measure the academic influence of a single paper, such as the citing number, paper quality index (Qiu et al., 2007), academic paper quality index based on citation strength (Wu, 2007), academic paper assessment in the same field based on principal component analysis (Long and Ge, 2007), single paper *h*-index (Schubert, 2009; Thor and Bornmann, 2011).

Sincethe paper quality is determined by multiple factors, a comprehensive evaluation index should be designed to integrate the advantages of various methods and avoid their disadvantages. Many suggestions have been proposed, for example, the integrated evaluation method based on indices such as periodical literature type, periodical influence, international influence, and fund assistance (Zhang et al., 2004); the integrated academic quality indices based on the periodical influence factor, the paper cited frequency and non-self-citation frequency (Guo, 2005); the comprehensive evaluation system based on non-self-citation amount and journal impact factor (Jin et al., 2010).

In spite oftheir special characteristics and respective advantages, a common problem occurs: emphasis is merely put on direct citation relationship among papers without considering various indirect citations and the value of references, and little attention is paid to citation context structure, namely a paper's position in the citation network. Taking direct and indirect references, direct and indirect citation, and citation network into consideration, the present research tries to establish a new evaluation index.

# 3 Method and Data

## 3.1 Research Method

Academic papers form a self-organized network by citing each other. Figure 1 is a simple citation network. In a citation network, some nodes or elements occupy central status because of their positional relationship, in which they play a key part in inheriting and diffusing the field knowledge. For a paper, except for its citation number, citation structure has provided important background information to present its influence. In Figure 1, if we simply use the citation number as a measuring standard, the citation number of node 2, 8, 10 all are 3, and their influence should be the same; but if we take the structure information into account, then the influence of node 8 and 10 may be greater than that of node 2. Particular attention should be paid to node 8 which is the bridge of the entire network and whose influence should be more greater. Therefore, the academic value of a paper is both associated with the backtracking depth of the references and the extending breadth of the cited papers. Backtracking depth and extending breadth is a comprehensive reflection of the quality effect of a paper in citation network.

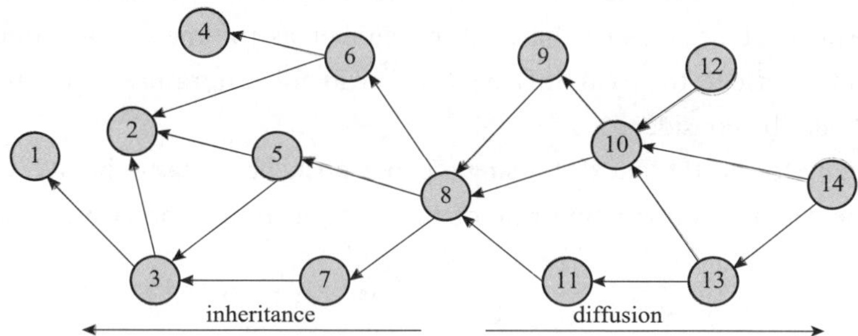


Figure 1　A simple citation network

If treating the nodes in each level equally (True a refiner processing requires that different weights should be put on the different elements structure and the different number relationship, but for the convenience of data processing, they are temporarily regarded as having the same weight), we can construct a simplified evaluation index to measure an academic paper influence as follows:

$$\mathrm{ID} = \sum_i \frac{r_i}{i^{\alpha}} + \sum_j \frac{c_j}{j^{\beta}} \quad (i=1,\ 2,\ \cdots,\ m;\ j=1,\ 2,\ \cdots,\ n) \tag{1}$$

In the index, $i$ indicates the backtracking depth of the cited papers of the assessed literature; $j$ indicates the extending breadth of the citing papers; $r_i$ indicates the cited number of the $i$-th backtracking level; $c_j$ indicates the citing number of the $j$-th extending level; and $\alpha$ and $\beta$ are the adjusting factors of the influence of references and citing papers. The calculation result is named ID, which is the comprehensive interacting effect of knowledge inheritance and diffusion of the assessed paper.

Generally, the closer to the measured paper, the greater the contribution of its influence is, so different weights should be given to different cited or citing levels. When $i=1$, $r_1$ is the direct references number of the assessed paper; when $i\geqslant 2$, $r_i$ is the indirect references; we use $1/i^{\alpha}$ as the adjustment factor to every backtracking level, which exemplifies out greater attention to short-distance effects. It is the same to $j$. Take node 8 in Figure 1 as an example. Its backtracking level is 3, and its extending breadth is 2. In terms of backtracking, there are 3, 3 and 1 cited papers on the first, second, and third level respectively. So far as extending breadth is concerned, there are both 3 citing papers in the first and second level. If let $\alpha = 2$ and $\beta = 2$, to node 8, $ID_8 = 7.61$. And similarly, $ID_2 = 4.01$ and $ID_{10} = 7.30$. Comparing their ID value, we can know that the influence of node 8 is greater than that of node 2 and 10, while the influence of node 10 is greater than node 2. The ID value is totally different from the result of measuring simply on the basis of citation number. This indicates that: even the citation number is completely the same, the paper influence is not the same because of the different location in citation network. Hence, using the citation number as the mere evaluation criterion may not fully reflect the real influence of academic literature, and the citation structure should be considered.

According to the ID index of a single paper influence measurement, we can also calculate the relative contribution rate of the references and the citing papers:

$$\mathrm{RID} = \frac{\sum_i \frac{r_i}{i^\alpha}}{\sum_j \frac{c_j}{j^\beta}} \tag{2}$$

If RID$\geqslant$1, it indicates that the contribution of knowledge inheritance factors is greater than that of knowledge diffusion factors, and vice versa. In Figure1, if let $\alpha = 2$ and $\beta = 2$, we can get the results: $RID_2 = 0$, $RID_8 = 1.03$, $RID_{10} = 1.43$. As can be seen, node 2 is without knowledge inheritance, which means this node is the source of the citation network and its main effect is knowledge diffusion. Knowledge inheritance contribution of node 8 is weaker than that of node 10. Of course, for those papers without citations, namely the sink of citation network, the relative contribution rate will be infinite with the indication that their full contribution is knowledge inheritance.

From the above, a problem will be raised naturally: whether the ID index can be used in the actual citation network? Whether the ID values reflect the importance of the node itself? In the following part, it will be verified by actual citation network of some sample field.

## 3.2 Data Collection

Sixteen kinds of core journals of Library and Information Science, collected by SCI and tabulated in Table 1 (Yan and Ding, 2009), are chosen in this research as a sample. The sample data was retrieved on the Web of Science on November 30, 2011. The document type of these data includes article, book, book chapter, book review, discussion, hardware review, letter, note, proceedings paper, review, software review. They come from SCI-expanded、SSCI、CPCI-S databases and their time range is "all year" (1900-2011).

**Table 1 Journal name of library and information science sample**

| Number | Journal | Number | Journal |
|---|---|---|---|
| 1 | *Annual Review of Information Science and Technology* | 9 | *Journal of the American Society for Information Science and Technology* |
| 2 | *Information Processing and Management* | 10 | *Information Society* |
| 3 | *Scientometrics* | 11 | *Online Information Review* |
| 4 | *College and Research Libraries* | 12 | *Library Quarterly* |
| 5 | *Journal of Documentation* | 13 | *Library Resources and Technical Services* |
| 6 | *Journal of Information Science* | 14 | *Journal of Academic Librarianship* |
| 7 | *Information Research* | 15 | *Library Trends* |
| 8 | *Library and Information Science Research* | 16 | *Reference and User Services Quarterly* |

First of all, main path analysis is used to sort all the nodes by traversal value. Secondly, some sample nodes in the main path are selected and their ID values are calculated. Finally, the correlation analysis between the traversal values of sample nodes and their ID values are done.

# 4 Research Results

## 4.1 The Main Path of Citation Network of Library and Information Science

The main path is a path from source to sink in acyclic network, whose arcs have the highest traversal values (De Nooy et al. , 2005). The main path analysis focuses on the connectivity of citation network (Hummon and Doreian, 1989). Main-path techniques examine connectivity in acyclic networks, and are especially interesting when nodes are time dependent, as it selects the most representative nodes at different moments of time.

Three models to identify the most important part of a citation network can be distinguished: the node-pair projection count, which accounts for the number of times each link is involved in connecting all node pairs; the search-path link count, which accounts for the number of all possible search paths through the network emanating from an origin; and the search-path node pair, which accounts for all connected vertex pairs along the paths (Hummon and Doreian, 1989). Of these three methods, the latter two algorithms are included in Pajek and a new algorithm, search path count, is designed (Batagelj, 2003). The search-path link count is the preferred algorithm for this analysis because all citation relations are taken into account. In this study, we use the search path count algorithm to get the main path of the sample (Figure 2).

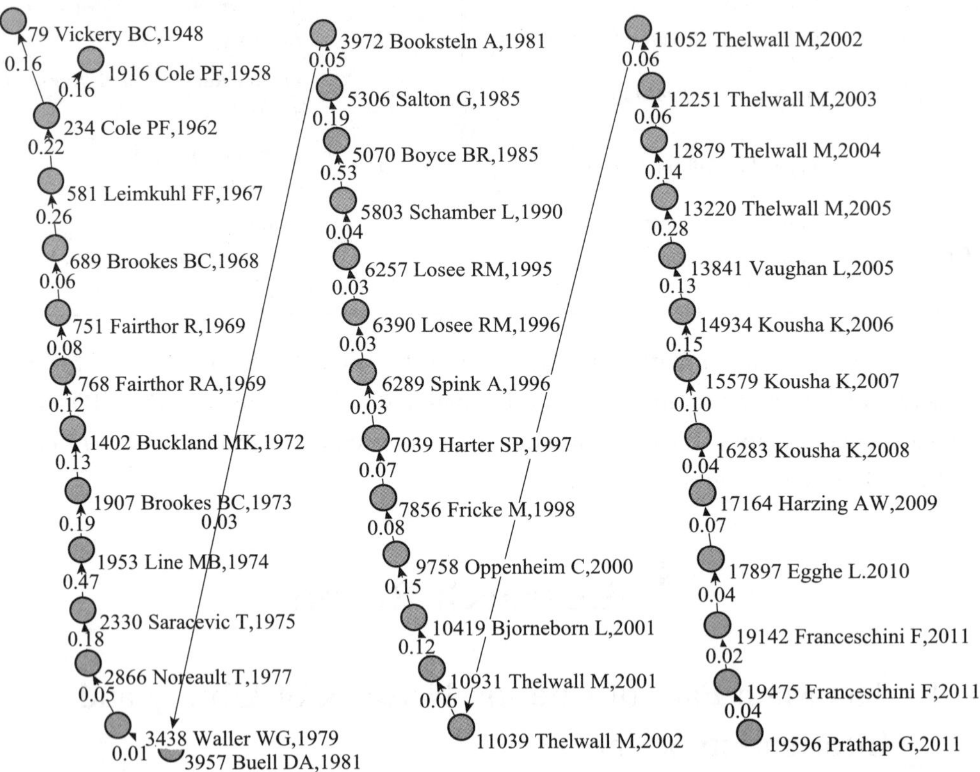


Figure 2　The main path and traversal values of library and information science sample

## 4.2　ID Value and RID Value of Sample Nodes

Traversal value reflects the degree of the importance of a node and the paper chooses those nodes with different traversal values to calculate ID value. ID value is also able to reflect the citation structure information of nodes. A higher linear relationship between traversal value and ID value means that the ID index is valid in theory.

12 representative nodes were chosen from the main path (Table 2). These nodes were analyzed with Pajek to achieve $k$ -out-neighbors ( $k$ -level backtracking references) and $k$ -

in-neighbors ( $k$ -level diffusion citation ) of these sample nodes.

**Table 2 Label and traversal value of sample nodes**

| Label | Traversal value | Label | Traversal value |
|---|---|---|---|
| 234 Cole PF, 1962 | 0. 22 | 581 Leimkuhl FF, 1967 | 0. 26 |
| 1907 Brookes BC, 1973 | 0. 19 | 1953 Line MB, 1974 | 0. 47 |
| 5306 Salton G, 1985 | 0. 19 | 7856 Fricke M, 1998 | 0. 08 |
| 9758 Oppenheim C, 2000 | 0. 15 | 10419 Bjorneborn L, 2001 | 0. 12 |
| 12251 Thelwall M, 2003 | 0. 06 | 16283 Kousha K, 2008 | 0. 04 |
| 17897 Egghe L, 2010 | 0. 04 | 19142 Franceschini F, 2011 | 0. 02 |

In order to display the composition of ID value in details, a bigger neighbor level ( $k = 10$) is chosen in the present research. To the sample nodes, the level numbers of neighbor, namely out-degree (backtracking level) and in-degree (diffusion level), are from 1 to 10, and the $r_i$ and $r_j$ can be obtained. So the ID value and RID value can be calculated. Taking the node 581 as an example, we can see that the number of each level backtracking document is 2, 1, 0, 0, 0, 0, 0, 0, 0, 0, and the number of each level diffusion document is 24, 66, 156, 661, 1550, 907, 291, 75, 26, 3. If let $\alpha = 2$ and $\beta = 2$, thus the influence of node 581 can be calculated, $ID_{581} = 196.05$, and $RID_{581} = 0.011$, 61.

The next problem is how to set the value of $\alpha$ and $\beta$. If $\alpha = 1$ and $\beta = 1$, this means that the ID is the sum of cited and citing number of all levels, which can not manifest the different contribution in different levels, so it is not a meaningful selection. But more bigger the value of $\alpha$ and $\beta$ are, the greater reduction of influence of indirect cited and citing papers are. In this research, $\alpha = 2$ and $\beta = 2$, $\alpha = 3$ and $\beta = 3$, and $\alpha = 4$ and $\beta = 4$, are selected to calculate the ID and RID, and their detailed data in Table 3. Why did we let $\alpha = \beta$ and not other relationship? If $\alpha = \beta$, it can be seen that we take the cited paper and citing ones as the same position. If not, they are in different position and the different influences are focused, and it is not suitable the hypotheses in this research, which is the cited and citing papers, have similar influence contribution to the assessed document.

As can be seen from Table 3, ID value differs greatly from the direct reference number and citation number. Let's take the node 16283 as an example. When $\alpha = 2$ and $\beta = 2$, its ID value is equal to 131. 32, but the number of direct reference (of the paper published in 2008) is 14 ($r_i = 14$) and direct citation is 15 ($r_j = 15$), and the latter is dramatically smaller to the former.

Table 3 ID value and RID value of each sample node ($i = j = 10$)

| Label | $\alpha = 2, \beta = 2$ | $\alpha = 3, \beta = 3$ | $\alpha = 4, \beta = 4$ |
|---|---|---|---|
| 234 | 173. 717 7 | 49. 886 9 | 21. 188 2 |
| | 0. 011 65 | 0. 041 77 | 0. 104 23 |
| 581 | 196. 051 9 | 68. 113 5 | 38. 019 1 |
| | 0. 011 61 | 0. 032 2 | 0. 573 6 |
| 1907 | 239. 388 1 | 78. 377 9 | 34. 315 9 |
| | 0. 038 06 | 0. 102 07 | 0. 237 52 |
| 1953 | 304. 135 5 | 126. 524 | 74. 260 8 |
| | 0. 112 71 | 0. 294 01 | 0. 599 29 |
| 5306 | 185. 604 4 | 68. 759 5 | 40. 194 4 |
| | 0. 406 36 | 1. 169 07 | 3. 296 61 |
| 7856 | 119. 000 1 | 41. 672 9 | 19. 532 7 |
| | 1. 616 77 | 2. 442 03 | 3. 816 15 |
| 9758 | 149. 749 6 | 57. 813 2 | 28. 868 |
| | 0. 811 21 | 0. 723 86 | 0. 603 57 |
| 10419 | 200. 523 6 | 112. 089 | 82. 459 7 |
| | 0. 401 13 | 0. 265 63 | 0. 218 93 |
| 12251 | 137. 997 6 | 70. 496 9 | 50. 057 9 |
| | 1. 177 9 | 0. 927 9 | 0. 863 1 |
| 16283 | 131. 315 6 | 61. 546 1 | 40. 805 7 |
| | 4. 007 8 | 2. 059 6 | 1. 344 73 |
| 17897 | 172. 098 | 93. 933 6 | 71. 363 2 |
| | 15. 790 05 | 8. 759 34 | 6. 663 17 |
| 19142 | 99. 544 83 | 33. 957 | 17. 103 1 |
| | 43. 242 15 | 14. 979 8 | 7. 292 42 |

Generally citation number is taken as the only criterion to measure the academic influence. Thus, the academic influence of node 16 283 is more important than that of node 9758 whose direct citation is 9 ($r_j = 9$); and node 16 823 has a better quality than node 5306 whose direct citation is 2 ($r_j = 2$). On the contrary, in terms of ID values, node 5306 (ID = 185. 60) is more important than node 9758 (ID = 149. 75) and node 9758 is better than node 16283 (ID = 131. 32).

According to RID value, the values of the front sample nodes are nearly less than 1, which indicates knowledge diffusion factors have a larger contribution thanknowledge inheritance factors in these documents' influential elements. RID value is smaller, and the influence of knowledge diffusion is more profound. RID value of the back nodes is more than 1, which shows the greater contribution of knowledge inheritance factors than knowledge diffusion factors. The larger RID value is, the longer the history of knowledge inheritance is. In a further sense, the node in the front of the network has more knowledge diffusion elements, while the nodes at the back of the network have more knowledge inheritance elements.

## 4.3 Correlation Analysis of ID Value and Traversal Value

A positive correlation has been found between the competence of knowledge diffusion in citation network and traversal value. Hence, the document has great power of influence and is the core document in the course of discipline evolution; accordingly, the author has great influence in this discipline. As can be seen from Table2, node 1953 has the greatest influence (traversal value is 0.47) and node 19142 has the least influence in the network (traversal value is 0.02).

A contrast between traversal value of sample nodes and ID value reveals that: if traversal value is larger, ID value is greater, which indicates a high relevance between the two measured values. From another perspective, the constructed ID index can be proved available if relevance reaches a fairly high significance level. The research calculates and inspects the Pearson correlation coefficient of ID value and traversal value by using SPSS 13.0 (Table 4).

**Table 4 The pearson correlation coefficient of ID value and traversal value**

| | | $ID_1$ | TRAN |
|---|---|---|---|
| | Pearson correlation | 1 | 0.865** |
| $ID_1$ | Sig. (2-tailed) | | 0.000 |
| | *N* | 12 | 12 |
| | Pearson correlation | 0.865** | 1 |
| TRAN | Sig. (2-tailed) | 0.000 | |
| | *N* | 12 | 12 |
| | | $ID_2$ | TRAN |
| | Pearson correlation | 1 | 0.530** |
| $ID_2$ | Sig. (2-tailed) | | 0.076 |
| | *N* | 12 | 12 |
| | Pearson correlation | 0.530** | 1 |
| TRAN | Sig. (2-tailed) | 0.076 | |
| | *N* | 12 | 12 |
| | | $ID_3$ | TRAN |
| | Pearson correlation | 1 | 0.251** |
| $ID_3$ | Sig. (2-tailed) | | 0.431 |
| | *N* | 12 | 12 |
| | Pearson correlation | 0.251** | 1 |
| TRAN | Sig. (2-tailed) | 0.431 | |
| | *N* | 12 | 12 |

** means correlation is significant at the 0.01 level (2-tailed).

It can be found from the analysis (Table 4): when $\alpha = 2$ and $\beta = 2$, there is a fairly high positive linear correlation (correlation coefficient is 0.865) between ID value and traversal value. The two-tailed testing result at 0.01 confidence level is 0.000, far less than the critical value of 0.01. The significant linear correlation is established in between. Thus, the constructed ID index scientifically reflects the citation network's structural features and the importance of the measured object. Values reflect the importance and influence of the measured document and they can be used to measure the influence of a single academic paper in the context of citation network. But when $\alpha = 3$ and $\beta = 3$, and $\alpha = 4$ and $\beta = 4$, the two-tailed significance is 0.076 and 0.431 respectively, so the linear correlation between ID value and traversal value is not significant. And the Pearson correlation coefficients are lower, 0.530 and 0.251 respectively. So we can get the conclusion: $\alpha = 2$ and $\beta = 2$ are the best adjusting factor to calculate ID index.

## 5 Conclusions and Discussions

The influence of an academic paper is rooted in the citation network structure. The number of direct citation and direct references cannot fully reflect its influence, and indirect citation and reference in the citation network also contribute to different degrees. Among them, all of the citations, direct and indirect, show the ability to diffuse knowledge, which manifests the penetrating effect of the influence. All the references, direct and indirect, reflect the competence of knowledge inheritance, which manifests the accumulative effect of the influence.

In this research, ID index based on the penetrating effect and accumulative effect, and RID reflecting constituent ratios of the two effects are constructed. The index not only changes the simple way to evaluate the academic influence of literature by means of the mere use of direct reference and direct citation, but also reveals the comprehensive and in-depth structural information of an academic paper in the whole citation network. It is a more comprehensive and influential evaluation index.

The research chose 12 nodes, which came from the 16 Library and Information Science journals collected by Web of Science, to calculate ID index and RID index based on the traversal value of the main path nodes with the result of an indication that ID value grows with traversal value. Pearson correlation coefficient test shows that there is a high positive linear correlation between ID value and traversal value at a strict testing level ($\alpha = 0.01$) when $\alpha = 2$ and $\beta = 2$. All the results have shown ID index has a relatively high effectiveness and reliability. But the other value of $\alpha$ and $\beta$ did not get satisfied testing significance. So $\alpha = 2$ and $\beta = 2$ are the best adjusting factor to calculate ID index.

Though it could fully reflect inheriting and diffusing characteristics of the measured object, the calculation of the ID index is relatively complex and must be with the aid of certain software. Thus, the practical application of the ID index would be somewhat limited. In addition, how to determine the adjustment factors of different levels (i. e., weight) is an issue for further study.

## References

Batagelj V. 2003. Efficient algorithms for citation network analysis. http://www.imfm.si/preprinti/PDF/00897.pdf [2012-11-15].

Bornmann L, Leydesdorff L. 2013. The validation of (advanced) bibliometric indicators through peer assessments: a comparative study using data from InCites and F1000. *Journal of Informetrics*, 7 (2): 286-291.

De Nooy W, Mrvar A, Batagelj V. 2005. *Exploratory Social Network Analysis with Pajek*. New York: Cambridge University Press.

Garfield E. 2006. The history and meaning of the journal impact factor. *The Journal of American Medical Association*, 295 (1): 90-93.

Geisler R. 2000. *The Metrics of Science and Technology*. New York: Greenwood Publishing Group Inc.

Guo L F. 2005. Research on bibliometric indicator to assess the quality of academic papers. *Modern Intelligence*, (3): 11-12 (in Chinese).

Hoeffel C. 1998. Journal impact factor. *Allergy*, 53 (12): 1225.

Hummon N P, Doreian P. 1989. Connectivity in a citation network: the development of DNA theory. *Social Networks*, 11 (1): 39-63.

Jin J, He M, Wang X N, et al. 2010. Feasibility research of evaluation and comparison of natural science papers in different fields. *Science and Technology Management Research*, 30 (14): 279-284 (in Chinese).

Long S, Ge X Q. 2007. Evaluation of academic level of scientific papers. *Science-Technology and Management*, (1): 133-135, 138 (in Chinese).

Qiu J P, Ma R M, Cheng N. 2007. New approaches for evaluation of scientific researches with SCI. *Journal of Library Science in China*, (4): 11-16 (in Chinese).

Schubert, A. 2009. Using the $h$-index for assessing single publications. *Scientometrics*, 78 (3): 559-565.

Sombatsompop N, Markpin T, Yochai W, et al. 2005. An evaluation of research performance for different subject categories using impact factor point average (IFPA) index: Thailand case study. *Scientometrics*, 65 (3): 293-305.

Thor A, Bornmann L. 2011. The calculation of the single publication $h$-index and related performance measures: a web application based on Google Scholar data. *Online Information Review*, 35 (2): 291-300.

Wu Q. 2007. Research on quality evaluation in the academic articles based on the intensity of citation. *Journal of China Society for Scientific and Technical Information*, 26 (4): 522-

526（in Chinese）.

Yan E，Ding Y. 2009. Applying centrality measures to impact analysis：a co-authorship network analysis. *Journal of the American Society for Information Science and Technology*，60（10）:2107-2118.

Zhang Y H，Pan Y T，Ma Z. 2004. Evaluation methods for scientific papers. *Acta Editologica*，（8）：243-244（in Chinese）.

# 四、科学计量与大学评估

# 4-1 Analysis of the Results and Innovation of the Idea and Method of Chinese University Evaluation in 2013

Qiu Junping①, Li Xiaotao, Dong Ke

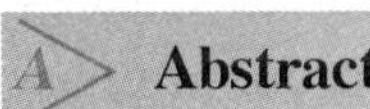

## Abstract

Chinese university evaluation is discussed from the aspects of the innovation (idea, method, technology), results and analysis, interpretation of the trend, focus on the analysis of the innovation of ideas, method and technology about university evaluation, and interpretation of the trend from the data of 2013's evaluation. Firstly, official introduced university web influence index which represents the value of the university's brand, performance index (comprehensive reflection of the school benefit and efficiency), and innovation index (reflecting university innovation ability and level), and the development of the corresponding information system is made up of all the evaluation activity automation and intelligence. From the result of Chinese university and subject evaluation in 2013, it is concluded that the Chinese university brand value (including the network brand, special major setting and the discipline construction, etc.), science and technology innovation level, and performance level have a great influence on university comprehensive competitiveness.

**Keywords**: Chinese universities; university evaluation; innovation; correlation methods; indicators

## 1 Introduction

Undergraduate education is the most important fundamental part in the higher

① SResearch Center for Chinese Science Evaluation, Wuhan University, Wuhan, China, jpqiu@whu. edu. cn.

education system, and how to evaluate the overall competitiveness of the institutions of higher learning and the quality and level of the construction of subjects and specialties, and used scientific, rational, objective and fair methods, is widely concerned by the community. Research Center of Chinese Scientific Evaluation of Wuhan University (RCCSE) follows the principle of classification of evaluation, and the intelligence services, and evaluates the colleges and universities according to different levels, different types, different disciplines and professionals, meanwhile uses of a variety of information technology to achieve deep excavation of evaluation information resources and intelligent services. Through the construction of a web-based information service platform, maximize the value of higher information resources of education evaluation (Qiu and Wen, 2010). Recently, RCCSE completed the 2013-2014 evaluation report of Chinese universities and disciplines, concluded 555 rankings, including: the education regional competitiveness ranking of the universities in China, China's top university rankings, the competitiveness ranking of China's key universities, the competitiveness ranking of science and technology innovation of Chinese universities, the competitiveness ranking of humanities and social sciences innovation of Chinese universities, the competitiveness ranking of classification of Chinese universities, the competitiveness ranking of fields of discipline of Chinese Universities, the competitiveness ranking of big categories of specialties, etc.

# 2 Innovation of the Philosophy and Method of Chinese University Evaluation of 2013

## 2. 1 Brand Value of Network of the University Becomes an Important Factor of Measuring the Overall Competitiveness

With the further improvement of our country network infrastructure, the Internet life of the public is increasingly rich. In a network environment, how to make full use of the network to expand its influence to further serve the public is an important thing to theuniversities (Qiu, 2006). Undergraduate colleges and universities are an important position of state higher education, and the degree of its network influence will directly affect the efficiency of the dissemination and popularization of scientific and technical information of our country. Therefore, we need to research the network influence of the Chinese university. In this university evaluation, we added the indicators of the university network reputation survey and network impact indicators. Theories and methods of the web impact factor, link analysis, we learn from the theory of network marketing brand value and webometrics, concluded the five aspects to evaluation the web influence of university, including the university website content quality, scale, search engine friendliness, number of inbound links to the website (inlinks, referenced by external sites, can be considered to represent the recognized and recommended), and the network academic influence (web citation analysis). Based on the above research, the system of the network influence evaluation was developed (Qiu and Cheng, 2009).

## 2.2 Pay More Attention to the Performance Indicators of Universities

In this evaluation, we pay more attention to the performance of universities, because the performance more fully reflects the ability of universities on the social, economic, and cultural services. *National Long-term Education Reform and Development Planning (2010-2020)* (*Planning (2010-2020)* for short) clearly stated: to improve the management mode, to introduce competition mechanism, and to implement the performance evaluation and the dynamic management. At the same time, the *Planning (2010-2020)* also pointed out: to establish the allocation of resources and academic development model oriented by academic and innovative performance, and to implement the performance appraisal of personnel of the institutions and the introduction of talent performance evaluation. Therefore, the evaluation of institutions of higher learning by the "quantity" measure to change the standard of the "effect" . The further development of the university should use the quality and effectiveness as measure factors. This is the future direction and focus of the development of higher education evaluation. In order to implement the *Planning (2010-2020)* in the performance evaluation requirements and tasks, on the basis of eight consecutive years released ranking of the universities, evaluation of the school performance of China's key universities was explored. The performance evaluation to the division are performance standards reflect the "quality" to "effect" will focus on the university evaluation of the "quantity" to the "quality" of the "quantity" to take into account the comprehensive scientific evaluation.

## 2.3 Pay More Attention to the Innovative Capacity Index of Universities

In this university evaluation we first proposed the "innovation index" refers to the scores in the evaluation of innovation capability, innovation capability of a quantitative expression. To emphasize innovative focus and the main part is the innovation of the basic science and scientific knowledge, involving the entire innovation chain, including universities, enterprises, government, finance, intermediary organizations, and other related support and auxiliary institutions, university innovation in the upstream in the National Innovation System has an important basic role and strategic role. Evaluation index system including a platform for innovation, innovators and innovation achievement three first level indicators, the university innovation index ranking will be issued as a special report.

## 2.4 A Multidisciplinary Approach Integration

Adopted a multidisciplinary approach to further integration in the evaluation process, the theoretical framework of evaluation has the interdisciplinary characteristics, the disciplines involved in decision science, management science, policy science, behavioral science, economics, metrology, engineering, social science, psychology and logic etc. Design and implementation of scientific evaluation activities require an interdisciplinary approach requires integration and application of the theory in different areas of expertise. Our university evaluation of the integrated use includes the following subject areas: comprehensive evaluation methods (such as analytic hierarchy process, fuzzy comprehensive evaluation method), statistical methods, bibliometrics methods (citation analysis), webometrics methods (such as link analysis, network information acquisition), and artificial intelligence (machine learning, neural networks, etc.).

# 3 Objects and Practices of the 2013 University Evaluation

## 3.1 The Object of Evaluation

According to the 2012 regular higher academic education enrollment eligibility colleges and universities list annouced by the Ministry of Education, as of May 22, 2012, a total of 844 colleges and universities. University evaluation report released this year in the name of the university is in the list of the Ministry of Education. In addition, we discarded not have an undergraduate vocational college, and universities by the Ministry of Education canceled undergraduate enrollment eligibility, and ultimately into the evaluation the university a total of 2, 742 (excluding military class institutions and Hong Kong, Macao and Taiwan areas colleges and universities), 131 key universities, 621 general universities, 403 private colleges. List of the key universities to increase 6 good local colleges of the Provincial-Ministry building universities and has a good social reputation and perform well in our rankings in the past few years, namely the Hebei University, Shanxi University, Zhejiang University of Technology, Henan University, Northwest Normal University and Heilongjiang University.

Professional directory of the university evaluationreferred to the Ministry of Education announced on September 14, 2011, *University Undergraduate Catalog* (Amendment Draft, second edition) were merged and the corresponding benchmark, and our previous 192 undergraduate evaluation of professional. According to the Ministry of Education's undergraduate catalog, a new category of art was added on the basis of the original 11 disciplines. According to principle of emphasis on basis and wide caliber, the actual undergraduate enrollment greater use of specialty big categories, we combine the latest professional catalog of the Ministry of Education, and will execute evaluation in accordance with the 94 specialty big categories.

## 3. 2 Data Sources and Processing Methods

In the evaluation of undergraduate education, the workload for primary data collection is very large, so we put a lot of manpower and material resources. After years of practice and exploration, We have established relatively stable, reliable data sources. The raw data is mainly from five aspects:

(1) Government departments, the relevant statistical data (including the yearbooks, reports, etc.);

(2) The database at home and abroad;

(3) Website of the relevant government departments and universities;

(4) The relevant national publications, books, newspapers, and internal information;

(5) The basic database platform of RCCSE.

RCCSE has independently developed the network reputation survey system, the evaluating raw data network acquisition system, the paper automatic classification system and information management system of Chinese university. Achievements of the automation and intelligent processing on data acquisition, statistics, computing, sorting diagnostic analysis, have greatly improved the efficiency and scientificality of the university of evaluation work.

## 3. 3 Indicator System of Evaluation

The University evaluation remained stable, and only a slight adjustment has executed in a few indicators and methods of data collection and calculation, in order to better reflect the level of technological innovation and scientific research quality of universities, for example, patent applications and license number change the number of patent licensing (Table 1).

**Table 1 Indicators and weights of Chinese universities' competitiveness evaluation**

| First level indicators | Second level indicators | Third level indicators |
|---|---|---|
| Educational resources | The basic conditions | 1. Total area of premises<br>2. Campus area per students<br>3. Total equipment<br>4. Total equipment per student<br>5. Sum of books<br>6. Sum of books per student |
| | Funding for education | 7. Sum of education expenditure<br>8. Sum of education expenditure per student |
| | Teachers | 9. The number of academicians of Chinese Academy of Sciences and Chinese Academy of Engineering<br>10. Outstanding People (Cheung Kong Scholars, Trans-century Talents, University Award for Young Teachers)<br>11. The number of PhD supervisor<br>12. Proportion of senior professional titles<br>13. Student-teacher ratio |
| | Advantage of subject | 14. The number of doctorate units<br>15. The number of master's degree units<br>16. The number of national key disciplines<br>17. The number of distinctive specialty |
| Standard of teaching | Students and graduates | 18. Average scores of entrance<br>19. The number of PhD graduates<br>20. The number of master graduates<br>21. The number of graduates<br>22. One-time employment rate of graduates |
| | Graduates and international students | 23. The ratio of graduate and undergraduate students<br>24. The ratio of International students and undergraduate students |
| | Teaching achievements | 25. Ministry of Education Outstanding Teaching Award<br>26. Ministry of Education quality courses<br>27. Well-known Teacher of Teaching<br>28. Top 100 Authors of National Excellent Doctoral Dissertation<br>29. The number of various international and national contest winners |
| Scientific research | Research team and base | 30. Team of outstanding S &T innovation<br>31. National key laboratories, research centers, scientific research base<br>32. Proportion of R & D full staff |
| | Output of S & R | 33. The number of patent licensing<br>34. The number of included papers of SCI, SSCI and A & HCI<br>35. The number of included papers of EI, ISTP and ISSHP<br>36. The number of included papers of CSTPC, CSSCI<br>37. The number of social science monograph |

Continued

| First level indicators | Second level indicators | Third level indicators |
|---|---|---|
| Scientific research | Quality of achievements | 38. The nation's highest science, invention, progress, and Ministry of Education, Humanities and Social Sciences Awards |
| | | 39. The number of outstanding papers of *Science*, *Nature* and ESI |
| | | 40. Symbolic quality achievements |
| | | 41. The number of citation of SCI, SSCI and A & HCI |
| | | 42. The number of citation of CSTPC and CSSCI |
| | Projects and funds of S & R | 43. The number of National Natural Science Fund projects |
| | | 44. The number of National Social Science Fund projects |
| | | 45. Total of projects of S&R |
| | | 46. Research funding that year |
| | Efficiency and effectiveness | 47. Output rate per capita |
| | | 48. Output rate per ten thousand *yuan* |
| Reputation | Reputation of universities | 49. Academic reputation |
| | | 50. Network influence |

# 4 Analysis of the Results of Chinese University Evaluation of 2013

## 4.1 Competitiveness Ranking of Top Chinese Universities

According to the characteristics and standards of world-class universities, and the actual situation of China's higher education, the rankings of Chinese top university were released. The standard of the first-class universities—first-rate scholars, first-class disciplines, first-class results, first-class efficiency, first-class management and first-class university must be a research focus of the university (Qiu et al., 2011). According to the competitiveness ranking of 2013 key university, we obtained its comprehensive ranking of the top 20% (the former 26) key universities for the first-class universities in China. The specific list is shown in Table 2.

Table 2 2013 China's top university rankings

| 2013 ranking | Name of university | Score | ER ranking | ST ranking | SR ranking | R ranking | Area ranking | | Type ranking | | |
|---|---|---|---|---|---|---|---|---|---|---|---|
| 1 | Peking University | 100 | 2 | 1 | 1 | 1 | Beijing | 1 | C | 1 |
| 2 | Tsinghua University | 95. 15 | 1 | 4 | 2 | 2 | Beijing | 2 | P | 1 |
| 3 | Zhejiang University | 91. 95 | 3 | 2 | 3 | 3 | Zhejiang | 1 | C | 2 |
| 4 | Fudan University | 81. 99 | 5 | 5 | 4 | 5 | Shanghai | 1 | C | 3 |
| 5 | Shanghai Jiaotong University | 81. 98 | 4 | 7 | 5 | 4 | Shanghai | 2 | P | 2 |
| 6 | Wuhan University | 81. 32 | 7 | 3 | 6 | 6 | Hubei | 1 | C | 4 |
| 7 | Nanjing University | 74. 54 | 9 | 9 | 7 | 16 | Jiangsu | 1 | C | 5 |
| 8 | Huazhong University of Science and Technology | 73. 29 | 10 | 8 | 9 | 10 | Hubei | 2 | P | 3 |
| 9 | Sichuan University | 71. 52 | 6 | 13 | 11 | 12 | Sichuan | 1 | C | 6 |
| 10 | Sun Yat-sen University | 70. 36 | 11 | 17 | 8 | 15 | Guangdong | 1 | C | 7 |
| 11 | Harbin Institute of Technology | 69. 65 | 15 | 14 | 19 | 24 | Heilong jiang | 1 | P | 4 |
| 12 | Renmin University of China | 69. 08 | 23 | 6 | 13 | 7 | Beijing | 3 | HL | 1 |
| 13 | Jilin University | 68. 76 | 8 | 18 | 15 | 8 | Jilin | 1 | C | 8 |
| 14 | Nankai University | 68. 36 | 17 | 10 | 14 | 69 | Tianjin | 1 | C | 9 |
| 15 | Beijing Normal University | 68. 33 | 19 | 12 | 12 | 9 | Beijing | 4 | N | 1 |
| 16 | University of Science and Technology of China | 67. 30 | 18 | 21 | 10 | 21 | Anhui | 1 | P | 5 |
| 17 | Central South University | 66. 90 | 12 | 15 | 18 | 23 | Hunan | 1 | P | 6 |
| 18 | Xi'an Jiaotong University | 66. 01 | 14 | 16 | 17 | 22 | Shaanxi | 1 | P | 7 |
| 19 | Southeast University | 65. 52 | 28 | 11 | 16 | 64 | Jiangsu | 2 | P | 8 |
| 20 | Shandong University | 65. 47 | 13 | 19 | 20 | 11 | Shandong | 1 | C | 10 |
| 21 | Beihang University | 64. 52 | 20 | 20 | 22 | 65 | Beijing | 5 | P | 9 |
| 22 | Tongji University | 60. 94 | 16 | 23 | 25 | 27 | Shanghai | 3 | P | 10 |
| 23 | South China University of Technology | 60. 27 | 27 | 22 | 24 | 66 | Guangdong | 2 | P | 11 |
| 24 | Tianjin University | 59. 46 | 26 | 26 | 21 | 13 | Tianjin | 2 | P | 12 |
| 25 | Dalian University of Technology | 58. 30 | 24 | 25 | 27 | 25 | Liaoning | 1 | P | 13 |
| 26 | Northwestern Polytechnical University | 57. 48 | 36 | 24 | 33 | 44 | Shaanxi | 2 | P | 14 |

Notes：ER（educational resources），ST（standard of teaching），SR（scientific research），R（reputation），C (comprehensive)，P (polytechnic)，HL (humanities and law)，N (normal).

Compared with the rankings of 2012，the overall competitiveness of China's top universities ranked relatively stable. Besides Nankai University climbed 4 places，Shandong University fell 8 places，Xian Jiaotong University fell 5 places，and other universities' ranking fluctuations are less than three. In the first-class universities，the three aspects of educational resources，scientific research，and teaching have not changed much compared with each year，and the differences of network brand value of the universities，to some extent，affect their competitiveness.

## 4.2 Competitiveness Ranking of Science and Technology Innovation of Chinese University

Chinese universities undertake personnel training, scientific research and social services for three main functions, in particular, scientific research in the three major functions occupy a significant proportion, especially for the key universities, scientific research is to the top in all its functions. Therefore, the evaluation of Chinese university research competitiveness is the main content and an important part of the evaluation of Chinese universities (Qiu and Zhao, 2007). Beginning in 2004, we follow the idea of "input-output-efficiency", the indicators of the absolute amount and the relative amount were combined to use in the evaluation of research competitiveness of the key universities and the general university, and the evaluation was executed from the innovation of science and technology and the innovation of humanities and social science. Table 3 shows the top 10 in this ranking.

**Table 3 Top 10 of the competitiveness ranking of S&T innovation of Chinese university**

| 2013 ranking | Name of university | Score | Area ranking | | Type ranking | |
|---|---|---|---|---|---|---|
| 1 | Tsinghua University | 100 | Beijing | 1 | P | 1 |
| 2 | Peking University | 97.92 | Beijing | 2 | C | 1 |
| 3 | Zhejiang University | 89.92 | Zhejiang | 1 | C | 2 |
| 4 | Shanghai Jiaotong University | 76.13 | Shanghai | 1 | P | 2 |
| 5 | Fudan University | 75.12 | Shanghai | 2 | C | 3 |
| 6 | Nanjing University | 64.50 | Jiangsu | 1 | C | 4 |
| 7 | University of Science and Technology of China | 63.95 | Anhui | 1 | P | 3 |
| 8 | Harbin Institute of Technology | 62.56 | Heilongjiang | 1 | P | 4 |
| 9 | Wuhan University | 61.40 | Hubei | 1 | C | 5 |
| 10 | Huazhong University of Science and Technology | 60.48 | Hubei | 2 | P | 5 |

Notes: C (comprehensive), P (polytechnic).

Top 10 of the competitiveness ranking of S&T Innovation of Chinese University remains stable. Wuhan University has increased from 10 to 9. In the top 20 of the competitiveness ranking of S&T Innovation of Chinese University, the fastest rising is the increase of 6 places of Harbin Institute of Technology, the fastest decline in ranking is Tianjin University, a decrease of 5 from last year's 12 to 17. Fluctuations of the other universities ranked in less than three, and remained stable.

## 4.3 Competitiveness Ranking of Humanities and Social Science Innovation of Chinese University

Table 4 shows the top 10 in competitiveness ranking of humanities and social sciences innovation of Chinese university.

**Table 4 Top 10 of the competitiveness ranking of humanities and social sciences innovation of Chinese university**

| 2013 ranking | Name of university | Score | Area ranking | | Type ranking | |
|---|---|---|---|---|---|---|
| 1 | Peking University | 100 | Beijing | 1 | C | 1 |
| 2 | Renmin University of China | 84. 52 | Beijing | 2 | HL | 1 |
| 3 | Fudan University | 78. 15 | Shanghai | 1 | C | 2 |
| 4 | Wuhan University | 74. 64 | Hubei | 1 | C | 3 |
| 5 | Beijing Normal University | 71. 44 | Beijing | 3 | N | 1 |
| 6 | Sun Yat-sen University | 69. 78 | Guangdong | 1 | C | 4 |
| 7 | Nankai University | 68. 26 | Tianjin | 1 | C | 5 |
| 8 | Nanjing University | 67. 77 | Jiangsu | 1 | C | 6 |
| 9 | Sichuan University | 66. 66 | Sichuan | 1 | C | 7 |
| 10 | Jilin University | 65. 97 | Beijing | 1 | P | 8 |

Notes：C（comprehensive），P（polytechnic），HL（humanities and law），N（normal）.

The competitiveness ranking of humanities and social sciences' innovation of Chinese university is relatively stable. Jilin University has increased by 3 to enter the top 10. In the top 20 of the competitiveness ranking of humanities and social sciences innovation of Chinese university，the fastest rising is the increase of 5 of Nanjing Normal University. Fluctuations of other universities ranked in less than three，and remained stable.

# 5 Evaluation Conclusions and Implications

The 2013 evaluation work has accumulated a large number of classification and evaluation data，from which we found a lot of valuable conclusions and inspiration. Concrete and vivid figures also make a better understanding of the characteristics and laws of undergraduate education development in China.

First，professional settings and subject development trends are characteristic and focused. With the release of the latest version of the undergraduate profession catalog of the 2011 by the Ministry of Education，how touse their own advantages and the external environment to adjust the subject development planning and professional settings，become the focus of attention of the managers of all colleges and universities. The key universities should be more international-oriented，and getting into the ranks of world-class universities are their target. As for the universities in general，they should study how to concentrate on the use of limited resources to do a few characteristics of the famous professional. The degree of public acceptance of private universities is very low, and their rankings often change radically. Therefore, the private universities should have a clear mission, to serve the local economic development, and combine industries with local characteristics to cultivate the characteristics of professional and expertise talent (Qiu and Luo, 2009). From this year's professional evaluation of the data, we found that universities in accordance with their own school type and school level, closely linked to the location conditions, like the natural environment, human environment and industrial structure features to develop the advantage profession. Reasonable professional sets of features can give full play to the regional economic service capabilities of universities that reflect the social value of universities.

Second, comprehensive development of key university's disciplines was the trend. From the Chinese university of humanities and social sciences competitiveness ranking of last few years, we can find that the polytechnic universities due to good conditions have made tremendous progress in the humanities and social sciences competitiveness (Qiu, 2009). The polytechnic universities have the more powerful educational resources and capital, so they can improve the teaching conditions in the short term, and the introduction of outstanding humanities and social sciences talent can boost its competitiveness in the humanities and social sciences quickly.

Third, more attention was paid to technological innovation and higher education performance. In the university evaluation of 2013, we pay more attention to the performance of universities, because the performance more fully reflects the university's service capabilities of social economy and culture. In order to reflect the level of university performance management, we first introduced performance ranking on the basis of the university competitiveness evaluation (will be published in addition). In order to reflect the science and technology innovation, this year we increased the weights of some indicators such as research output, innovation (such as patents) etc. Evaluation index system includes three first level indicators: innovation platforms, innovation talents and innovation achievements. The university innovation index ranking will be issued in addition.

Fourth, the process of internationalization of universities was speeding up. Combined with the data accumulated by RCCSE in China and the world of university evaluation, we can find that in recent years, our key universities have improved steadily in three aspects of comprehensive competitiveness, competitiveness of research and innovation, competitiveness of discipline. But the quality is always the main factors which can enhance the international influence of Chinese university. The international level of our universities needs to be improved, mainly in three aspects:

(1) Our universities' awarding is relatively small in various international competitions.

(2) The level of coordination between our universities and foreign universities and research institutions was low.

(3) China's scientific research especially in the humanities and social sciences was not strictly in accordance with the norms of international research, resulting in that academic exchanges cannot be carried out. The fact can also be seen from our papers' low citation in foreign journals. The lack of international vision and standardization of academic norms seriously hindered the international development of humanities and social sciences.

Chinese key universities in the future development and construction must strengthen the international scientific research cooperation, personnel exchanges and visits between countries, while strengthening the in-depth study of the processes and methods of academic norms and international standards. On the other hand, the introduction of overseas talents, funds, and projects should be increased. This can shorten the gap with world-class universities, and continuously improve their international visibility and influence as soon as possible.

## References

Qiu J P, Cheng N. 2009. Study on evaluation of web influence of key university in China. *Studies in Science of Science*, 27 (2): 175, 190-195.

Qiu J Q. 2006. University evaluation report of China (2006-2007) —the concept and practice of evaluation of Chinese universities and disciplines. *Science & Technology Progress and Policy*, (7): 23-26.

Qiu J Q. 2009. Method, result and inspiration of evaluation of Chinese universities and disciplines. *Science & Technology Progress and Policy*, 26 (1): 118-123.

Qiu J Q, Luo L. 2009. Reforms of Chinese universities evaluation—analysis of chinese universities and majors ranking in 2009. *Higher Education Development and Evaluation*, 25 (3): 19-28.

Qiu J Q, Wen T X. 2010. *Evaluation Science Theory, Methods, Practice*. Beijing: Science Press.

Qiu J Q, Zhao R Y. 2007. *Report of Disciplinary Competitiveness of World-class Universities and Research Institutions in the Evaluation Studies*. Beijing: Science Press.

Qiu J Q, Zhao Y J, Luo L. 2011. Research on method of paper classification in scientific evaluation. *Journal of the China Society for Scientific and Technical Information*, 30 (5): 554-560.

# 4-2　SUR模型在大学评价中的应用

丁　宁[①]　王　培

**摘　要：**我国大学评价体系呈现多元化和多样化的特征，不同体系针对自己的读者群在指标体系和计算方法上各有侧重，难分优劣。为了尽可能有效地综合利用比较排名中蕴含的信息，本文尝试用计量经济学中的SUR模型来定量分析处理各评价体系产生的排名结果，并给出相应的参数估计方法与假设检验的渐近分布。
**关键词：**SUR模型；大学评价；计量经济学

作为高等学府，大学承担着为国家和社会培养高端人才和研发前沿学科知识的重任。在当今高度竞争的世界，高等教育的水平高低不仅衡量社会资源对知识和人才的投入产出效益，更是上升为一个国家的竞争战略，决定着未来的国运兴衰。近年来，在从大国迈向强国的战略目标指引下，我国对于高等教育的财政投入增长迅速，大学办学条件日新月异。为有效地评价大学的办学质量，自从1987年中央教育科学研究所蒋国华教授等学者在《科技日报》上发布了我国第一个大学排行榜，我国的一些媒体和官方民间组织效仿国外发达国家的做法，采用像流行歌曲排行榜一样的形式来给大学排座次，以增强大学之间的竞争。不仅教育主管部门在分拨教育经费时参考大学排行榜，家长在为子女填报高考志愿时也非常看重各校的最新排名。特别是最近几年学校自主招生的权限放大，付费上学成为趋势，大学排名更是成为各校招揽生源的一大卖点。但是，对大学进行正确的排名不是一件容易的任务。异质化的人才和知识不像同质化的实体资产那样有统一的计量标准，即使是同一学科的知识，同一专业的人才，其质量的高低也可能受到一些非大学本身可以控制的环境因素所决定。比如，一所大学毕业生的就业率和薪资水平比外地同类高校高，可能是因为当地大学生就业市场的相对景气，而非该大学办学更成功所致。为了有效地做出大学评价，市场上产生了各种不同的大学评价体系，比较权威的编制者有武书连领导的中国管理科学研究院课题组、武汉大学中国科学研究评价中心联合中国科教评价网组成的课题组，以及中国校友会网大学评价课题组。这些评价体系产生不同的排名结果，都在市场上占有一席之地，受到不同使用人群的追捧。比如，从

① 浙江大学宁波理工学院，地址：浙江省宁波市学府路1号；邮编：315100；电子邮箱：dingn@nit.zju.edu.cn。

2011 年到 2013 年，武书连的课题组连续三年把浙江大学评为全国冠军，受到了北京传统名校的非议。总体来讲，一个大学评价体系主要由四个要素决定：①评价指标的选取；②赋予各指标的权重；③学校数据的采样；④计算结果。评价指标主要围绕着大学的两大任务——人才培养和科学研究展开，有的还包括学校声誉和办学资源等。指标权重的设定则有可能大相径庭，有的是采用专家的主观意见，有的是用层次分析法等定量方法。对于评价体系的比较，文献中已经有很多的论述。本文的意图不是在于评估不同评价体系的优劣，相反，是基于现有所有评价体系，提出一种新的大学排名方法。这种方法尽可能完整地综合各种评价体系的排名，给使用者一个全面客观的展示。为达到这个目的，和文献中采用统计方法的做法不同，我们应用计量经济学中的似不相关回归（seemingly unrelated regression，SUR）模型来有效地处理排名数据中蕴含的排名信息。计量经济模型与统计模型最大的区别在于对待数据的观点。统计学者把每个学校的原始数据和每次排名的结果都看成是随机抽取的样本，而计量经济学者则把它们看成是给定的已知整体，即已经包含了所有的可能的样本，需要决定的是如何用模型来刻画不同变量之间真实的关系。之所以用 SUR 模型是因为表面来看不同评价体系是自成一体、各有侧重的，其实它们在评价大学这一共同目标上还是有着千丝万缕的关系的。比如，有些大学评价的指标（像生源所在中学的学术水平和所在地区的经济发展水平）因为不便度量而被排除在外，这些变量对所有评价体系产生的误差都会有着相似的作用，也就是说其实这些表面看似不同的评价体系是有相关因素的。具体来说，我们想要回答下列三个研究问题：第一，一个评价体系中的指标是否统计显著；第二，不同评价体系是否在统计意义上一致；第三，能否找到一个排名方法可以包含所有评价体系结果蕴含的信息。

## 1 SUR 模 型

SUR 模型是一种把多个回归方程放在一起回归的方法。它要求所有的被解释变量不能出现在其他方程的解释变量中，这是它与联立方程回归模型的区别。SUR 模型的一般形式为

$$\begin{cases} y_1 = x_1 \cdot \beta_1 + u_1 \\ y_2 = x_2 \cdot \beta_2 + u_2 \\ \quad \cdots\cdots \\ y_G = x_G \cdot \beta_G + u_G \end{cases}$$

这里，模型有 $G$ 个方程，每个方程代表一组数据，共有 $G$ 组数据。把 $G$ 组数据放在一起，形成一个共有 $N$ 个观察值的横截面数据。每个方程都有自己的斜率系数 $\beta_i$，即每个横截面个体的解释变量对被解释变量的影响是不随数据变化的确定性关系，这些系数恰好是我们所感兴趣的。不失一般有外生性假定：

$$E(u_g \mid x_1, x_2, \cdots, x_G) = 0 \qquad \forall g = 1, \cdots, G$$

这个假定要求我们对上面的 $G$ 个回归方程是正确识别的，即没有省略变量、观察误差等问题。这种 SUR 模型通常用于对同一个个体的多个不同指标 $y_1$，$y_2$，…，$y_G$ 同时进行分析。这时，$G$ 个不同的扰动项 $u_1,u_2,\cdots,u_G$ 虽然是不同的随机变量，但是它们都表示同一个个体的不可观察因素的整合，其中必有其公共因素。不同个体的扰动项相互之间是线性无关的，这个是横截面数据的一般特征，我们这里也这样假定：

$$\mathrm{Cov}\,(u_{it},\ u_{js})=\begin{cases}\sigma_{ij},\ t=s\\0,\ t\neq s\end{cases}$$

$$i,j\in\{1,\cdots,G\}\qquad t,s\in\{1,\cdots,n\}$$

由假设条件可知，各个回归方程之间实际上是有关联的。似不相关回归（SUR）容许各个回归方程的扰动项之间存在跨方程相关，方程中的 $u$ 在任何一个个体中不必相互独立，即不同方程的扰动项之间可以存在同一个个体内的相关关系。这样，SUR 估计程序就可以使用扰动项的相关来改善估计值。各个回归方程之间任何的相关都是有价值的信息，它可能是告诉我们某个个体中有一个共同的因素不可观测，而只能反映在扰动项中。

事实上，在经济活动中，有许多问题具有同期相关性，例如，在考察家庭消费时，对于同一个家庭样本，我们经常会同时考虑这个家庭的食品消费、服装消费等多个指标。由于这些指标属于同一家庭，它们必然受到一些共同的不可观察因素的影响，如生活习惯、宗教信仰等。因此，在研究这些问题时，就可将模型设定为 SUR 模型。

## 2 SUR 模型的参数估计

SUR 模型的参数估计按以下四个步骤进行。

第一步：用 OLS 法估计每个方程的 $\beta_i$：

$$\hat{\beta}_{\mathrm{OLS}}=\left(\sum_{i=1}^{N}X'_iX_i\right)^{-1}\cdot\left(\sum_{i=1}^{N}X'_iy_i\right)$$

第二步：计算并保存回归中得到的残差 $\hat{e}_{it}$。

第三步：用这些残差来估计扰动项方差和不同回归方程扰动项之间的协方差（covariance），得到扰动项协方差矩阵的估计值 $\hat{\Omega}$：

$$\hat{\Omega}=\frac{1}{N}\sum_{i=1}^{N}\hat{u}_i\cdot\hat{u}'_i,$$

这里 $\hat{u}_i=(\hat{e}_{i1},\hat{e}_{i2},\cdots,\hat{e}_{iG})'$。

第四步：上一步估计的矩阵 $\hat{\Omega}$ 被用于执行广义最小二乘法，得到各方程参数的 GLS 估计值：

$$\hat{\beta}_{\mathrm{GLS}}=\left(\sum_{i=1}^{N}X'_i\Omega^{-1}X_i\right)^{-1}\cdot\left(\sum_{i=1}^{N}X'_i\Omega^{-1}y_i\right)$$

SUR 得到的估计值具有一致的特性，也就是说当 $N$ 趋向于无穷大时，$\hat{\beta}_{\mathrm{GLS}}$ 在概

率上收敛于真实的 $\beta$。而且，重要的是 $\sqrt{N}(\hat{\beta}_{GLS}-\beta)$ 服从均值是零的正态分布，它的渐进方差是 $\hat{A}^{-1}\hat{B}A$ ，其中：

$$\hat{A}=\frac{1}{N}\sum_{i=1}^{N}X'_i\hat{\Omega}^{-1}X_i$$

$$B=\frac{1}{N}\sum_{i=1}^{N}X'_i\Omega^{-1}u_iu'_i\Omega^{-1}X_i$$

由于 SUR 充分利用了扰动项之间的相关关系，一般来说比单个方程的回归有效，具体表现在估计参数的方差较小，比较容易通过显著性检验，这一点在很多情况下是至关重要的。

## 3 建立大学排名 SUR 模型的示例

我们设 $y_1$，$y_2$，$y_3$ 分别为“全国大学综合评价指标排名”“全国大学校友会网排名”“武书连中国大学排名”的排名结果，回归变量可以大致包括以下几个方面。

（1）办学基本条件类（base）：①校舍总面积或学生人均校舍面积；②仪器设备总额或学生人均仪器设备额；③图书总量或学生人均图书量；④当年教育经费支出总额或当年生均教育经费支出额。

（2）教师队伍类（teach）：①中科院院士与工程院院士数；②杰出人才（长江学者、跨世纪人才、教学名师）；③博士生导师数；④高级职称教师占教师总数比例；⑤师生比。

（3）优势学科类（discip）：①博士点数；②硕士点数；③国家级重点学科数；④特色专业数。

（4）生源与毕业生（stud）：①新生入学平均分数；②毕业生一次就业率；③毕业生工作半年后工资水平；④杰出校友数。

（5）科研项目和产出（research）：①国家自然科学基金项目数；②国家社会科学基金项目数；③科研项目总数；④当年科研支出经费；⑤人均产出率；⑥专利申请与授权数；⑦高水平论文收录数；⑧科学专著数；⑨获国家最高科学技术奖、国家自然科学奖、国家技术发明奖、国家科学技术进步奖、教育部人文社会科学奖数。

这样我们的高校排名影响因素的 SUR 模型为

$$\begin{cases}y_1=\beta_{1,0}+\beta_{1,1}\cdot\text{base}+\beta_{1,2}\cdot\text{teach}+\beta_{1,3}\cdot\text{discip}+\beta_{1,4}\cdot\text{stud}+\beta_{1,5}\cdot\text{research}+u_1\\ y_2=\beta_{2,0}+\beta_{2,1}\cdot\text{base}+\beta_{2,2}\cdot\text{teach}+\beta_{2,3}\cdot\text{discip}+\beta_{2,4}\cdot\text{stud}+\beta_{2,5}\cdot\text{research}+u_2\\ y_3=\beta_{3,0}+\beta_{3,1}\cdot\text{base}+\beta_{3,2}\cdot\text{teach}+\beta_{3,3}\cdot\text{discip}+\beta_{3,4}\cdot\text{stud}+\beta_{3,5}\cdot\text{research}+u_3\end{cases}$$

## 4 大学排名 SUR 模型的假设检验

回到我们前面提到的三个问题，对于第一个问题，即评价体系中的指标对于排名是否统计显著，有了估计值的渐进方差，我们可以采用 $t$-检验来检验每个 $\beta$ 值等于 0 的零假设。

对于第二个问题，即各评价体系是否在统计意义上一致，我们需要检验 $\beta_i = \beta_j$ 的零假设。由于前述 $\beta$ 的估计值服从渐进正态分布，我们可以借用 Wald 检验方法来完成假设检验。Wald 统计量在下面给出，服从 $\chi^2(1)$ 分布。

$$\text{Wald} = (R\hat{\beta})'(R \cdot \text{Avar}(\hat{\beta}) \cdot R')(R\hat{\beta}) \sim \chi^2(1)$$

其中对于矩阵 $R$，当 $i=1, j=2$ 时，$R=(\vec{1}, -\vec{1}, 0)$，其他 $i$ 和 $j$ 的情况以此类推。

这里要求必须用方程组来做回归，单个回归方程无法完成这个功能。

最后，对于第三个问题，即能否找到一个模型可以包含所有评价体系蕴含的信息，我们只需要把所有评价体系的指标变量都包括，系数用前述的 $\beta$ 估计值即可。

## 5 小　结

本章基于现有的各种大学评价体系，采用计量经济学中的 SUR 模型把它们综合起来，提出了一种新的大学排名方法。这种方法的好处是最大限度地整合所有评价体系蕴含的信息，有效地回避了不同评价体系的指标体系和权重不一致的问题，并且采用假设检验的方法来检验指标的统计显著性并比较评价体系的差异，具有科学性、严密性、有效性、可操作性的特点。

# 4-3 大学排行的科学与文化
## ——再看大学排行榜的认识与误区

蒋国华[①]

**摘 要：**世界上第一个大学排行榜是由美国《美国新闻与世界报道》于1983年首创的。其实我国也有第一个大学排行榜，那是中国管理科学研究院科学学研究所于1987年首创的。大学排行榜之在美国始于市场，而中国始于科学；在西方责难表现为学术批评，在中国则往往表现为道德审判。大学排行榜之未来：1%研究型大学，主要采用科学计量学指标；99%非研究型大学，主要采用教育计量学和经济计量学指标。

**关键词：**大学排行榜；科学计量学；网络计量学；教育计量学

改革开放30余年来，我国教育界有几项论争是几乎贯通始终的，且至今尚无见到有趋于平息的迹象，诸如应试教育之争、高考改革之争、教育产业之争、教育经费之争、教育公平之争等。倘若其中还有亦堪位列前茅的，则必定是大学排行榜之争了。

有媒体朋友问：大学排行榜自发布以来，缘何饱受诟病，且连绵不断（李红颖，2012）？于2004年5月19日，在人民政协报教育之春沙龙第5期“关注教育评估”主题报告会上，笔者曾应邀做过一次题为“我看大学排行榜：认识与误区”的讲演。本文则打算从科学和文化的角度，再来谈谈有关大学排行榜的认识与误区。

### 1 大学排行榜之缘起：美国始于市场，中国始于科学

众所周知，迄今在国际上做世界大学排行榜且做出名声的，世称有三大全球大学排名机构，这就是中国上海交通大学高等教育研究院世界一流大学研究中心之“世界大学学术排名”，英国泰晤士高等教育杂志之“世界大学排行榜”，西班牙国家研究委员会网络计量学研究中心之“网络计量学大学排行榜”（该排行榜创始于2004年，每半年，即1月和7月，更新一次）。而在国内做中国大学排行榜且做出名声的，则有四大排名机构，这就是中国管理科学研究院武书连（广东深圳）之“中

① 蒋国华（1944— ），男，教授，江苏无锡人，研究方向为科学学、科学计量学、教育科学；中国民办教育协会网站总编辑、中国民办教育研究院副院长、北京吉利大学教育研究所名誉所长。

国大学评价"，广东深圳市网大教育服务有限公司（简称网大）之"中国大学排行榜"，中国校友会网（广东深圳）之"中国大学排行榜"，武汉大学中国科学评价研究中心之"中国大学评价报告"。

现在世所公认，世界上第一个大学排行榜乃是由美国《美国新闻与世界报道》于1983年首创的（Morse，2001）。从形式和现象上看，随后跟进的则有英国（1986年）、中国（1987年）、德国（1989年）、加拿大（1991年）、日本（1993年）、西班牙（2004年）、法国（2007年）、荷兰（2007年）……欧盟（2013年）（王寰鹰，2011）。在我们中国，第一个大学排行榜乃是由中国管理科学研究院科学学研究所于1987年首创的（中国管理科学研究院科学学研究所，1987）。随后跟进的则有广东武书连（1992年）、中国科技信息研究所（1993年）、香港《亚洲周刊》（1997年）、深圳网大（1999年）、中国校友会网（2002年）、上海交通大学（2003年）、武汉大学（2004年）、浙江大学（2006年）、台湾财团法人高等教育评鉴基金会（2007年）……中国人民大学（2008年）（必艰，2008）。

毫无疑问的是，上述国内外这两个大学排行序列一定会延伸下去，换句话说，鉴于在21世纪知识经济时代大学正在并将"在知识创新、知识传播和培养人才等方面承担更加重要的使命"（陈至立，2002），随着时间的推移，一定还有更多关注高等教育发展的国家、地区、大学和社会研究机构加入到世界、地区和本国的大学排名队伍中来。

本文首先关心的却是大学排名之始，她的首创及缘起。

美国著名诗人、作家，1987年诺贝尔文学奖得主约瑟夫·布罗茨基说过："熟悉的，就叫制造；不熟悉的或者前所未有的，就叫创造。"（冈特恩，1998）布罗茨基表达的正是著名华裔美国物理学家、1976年诺贝尔物理学奖得主丁肇中多次申述的"科学研究只有第一名，没有第二名"的著名原理（蒋廷玉，2012）。

因此，大学排名之第一归于美国，这没有问题，如同100多年来美国夺得了许多世界第一一样（石小玉和涂勤，1999）。但重要的问题还在于，这个第一为什么是美国？笔者在《人民政协报》教育之春沙龙第5期"关注教育评估"主题报告会上曾详细分析过大学排名为何首先在美国诞生的四大原因（蒋国华，2004）。现在看来，这四大原因没错，但还不够，经过8年的思考和比较，还应该有第五个原因，即文化的原因，确切地说，是美国的得天独厚的市场经济文化孕育和孵化了她。正如国际创造性和领导学基金会创始人戈特利布·冈特恩指出的：尽管"人类创造性是一种事实上不可耗尽的自然资源"，但是归根到底，"在个人的系统里，都有着创造性的三个主要资源：人脑、人格和系统文化"（冈特恩，1998）。由此可见，倘若忘却了作为创造性"三个主要资源"之一的人们所处的社会系统文化因素，或许也难以深刻理解为何大学排名之第一被美国夺得的更本质的缘由。质言之，如若没有美国的成熟的市场经济文化，也许就不会有大学排名之美国第一。

事实正是如此。正如我国台湾辅仁大学大学排名研究专家侯永琪教授指出的：从国际上看，"高等教育在20世纪末已正式地迈进市场化的新时代"（侯永琪，

2008)；她在另一个场合更是认为，高等教育市场化乃是三大"大学排名的理论基础"之一（侯永琪，2007)。2001 年 7 月笔者作为大会主席邀请美国《美国新闻与世界报道》数据研究部主任，亦即该刊"美国大学排行榜"的主要研制者和领导者罗伯特·莫斯先生，来华参加第三届科研评价与大学评价国际研讨会时，他送来的讲演稿就是是年 3 月在美国加州长滩一个全美年会上的稿子。他告诉我，他们杂志 1983 年推出的第一个排名就一个指标，即"学术声誉"，其数据来源也是靠对全美 1308 位大学校长做市场问卷调查得来的。他们把这第一次排名冠名为"美国最好大学"(National Best University)，前三甲为斯坦福大学、哈佛大学和耶鲁大学。也正是这篇讲稿中，莫斯先生在美国大学排名指南前言部分明确指出，本排名旨在帮助有志上大学的学生做出信息充分的选择，因为上大学不仅是一次职业生涯的决定，还是一笔重要而又不菲的投资；配合大学排名，该杂志社每年还推出基于排行榜的升学指南特刊，广为发行。简言之，美国大学排行榜之生与长，可以说均系市场所赐，为市场而生，因市场而长。

换句话说，美国大学排名是顺应了世界高等教育市场化的潮流的产物，是冲着"有助于消费者"去的，始于市场，长于市场，并且由此而进，作为一种新的思想和方法，同时作为美国软实力的一部分，渐次输出，影响远及国门以外，誉满全球。

近几年来，正当国内部分大学校长、教育专家和媒体记者还在同仇敌忾、兴致勃勃地着力组织大学排行榜批判大合唱之际，有三个标志性事件也许是不得不面对的：一是 2006 年联合国教育、科学及文化组织（简称联合国教科文组织）欧洲高等教育中心主任牵头成立的"国际排名专家小组"(International Ranking Expert Group，我国上海交通大学刘念才教授是该专家组副主席）在德国柏林召开第二次会议，并公布了"高等教育机构排名柏林原则"；二是 2008 年欧盟委员会发布了一项新型大学排名体系研究项目的招标启事，项目支持预设金额为 110 万欧元（刘念才教授是该项目 7 个专家顾问成员之一)；三是 2011 年联合国教科文组织在其巴黎总部召开大学排名专题会议（刘念才教授应邀作了大会报告)，联合国教科文组织副总干事唐虔在总结讲话时指出："大学排名已经成为世界范围内高等教育质量保障的重要工具之一，要合理应用，避免滥用。"他还呼吁尽快制定大学排名使用指南。这一方面表征美国大学排行榜已成为一种软实力输出的成就，另一方面也表明，大学排行榜正在成为一种颇有影响且受到社会欢迎的高等教育评估模式，一种推动和促进高校间竞争、进步和发展的杠杆。正如大文豪雨果所说："当一种新思想的时代到来之时，世界上任何力量都阻挡不了。"

值得指出的是，我国第一个大学排行榜是 1987 年 9 月 13 日由《科技日报》发布的，比美国晚了四年。从文献上可以看到，我们文章的署名是中国管理科学研究院科学学研究所（实际作者应是四位：赵红州、郑文艺、熊学敏、蒋国华)，赵红州时任所长，蒋国华时任副所长，文章的题目是"我国科学计量学指标的排序"。尤其需要补充指出的是，当初我们搞这个大学排行榜，完全是自身科学计量学研究的应用性副产品。当时我们可以说是一心扑在科学学和科学计量学的开垦和研究上，对教育特别是高等教育也没有什么研究和积累，再加上 20 世纪 80 年代中期的国家通信和国际科教情报交流状况导致的信息闭塞的原因，我们也无从知晓美国早在 1983 年就捷足先登，握有优先权了。不过，30 年后来评价我们当时的这项成果，一方面固然是科学发现同时性原理的一个小小例证，另一方面依然可以看到我们这项成果是不无遗憾和悲剧性的成分的。和《美国新闻与世界报道》推出的美国大学排行榜相比较来看，我们与之两者在思路、指标和方法是不同的：美国始于市场，中国始于科学。

历史已经证明，在我们之后整整 20 年，亦即 2007 年，我们的科学计量学同行友人、荷兰莱顿大学科学与技术政策中心才发布了基于科学计量学指标的世界大学排名榜（刘念才，2008）。当然，无论如何怎么解释，历史已成过去，我们失去了一次把我国有关大学排序的思想和方法使之成熟并走出国门，作为一项国家软实力展示和影响世界的机会。其根源显然缘于我们当时所处的社会系统文化因素，在计划经济一统天下的重压下，任何市场经济的幼苗是无从侥幸和存活的，有时科学幼苗也不例外。

当然这已经是 21 世纪的认识和分析了。“可堪回首”？当初我们也只能认识到社会包容精神之缺失及项目与经费支持体制之僵化。因为赵红州和笔者曾骑着自行车几乎是满京城地奔走呼号，向有关计划部门争取过哪怕给一个小小课题项目名义支持和哪怕一点点经费名义资助，可是，诺大的北京城竟然没有一个部门可立题，没有一元钱可资助。遗憾的是，这类崇尚计划、排斥市场，崇尚衙门、排斥民间的贻害，至今犹存，君不见诸如“武书连的排行榜也只是个民间行为，并不具有权威性，但我们仍然必须小心这种排行榜的严肃性，以及它可能给考生们造成的误导，甚至是对中国整个高等教育、学术风气的负面影响”（韩晓波，2012）的网文观点，不还在大行其道吗？

## 2　大学排行榜之遭遇：在西方责难表现为学术批评，在中国则往往表现为道德审判

乍一看来，或许好多人觉得，排行榜可能是一种十分“现代”的产物，其实，文献研究表明，“它的基本思想却几乎和人类的历史一样古老”（欧文・华莱士，1993）。然而，从高等教育历史发展上看，大学排行榜乃是当代的产物，是当代大学评估，或者更规范些说，是高教评估的副产品；进入 21 世纪以来，则正在成为高教研究与评价的一部分。因此，对大学排行榜大体可以从两个方面看：一是学术研究及其成果，二是其研究成果的社会应用。

积 20 余年的经验，从我国现有对大学排行榜的批判、指责或批评看，大多集中在后者，即有关大学排行榜应用而起的社会反响、社会影响与社会观感。比如，北大原校长许智宏就说过“排行榜误导公众”（许智宏，2006），复旦大学校长杨玉良则更认为大学排行榜“污染学术空气”（赵婀娜，2009），甚至也有国外大学校长来华访问，如英国剑桥大学校长艾莉森・理查德，当国内媒体记者追问到她时也说过“大学排名本身很有可能毫无意义”的话；还有诸如“质疑‘中国大学排行榜’”（杜

星，2009)、"一流大学不是世界排名榜排出来的"（仇方迎，2003)、"陷入误区的中国大学排名"（顾海兵，2003)、"政协委员黄因慧指出，大学排名，误导学生，禁止民间组织进行大学排名"（袁祖君，2003)，甚至还有"一位名叫郭军辉的中国人民大学毕业生，在2000年因不满'网大'将其母校排在重点大学综合排行榜第25位，以侵害名誉权为由，一纸诉状把深圳'网大'公司告上法庭"的（凌云，2000)。这也就是当初我们在《科技日报》发布首个中国大学排行榜时，仅仅署了研究所的名称的原因。因为我们估计到大学排行榜一旦公诸于众，是免不了会引发地震式的攻击、谩骂和骚扰的，为此，我们曾就榜单署名费了一番心思：一是不在标题里出现"大学"二字，二是署名不出现作者名字。可是，尽管标题没写"大学"，但是还是受到了主管部门有关司局的关注并随即召见了我们，认为是"有什么科学根据""大学之间不可比，怎么排名""瞎胡闹""不要搞了"。

作为该排行榜的研究者制作者，我们亲历亲为并目睹了大学排行榜在我国遭遇风风雨雨的全过程，但是毋庸讳言的是，20余年来林林总总对于大学排行榜的责难，基本上隶属于自说自话的"道德审判"的范畴。在高举着的科学性、公平性的旗帜下，最常见、最有代表性的是这样的观点："每年都有一些大学排行榜出炉，吸引了不少眼球。除了一些民间商业机构受利益驱使，发布形形色色的榜单外，一些大学自己的高等教育研究所，也热衷于向媒体发布大学排行榜之类的新闻，甚至还煞有介事地'定量'计算出我们的大学与世界一流大学的距离。"（赵婀娜，2009）不用细说，这么多字的表达，其核心则是质疑大学排行榜中的"利益为先"问题。还有一个典型例子是三年前有人撰文《人民日报》，指称我国大学排行榜知名研究者制作者武书连研究员是按收取大学赞助费而使该校排名得到"飙升"的（赵亚辉和张炜，2009)，但从通篇报道看，文章作者疑似想当然有余，而科学性不足。其实，大凡排行榜皆要依靠数据说话，没有对个案指标体系及具体数据做深入的调查研究就大胆推断是因"利益为先"所致，显然是欠妥的。遗憾的是，人民日报的这篇文章直到近几个月还在《天津网》上被人津津乐道地引证（韩晓波，2012)，他们的逻辑很简单：因为武书连"利益为先"，所以这个"前科"就决定了他今年发布的2012年我国大学排行榜也必定"具有误导作用"！试问，这是哪家的科学证伪逻辑呢？难道天下科学研究必定是"免费的午餐"吗？

和任何一件新事物诞生之初的遭遇一样，大学排行榜也可以说是"在骂声中成长"的。这在美国也不例外，但被"骂"的内容是有差异的。如果说在在中国往往表现为自说自话的道德审判，那么，在美国为代表的西方诸国对大学排行批评责难基本上表现为严谨理性的学术批评。文献研究表明，《美国新闻与世界报道》于1983年推出世界上第一个美国大学排行榜，但其在以后的若干年里，同样受到美国教育界内外广泛的批评与诘难，但主流是与排名有关的数据、指标、权重、分类等之可靠性科学性的批评或探讨，据说，国际驰名的高教权威专家菲利普·阿尔特巴赫（Philip G. Altbach)、马丁·特罗（Matin Trow）等都曾对排名指标的科学性发表过意见。

尤其具有相映成趣意味的是，自2003年起，上海交通大学高等教育研究院院长刘念才教授领导并发布“世界大学学术排名”以来，的确在国际上反响硕大，大到有的国家外交部因上海交通大学排名涉及该国大学名次而出面与我国外交部交涉的程度；甚至，国际著名杂志《科学》和《自然》也加入进来，对上海交通大学排名榜依据的数据和指标提出质疑。但是，笔者初略的文献调研表明，尚未发现有国际高教界、学术界和新闻界有对刘念才领导的“世界大学学术排名”及其团队进行无休止的“利益驱使”“利益为先”“污染空气”之类的抨击文字，更没有挖祖坟式的人身攻击，恰恰相反，均属于严谨、理性的学术批评与论争的范畴之内。比如《科学》杂志，2007 年 8 月 24 日，发表了该刊资深记者恩森林克（Martin Enserink）题为“谁能给大学排名”的文章。文章一开头就对上海交通大学排名存在的问题提出商榷，比如，像诺贝尔奖得主这样的著名科学家如何计入？诺贝尔奖得主归属之争（爱因斯坦归属柏林洪堡大学？还是柏林自由大学？）该如何处置？所占分值又该几何？人文社科类学科在此指标体系下处于劣势问题、加权系数差异、大学提供数据作弊、问卷调查中同行评议的可靠性问题又该如何？等等，而所有这些因素均会极大影响到学校的“名次”（Enserink，2007）。再比如《自然》杂志，2010 年 3 月 4 日，发表了巴特勒（Declan Butler）题为“大学排名正阔步前进”的特稿。文章开言第一句话便指出：“每年秋天，政治家、大学校长、高教基金会官员，以及数不清的莘莘学子，都会耐心等待泰晤士高教杂志社（THE）制作的‘世界大学排行榜’。”（Butler，2010）但话锋马上转为对这类排行榜的问题与商榷，诸如排名方法和数据有问题；部分大学名次涨落无道理；THE 赋予学术声誉调查权重过大；调查中语言倾斜问题；法国抱怨科学家兼署了国立科研单位的名，使相关高校相应得分被稀释了；等等。文章还特别提到，欧盟委员会已经下令建立“全球多维大学数据库”（the Multi-dimensional Global Ranking of Universities，U-Multirank），以便研制新型大学排名体系；作为拥有 800 多所大学会员的欧洲大学协会，则计划每年出版一本全球大学排名述评报告，以评估其方法，仔细研讨诸大学排名的涨落。

顺便提一下，我国大学排行榜多年来的诸般遭遇中，在今年五月第一次出现了一个真正的反例：不再纠缠为道德审判，而是严肃的学术批评。这就是中国科学技术大学大学评价课题组（简称中科大课题组）在《中国高教研究》杂志 2012 年第 5 期第 5 页至第 11 页，发表了题为“基于公信力视角的大学排名研究——对《2010 中国大学评价》指标体系及算法的质疑”论文（中国科学技术大学大学评价课题组，2012）。在中国大学排行研究界出现严肃的学术批评的文章，非常值得肯定，因为它开辟了新鲜风气。然而，从文章本身及网上公布的武书连研究员的反批评看（武书连，2012），中科大课题组文章确乎存在弱点，有的还是致命的；另外，该中科大课题组不知出于何种目的，匆匆忙忙几乎一夜之间把他们得到的某些结论拱手送人，让人署名发文到了《人民日报》《中国青年报》《中国科学报》上，随后几乎迅速占领了包括《新华网》在内的无数网站，声称“2011 年在英国《泰晤士报高等教育副刊》全球大学排名中名列第一的美国加州理工学院，其“人才培养”得分竟然无缘进入武书连排行榜单的前 500 名；巴黎高师勉强进入前 500 名，和咸宁学院、宜春学院的排名大体相当（刘道彩，2012；何聪等，2012；蒋家平，2012）。搞出如此耸人听闻的大批判式新闻，把本来严肃的学术批评新闻化公众化娱乐化，到头来砸了谁家的脚，让我们拭目以待吧。

即便如此，笔者依然认为，所有这些皆不足为怪，并且深信，在学术的批评和反批评中得到升华的必将是更加科学公正公信的中国大学排行榜。

## 3 大学排行榜之未来：1％研究型大学，主要采用科学计量学指标；99％非研究型大学，主要采用教育计量学和经济计量学指标

大学评估与排行的出现并迅速发展，本质上是由社会发展之需要及高等教育发展与研究之需要推动的。比如，上海交通大学刘念才教授领导的世界大学学术排行榜，如今其影响远及国门以外。正如美国某名校校长指出的："刘念才的'世界大学学术排行榜'（ARWU）是一项在世界上产生了广泛影响的学术成果，值得中国人骄傲。"（刘念才，2008）同样毋庸讳言的是，就排行榜的国内处境来说，这与其说是为祖国为上海交通大学赢得了荣誉和光彩，还不如说是国际高教界与部分国家政府因重视他的排行榜而给予了刘教授及其团队以极大的支持和推动，否则来自国内方方面面的重压与阻击早就把他们摧跨了。至于"大学是一个社会复杂综合体，不能进行量化""学术是不能量化的"等这类观点的持有者，应该首先重温马克思"科学只有她成功地应用数学的时候，才算达到了完善的地步"的这句名言。试问，谁敢下断言和提出"大学不能作为量化研究的对象"这么一个命题呢？

还有人说，一个大学如果仅仅用各种数据去衡量会不会有失公允？一个大学的精神与气质又该如何用数字去体现呢？试问，谁说过可以"仅仅用各种数据去衡量"一个大学呢？这大概是有些人为制造新闻而想当然编造出来的这么一个伪命题。这与前几年媒体上流传炒作的所谓"GDP 崇拜"的说法，可以说基本上是如出一辙的。作为一个科学计量学研究者，我们始终秉持的精神是国际科学计量学与信息计量学学会（International Society for Scientometrics and Informetrics，ISSI）现任会长、国际普赖斯科学计量学奖章获得者鲁索（Ronald Rousseau）教授的权威表达："科学计量学指标并不是要取代专家，而是为了能够对研究工作进行观察和评论，从而使专家能掌握足够的信息，形成根据更充分的意见，并在更高信息集成水平上更具权威性。"（鲁索，2000）

据笔者所知，世界各国的大学排行榜研制者或大学量化评估者，包括笔者的同行朋友西班牙"网络计量学大学排行榜"研制者在内，大家恪守的也是这个原则。我们当初研究制作我国第一个大学排行榜的时候，也是这样，我们不过是希望通过我们的大学排序向社会公众提供一个"观察和评论"中国大学状况的新的方法和视角而已。

大学排行榜本质上说是一个宏观指标系统，换言之，样本越大，其评价的结果

也就越可靠越可比。反之，其可靠性可比性会急剧下降，甚至不仅是可靠性可比性高低的问题而是适得其反的大问题。因此，有关大学排行榜改进方向应当把主要精力投放到大样本的宏观层面上，而不宜在大学各级各类基础个案层面及求全求细的方向上耗费智力与功夫。如果一味朝着评价的指标和要求太烦琐、太复杂的方向走，如果只知道一味地应用与扩大效果而不探究其中的原理，那么这样的大学排行榜必定会走向反面。可以预料，当代正处在高等教育走向后大众化、后普及化的伟大时代（蒋国华，2010），也一定会诞生与以往相较全新而又更切合实际更受市场欢迎的大学排行指标体系。

这是因为，今天世界正处在一个拥有 20 000 余所大学的高等教育鼎盛时代。自 20 世纪下半叶后期开始，世界高等教育出现了走到一个新的十字路口的明显迹象，其主要标志就是在这 20 000 所大学出现了“1∶99 现象”。这就是研究型大学只占其中的 1%弱，其余 99%则均非研究型大学，或许可用“职业培训站”来表征之（蒋国华，2010）。对大学排行来说，也必须依此“1∶99 现象”作相应的分类和调整，即 1%研究型大学，以学术排名为主，主要采用科学计量学指标；99%非研究型大学，则以教育教学和学生学习就业状况排名为主，主要采用教育计量学和经济计量学指标（王战军和蒋国华，2002）。

只要科学在人类的生活和工作中不可离开须臾，只要选择在人类的生活和工作中不可离开须臾，那么，大学排行榜将和世界上其他众多的排行榜一样，一定会永远存在下去。

## 参考文献

必艰．2008-05-09. 排行榜掺杂利益因素 大学给大学排名难言公正．东方早报．

陈至立．2002. 在中外大学校长论坛开幕式上的致辞//中外大学校长论坛文集．北京：高等教育出版社：2.

杜星．2003-03-10. 质疑“中国大学排行榜”．科技日报．

冈特恩．1998. 创造性领导的挑战．郑泉水，等译．北京：清华大学出版社．

顾海兵．2003-04-08. 陷入误区的中国大学排名（上）．科技日报．

顾海兵．2003-04-15. 陷入误区的中国大学排名（下）．科技日报．

韩晓波．2012. 评论：“武书连排行榜”的误导作用．http：//www. tianjinwe. com/tianjin/tjxwcg/201203/t20120321 _ 5391250. html [208-03-21]．

何聪，姜泓冰，陈星星．2012-05-24. 大学排行榜，你还信吗？人民日报．

侯永琪．2007. 高等教育大学排名现况与未来．http：//max. book118. com/html/2011/1223/892785. shtm [2011-12-23]．

侯永琪．2008. 全球与各国大学排名问题之探讨，在台湾“教育政策与评鉴新议题：对新政府的教育期许”研讨会上的演讲．

剑桥大学首位女校长艾莉森·理查德：大学排名很可能毫无意义．http：//news. sohu. com/20070405/n249233254. shtml [2007-04-05]．

蒋国华 . 2004. 我看大学排行榜：认识与误区，在人民政协报教育之春沙龙第 5 期“关注教育评估”主题报告会上的演讲（北京师范大学英东会堂）.

蒋国华 . 2010-02-26. “后大众化”时代的高等教育：1∶99 现象 . 科学时报 .

蒋家平 . 2012. 大学评价：质量优先还是数量制胜 . http://news. sciencenet. cn/sbhtmlnews/2012/5/257790. shtm [2012-05-09] .

蒋廷玉 . 2012-06-27. 科学研究，只有第一没有第二 . 新华日报，http://www. qstheory. cn/wz/jiangt/201206/t20120627_166500. htm.

李红颖 .《高考七日谈》书面采访 . http://edu. 163. com/special/2012_7days_06/.

凌云 . 2000-07-11. “人大”学子状告“网大”. 羊城晚报 .

刘道彩 . 2012-05-25. 大学排行越排越乱 . 中国青年报 .

刘念才 . 2008. 大学排名与世界一流大学建设 . http://wenku. baidu. com/view/95c27d8c84868762caaed5c5. html [2008-09] .

鲁索 . 2000. 评估科研机构的文献计量学和经济计量学指标 . 董晋曦译（参见：蒋国华 . 科研评价与指标 . 北京：红旗出版社：17.）

欧文 · 华莱士 . 1993. 世界排行榜 . 孟学雷，袁嘉谋，周典霞，等译 . 上海：上海人民出版社 .

仇方迎 . 2003-04-08. 一流大学不是世界排名榜排出来的 . 科技日报 .

石小玉，涂勤 . 1999. 美国经济实力分析 . 北京：民族出版社 .

王寰鹰 . 2011. 欧盟欲推出“欧标”世界大学排名榜 . http://www. edu. cn/gao_jiao_news_367/20110922/t20110922_687160. shtml [2011-09-22] .

王战军，蒋国华 . 2002. 科研评价与大学评价 . 北京：红旗出版社 .

武书连 . 2012. 加州理工学院在中国能排第几名——答复中国科学技术大学大学评价课题组 . http://blog. sina. com. cn/s/blog_4b2cb00e0102dyec. html? tj=1 [2012-6-4] .

许智宏 . 2006. 如果追求大学排行榜会产生误导 . http://edu. qq. com/a/20061024/000208. htm [2006-10-24] .

袁祖君 . 2003-03-10. 禁止民间组织进行大学排名 . 北京青年报 .

赵婀娜 . 2009-03-17. 复旦大学校长认为大学排行榜污染“学术空气”. 人民日报 .

赵亚辉，张炜 . 2009-05-05. 大学排行榜真有“潜规则”? 人民日报，http://edu. people. com. cn/GB/79457/9236770. html.

中国管理科学研究院科学学研究所 . 1987-09-13. 我国科学计量指标的排序 . 科技日报 .

中国科学技术大学大学评价课题组 . 2012. 基于公信力视角的大学排名研究——对《2010 中国大学评价》指标体系及算法的质疑 . 中国高教研究杂志，(5)：5-11.

Воспомиhahия о Марксе. 1956. Маckba：Госполитиздат.

Enserink M. 2007. Who can make university ranking? Science，317 (24)：1026-1027.

Morse R J. 2001. U. S. News & World Report's College Rankings. AIR annual meeting，Long Beach，CA. June 3-6.

# 4-4 中外合作办学人才培养质量内部监控体系的构建

李荷迪[①] 樊丽淑[①]

**摘 要：**人才培养质量是高等教育的生命线。中外合作办学由于其特殊性，人才培养质量也呈现多样性，急需对其进行监控。针对中外合作办学的特性，参照国外经验，结合国内当前实际，中外合作办学人才培养质量内部监控体系主要由决策与监控的组织系统、人才培养质量标准系统、检查与督导系统，以及信息系统四部分组成。

**关键词：**中外合作办学；人才培养质量；监控体系

中外合作办学通常是指外国教育机构同中国教育机构在中国境内合作举办的以中国公民为主要招生对象的教育机构（李盛兵和王志强，2009）。伴随着改革开放，中外合作办学已经走过了30多年，成为高等教育国际化的重要形式，对于引进国外先进办学理念和优质教育资源，推动我国高等教育办学模式改革和进步具有重要意义。

人才培养质量作为高等教育的生命线，已成为高等学校办学的共识（沈健，2009）。中外合作办学由于其特殊性，人才培养质量也呈现多样性，急需对其进行监控。目前，我国部分高校开始探索人才培养质量的评价与监控工作，但这项工作还处于初级阶段，尤其是专门针对中外合作办学人才培养质量的监控工作基本处于空白（朱苏飞，2006；孙泽敏等，2011）。中外合作办学人才培养质量监控体系包括外部监控体系和内部监控系统，其中外部监控系统主要指外部环境对中外合作办学人才培养质量的监督、控制和管理，具体包括教育行政部门的评估、第三方中介机构的评价，以及社会公众的影响等方面，而内部监控是指中外合作办学双方学校对办学项目的自我监控和内部管理。相比之下，内部监控显得更为重要。

## 1 人才培养质量内部监控面临的问题

经过不断的探索发展，中外合作办学在理念创新、资源引进、学科建设和提升国际化等方面发挥了积极作用，但在人才培养质量的内部监控方面，还远没有达到

① 浙江大学宁波理工学院，浙江宁波，315100。

科学化和规范化的程度，面临着一些迫切需要解决的问题。

### 1.1 认识不足

中外合作办学项目一般多重视短期投资收益，较少立足于长远发展。国家对中外合作办学也没有提出相关监控办法和指标要求，各层面都存在认识不足的问题。因此，一些办学机构往往对上级部门的评估等外部质量评价工作非常重视，以期通过项目评估获得继续办学的资格，但对于内部人才培养质量的监控工作则觉得可有可无，往往会忽视。

### 1.2 机构人员缺乏

中外合作办学机构一般没有设置独立的人才培养质量监控机构，常挂靠在教务部门，不能独立实施监控。少数虽有形式上的质量监控与评价部门，但没有配备具有相关业务知识的工作人员，人才培养质量监控工作常流于形式，不能真正起到监控作用。

### 1.3 体系不完善

目前，国内很少有学者涉足人才培养质量的内部监控体系研究，现行也没有一套适合中外合作办学的内部人才质量监控体系。很多机构往往简单照搬国内普通高等教育的监控模式，不仅不适应中外合作办学的性质和特色，也缺乏自我约束与完善的机制。

## 2 人才培养质量内部监控体系构建的原则

中外合作办学人才培养质量的内部监控，必须充分考虑中外合作办学的性质和特征，形成一套具有科学合理性、可行可操作性，以及目标导向性的监控体系。

### 2.1 科学性原则

科学性原则要求制定的监控体系、建立的质量标准，以及开展的监控工作都必须符合中外合作办学特性和人才培养质量监控的客观规律。监控指标应全面再现监控对象的培养目标，也要保证同一层次的各指标相互不重叠。

### 2.2 可行性原则

可行性原则体现在监控对象之间的可比性、监控指标的可测性和监控工作的可操作上。可行性主要涉及两个方面：首先应从实际出发得出监控指标，使其符合我国人才培养情况的实际；其次，按指标进行监控应该是切实可操作的。人才培养质量监控的可行性将决定人才培养质量监控能否在更大的范围内开展起来。

### 2.3 导向性原则

导向性原则是指人才培养质量监控应有正确的目标导向。在实施人才培养质量监控时，学生如何被评价是影响他们目标取向、成就感和实际行动的重要因素。如果监控和评价的导向性发生偏差，学生就有可能把注意力主要集中到评价结果、评价等级或评价所暗示的发展方向上，而忽视个人的能力培养和全面发展等更为本质的东西。

## 3 人才培养质量内部监控体系的构建

由于中外合作办学的性质不同于普通高等教育办学，应该在开放办学思路和国际对接教学理念的指导下，设立独立的监控部门，安排专业人员，制定监控制度，对人才培养的各个环节定期进行检查、督导、评价和反馈。参照国外经验，结合国内当前实际，中外合作办学人才培养质量内部监控体系主要由决策与监控的组织系统、人才培养质量标准系统、检查与督导系统，以及信息系统四部分组成。

### 3.1 决策与监控的组织系统

目前，中外合作办学的人才培养质量监控工作大多由教务部门组织实施，很少有专门的监控机构，这是一种不合理的现象。中外合作办学机构应该在中外合作双方的共同指导下，由分管教学工作的院长主持工作，建立专门的监控机构，规划监控工作目标，确定监控具体内容。机构内应配备专业人员具体协调实施监控工作，同时成立由教学委员会、专职质量监控机构等为主体组成的人才培养质量实施组织保障系统，借用学术力量通过教学委员会、教学督导专家组对各人才培养单位的人才培养质量进行监控，定期组织检查、督导、评价和发起反馈。形成人才培养监控的相关制度，定期召开人才培养工作会议，对人才培养工作进行研究和决策，并做出调整，形成人才培养质量监控的长效机制，从而保证机构人才培养质量监控工作能在一个较高层次上运行和发展。

### 3.2 人才培养质量标准系统

中外合作办学机构应根据双方学校的办学理念，定位人才培养目标，制订人才培养方案。中外合作办学项目的人才培养理念应体现双方高校的办学思想、理念和特色，既能引进国外先进办学理念又能适应国内办学环境。项目的人才培养目标应该是培养出既能服务于地方经济社会发展，又具有国际化背景和视野的外向型人才。在培养理念的指导下，通过质量功能分解的方法，将人才培养目标转化为人才培养方案和各培养环节的质量标准，并形成人才质量标准系统。标准体系的制定，既不能照抄国内普通高等院校的标准，也不能照搬国外大学的标准。标准体系应包括人才培养质量标准、专业建设质量标准、课程建设质量标准、实践质量标准，以及毕业设计（论文）质量标准等。这些标准应具有一定的稳定性，形成文件，作为共同遵循的准则。同时，标准系统应能够全程监控和实际测量，避免标准缺口和质量真空，并能与国际标准相对接。

### 3.3 检查与督导系统

中外合作办学机构应建立和完善人才培养的检查与督导系统。成立督导组，督导组成员应为理论水平高、责任心强、教学经验丰富的资深教师，也可吸收在岗骨干教师做兼职督导员。检查与督导的对象为学生培养的各个环节，根据时间划分为培养前、培养过程和培养结果三个阶段。

(1) 培养前的监控对象是生源质量，主要涉及招生工作。由于学生是学习的主体，其所具备的基础、素质和能力最终决定了教育的可塑性。通常中外合作办学的学费比普通高等教育要高，学生需要接受外语或双语课程教学，还需要强化外语学习，使得中外合作办学的学生不仅需要有较好的经济实力做后盾，也要有较好的学习基础和能力。由于学费较高，对家庭经济一般的学生具有一定阻碍。不断加强招生工作，改善和提高中外合作办学的生源质量，是保证人才培养质量的前提。

(2) 培养过程的监控对象主要由教学、管理和服务三类工作组成。除了教学，管理和服务也是育人的重要手段，也应纳入监控范围。但是最主要的人才培养工作是教学工作，对教学工作的监控主要涉及教学大纲、课堂教学、实践教学和毕业设计（论文）的监控。教学大纲是以系统和连贯的形式，按章节、课题和条目叙述该学科的主要内容的教学指导文件。它不仅是对人才培养目标的分解和分项落实，也是指导教师教学的书面文件，是一个重要的监控环节。对教学大纲的监控应该由研究所（教研室）、二级院系、教务处和教学委员会共同负责，监控内容是保证研究所及时制定与人才培养目标相符合的教学大纲，并根据学科发展新趋势和社会对人才新需求做出调整。课堂教学是教学工作中最重要的组成部分，因此对课堂教学的监控对于提高人才培养质量具有重要意义。课堂教学的监控包括教学前准备工作监控、教学实施过程中监控和教学完成后监控三个组成部分。课堂教学中，师资力量是重点监控的内容，中外合作办学的师资应该由中外双方选派有经验的教师，并且包含一定量的外语课程和双语课程。实践教学既是培养学生运用理论知识分析和解决实际问题的重要环节，也是对应用型人才培养目标的具体落实与贯彻，对于提高学生的实际业务操作能力具有重要意义。实践环节监控主要内容为监控人才培养方案中所确定的社会实践环节和专业实践环节能否得到有效的贯彻、落实和实施。毕业设计（论文）是对学生基本知识、基本理论和基本技能掌握与提高程度的一次总测试，同时通过毕业设计（论文）对学生进行科学研究基本功的训练，培养学生综合运用所学知识独立地分析问题和解决问题的能力。对于毕业设计（论文）的监控可以反映出人才培养质量的高低。

(3) 培养结果的监控主要是毕业生的就业情况，包括就业、升学和出国等方面，涉及就业指导工作。随着教育观念的转变，毕业生去向的好坏将直接决定中外合作办学项目未来的发展。那些专业设计合理，人才培养质量高，毕业生就业好的项目将继续得到政府、社会和学校的支持；而那些专业设计不符合社会需求，人才培养质量较差，毕业生就业差的项目必然面临着停止或压缩招生的命运。因此，迫切需要建立中外合作办学项目毕业生就业的监控体系。这项工作通常由学校就业处和学生处共同负责，通过定期和毕业生及用人单位沟通，掌握毕业生就业率和就业质量，并及时根据反馈信息，调整和优化专业培养方案，使学生知识结构方面的缺陷能够得到及时的弥补。

### 3.4 信息系统

建立一个高效的信息系统是中外合作办学人才培养质量监控体系运行的先决条件。信息系统包括信息的采集、处理和反馈三方面。信息采集主要集中在检查与督导系统的各个人才培养环节。对采集的信息需及时汇总并加以分类整理，并进行多参数分析和符合度检验，找出问题的轻重程度、存在范围及偏离原因。信息的处理和分析应注重全面、公正和合理，对于一些重点问题要有足够长时间的标本量。信息反馈是人才培养质量监控体系形成闭环的重要保证，将结果反馈给被评者和中外合作双方管理者，要有信度、效度和针对性，并注意信息的通畅。同时，也要将分析整理好的结果准确地上传到决策与监控系统，以便及时进行调控，为研究和部署下一步质量监控工作提供科学依据。信息采集、处理和反馈形成良性互动，才能不断提高中外合作办学人才培养质量监控的管理水平。

## 4 小　结

中外合作办学人才培养内部质量监控体系的构建是一项复杂的系统工程，必须要随着中外合作办学的发展和变化不断探索、不断推进、不断创新，制定出更加科学合理、可行可操作、目标导向的监控体系，并确保体系长期有序运行，从而不断提高中外合作办学人才培养质量，推动教育国际化水平提升和中外合作办学健康发展。

## 参考文献

李盛兵，王志强 . 2009. 中外合作办学 30 年——基于 11 省市中外合作办学分析 . 华南师范大学学报（社会科学版），(2)：96-99.

沈健 . 2009. 当前国际高等教育改革发展的新趋势 . 江苏高教，(2)：1-4.

孙泽文，叶敏，刘俊平 . 2011. 中外合作办学教学质量内部监控及其体系构建 . 教育与教学研究，25 (1)：64-68.

朱苏飞 . 2006. 保障中外合作办学质量的实践和思考 . 高等教育研究，(2)：30-33.

# 4-5 世界一流大学与科研机构竞争力评价的做法、特色与结果分析

邱均平[①] 赵蓉英[①] 楼 雯[①] 吴胜男[①] 余厚强[①]

**摘 要：** 本研究从美国汤森路透科技信息集团出版的《基本科学指标》(ESI) 数据库获取原始数据，对世界一流大学与科研机构的竞争力进行了科学、合理、客观、公正的评价研究和综合分析，最终得到了5类共37个排行榜，它们分别是：《2013年世界国家或地区科研竞争力排行榜》、《2013年世界大学科研竞争力排行榜》、《2013年世界大学与科研机构学科竞争力排行榜》(分22个学科)、《2013年世界大学科研竞争力一级指标排行榜》(分5个一级指标) 和《2013年世界大学科研竞争力基本指标排行榜》(分8个基本指标)。2013年的评价结果显示，中国的科研竞争力排名有所上升，由2012年的第六位上升为2013年的第五位，而且2013年中国进入世界一流大学排行榜的学校无论从数量上还是从排名上，都有了显著提高。这在一定程度上表明，我国在建设世界一流大学道路上取得了可喜成绩，但通过仔细比较和分析，可以看出我国大学整体排名还比较靠后，说明我国大学离世界一流大学还有相当差距，建设世界一流大学的任务仍很艰巨，特别是在建设世界高水平的一流大学和科研机构以及创新型成果方面差距还很大。从自身来看，与2012年相比，我国大学与科研机构不论在数量上还是质量上都有明显进步，虽然各项指标与世界科研强国相比还有很大差距，但是这些成绩的取得会给我们增加信心，相信不久的将来中国一定会迈入世界科研强国的行列。

**关键词：** 基本科学指标；ESI；大学评价；学科建设；科研机构；科研竞争力

## 1 引 言

世界一流大学是一个国家高等教育发展水平的标志，是综合国力的集中体现，更是一个国家经济、科技、社会发展到一定阶段后的需要。近年来，对世界一流大学进行评价受到越来越多的关注，各种大学排行榜如雨后春笋般涌现，受到各国学子与家长的热烈追捧。大学排行榜的应运而生，显示了在高等教育逐步迈向高度普及化的进程中，社会各阶层对有关大学的各种信息的知情权的需求不断攀升，是经济增长与社会发展的标志（邱均平等，2013）。

① 武汉大学中国科学评价研究中心，武汉，430072。

在前几次世界大学科研竞争力评价的基础上，由武汉大学中国科学评价研究中心、中国科教评价网和武汉大学中国教育质量评价中心共同完成的2013年有关世界大学科研竞争力的排行榜也应运而生。在此次评价中，我们同样认为世界大学与科研机构学科竞争力的评价指标主要由科研生产力、科研影响力、科研创新力、科研发展力这四个部分组成，但是，针对大学科研竞争力评价，我们在此次评价中创新性地引入了网络影响力指标。用这一指标来进一步考察各学校的声望情况、科研成果的开放获取程度来作为Web环境下的科研影响力评价的补充，以达到从科研产出到现实影响再到网络影响的综合实力评价。

我们从2013年6月开始，第六次利用ESI和DII这两种权威工具作为数据来源，集中科研力量对世界大学与科研机构学科竞争力评价进行了较为系统和深入的研究，并且研发了《2013年世界国家或地区科研竞争力排行榜》、《2013年世界大学科研竞争力排行榜》、《2013年世界大学与科研机构学科竞争力排行榜》（分22个学科）、《2013年世界大学科研竞争力一级指标排行榜》（分5个一级指标）和《2013年世界大学科研竞争力基本指标排行榜》（分8个基本指标）。从评价结果可以看出，我国大学近两年进步很快，取得了可喜的成绩，但与世界一流大学相比，差距还很大，建设世界一流大学的路还很长，任务十分艰巨，特别是在前沿学科的高水平研究成果和国际竞争力、影响力方面存在着较大的差距。这些鲜为人知的排名结果和评价结论，为我国各个大学、科研院所、政府管理部门、相关研究人员、欲出国求学的学子，以及其他社会各界人士提供了一份较为全面、详细、有特色的评价报告。这对于我们清楚地认识国内大学在世界上所处的位置，从而提高我国大学的国际竞争力具有重要的指导意义和参考价值。

## 2 世界一流大学的研究现状与评价标准

### 2.1 国外研究现状

目前，国外对世界一流大学研究的侧重点各不相同，其影响力比较大的有以下几个。

#### 2.1.1 《美国新闻与世界报道》

早在1983年，《美国新闻与世界报道》（*US News & World Report*）率先推出全美大学排名，在全美本科院校每两年评选一次。1987年，《美国新闻与世界报道》开始面向研究生教育，改为每年评选一次。这种最初的排名主要是为了给学生和家长在选择高校时提供一些参考数据。它每年春季都公布最新的“全球大学排行榜”（Global Universities Ranking）主要供秋季新生入学参考。《美国新闻与世界报道》对高校进行排行是依据卡内基教学促进基金会公布的高等学校分类法，先将高校进行分类，然后在同类之间进行评比，由于它的调查过程科学严谨，所以具有权威性。

《美国新闻与世界报道》的评估主要基于两个原则展开：其一是根据专家确定的标志学术质量的定量指标；其二是根据他们作为局外人对有关教育质量的认识。它利用的重要数据源之一是来自大学董事会、彼得森公司、《美国新闻与世界报道》联合组成的数据中心。

《美国新闻与世界报道》对大学评价的指标体系，主要包括同行评议、教师资源、财政资源、学生保持率、招生选拔、毕业率表现、校友捐赠率等方面内容，但对于全国性大学、美国文理学院和地区性大学及学院这两类大学的评价指标权重是有所区别的（Morse，2013）。

1）本科学术声誉（undergraduate academic reputation）

《美国新闻和世界报道》对学术声誉赋予最大的权重，在地区性大学及学院评价中权重达到25%，在美国国立大学、美国文理学院评价中的权重为22.5%。在前者的评价中，声誉评价主要集中在同行评估（peer assessment survey）这一方面。在后者的评价中，还引入了高中辅导员的评定（high school counselor's rating），占到声誉指标权重的1/3。

2）学生选择（student selectivity）

学生选择主要指入学录取标准，这是衡量学生素质的重要尺度，它在两类大学的评价中都占到总分的15%。其中，学生的考试成绩——学生入学的SAT或ACT，即Scholastic Aptitude Test/American College Testing（学术能力测验/美国高校测验）的平均分数占学生选择的50%；学生在高中班级中的名次也需要考虑，包括入学的新生中占高中时班上最好10%的比率（国立大学评价）和25%的比率（地区性大学评价），高中班级名次占学生选择的40%；录取率占学生选择的10%，指的是录取学生数目与申请学生数目的比率。

3）师资情况（saculty resources）

师资情况通过班级规模、师资薪酬、师资学位、师生比、全日制师资比等指标来反映，在评估中的权重均为20%。其中，小班教学（班级人数低于20人）所占权重为30%，大班教学（班级人数多于50人）所占权重为10%。好学校愿意并且能够高薪聘请优秀的教授，所以要考虑教师的收入，教师薪酬所占权重为35%，具有博士学位或该学科最高学位的教授比率权重为15%，教师与学生数目比率权重为5%，专职教师比率的权重为5%。

4）毕业和保持率（graduation and retention rate）

这项指标的权重在国立大学评价中为20%，在地区性大学评价中为25%，包括两个方面：平均毕业率和平均大一新生持续注册率，即保持率。其中六年期间的平均毕业率占80%。保持率指的是一年级新生第二年继续返校注册的比率，保持率在该项指标中的比重为20%。毕业和保持率得分越高，就表明学校的课程及相关服务越能够满足学生的需求。

5）财务资源（financial resources）

财务资源主要是指学校每年在每个学生的教学、研究、服务及其他教育开销的费用，学校花在学生身上的钱越多，就越能提供更好的服务。该项的权重在两类大学中均为10%。财务资源是以教育费用和其他费用等要素来综合评定的，其中教育费用权重为80%，而其他费用的权重为20%。

6）毕业率履行情况（graduation rate performance）

毕业率履行情况在国立大学评估中的权重是5%，在地区性大学评价中为0%。它是学校基于入学学生的入学成绩而做出预期的6年学生毕业率，如果实际毕业率高于该预期，则表明学校取得了进步，作为“加分”。

7）校友捐赠（alumni giving）

校友捐赠可以间接反映校友对母校的满意程度，在两类大学中均占总分的5%。

《美国新闻和世界报道》首先计算各标准分的加权平均分，然后进行必要的调整，最后以最佳学校为100分，对所有学校评分按比例归一化，四舍五入，然后按顺序排行。《美国新闻和世界报道》对国家级大学、国家级文理学院，地区级大学和学院、地区级文理学院、专业院校等分别排行。

### 2.1.2 英国《泰晤士报高等教育》

英国《泰晤士报高等教育》（*The Times Higher Education*，THE）是由公司TSL教育公司出版的周刊，在世界范围内有较大影响。从2010年起，《泰晤士报高等教育》与世界首屈一指的数据公司美国汤森路透科技信息集团合作，由美国汤森路透科技信息集团负责收集和分析所有的与排名相关的数据。《泰晤士报高等教育》还将采用新的评价标准和方法。在新的世界大学排名标准中，《泰晤士报高等教育》保留了“同行评议”这一指标，由民意调查公司Ipsos Mori接手声望调查工作，并采用一种更为谨慎的抽样调查方式，在公信力方面将有较大的改善（Baty，2013）。

2010年的评价方法新增了经济活动/创新这一级指标，二级指标也由2009年的6个增加为13个，改变了往年一个二级指标代理一个一级指标的较为单一的评价方式。舍弃了雇主调查这一定性指标，并且在学术声誉调查这一部分做了较大变动，声誉调查的规模更为扩大，更具严密性和代表性；使其从一级指标降为两个二级指标，将教学相关和研究相关调查结果分列在教学指标和研究指标中，成为教学指标和研究指标的支撑，降低了其独立性；从比重上来看，同行评议的比例由之前的高达40%的比重降到20%，使世界大学排行榜的主观指标降低了至少20%，大大增加了量化指标的比重（Methodology，2012）。总体而言，2010年新的指标体系设置一级指标5个，即工业收入（industry income）所占权重为2.5%、国际师资和学生（international mix-staff and students）占7.5%、授课（teaching—the learning environment）占30%、研究（research—volume，income and reputation）占30%、论文引用影响（citations—research influence）占30%（Baty，2013）。

### 2.1.3 《新闻周刊》

《新闻周刊》（*Newsweek*）是一份在纽约出版，在美国和加拿大发行的新闻类周刊。《新闻周刊》的世界大学排名（Top 100 Global Universities）是以上海交通大学和英国《泰晤士报高等教育》为基础进行排名的。

《新闻周刊》借鉴上海交通大学所使用的教师素质、研究成果等指标来衡量大学的学术情况。其中，要考虑发表于《自然》与《科学》杂志的论文数目和收录于科学引文索引扩展版和社会科学引文索引的论文篇数。这部分得分的权重为50%。

《新闻周刊》借鉴英国《泰晤士报》的高等教育增刊对大学进行评价中所使用的海外教师比例、海外学生比例、论文引用比例、师生比等指标对大学进行评估。这部分的权重为40%。

此外，《新闻周刊》还考虑了大学图书馆的藏书量，来反映大学的资源拥有情况。这部分的权重为10%。

### 2.1.4 西班牙网络计量实验

西班牙人文与社会科学研究中心（Centro de Ciencias Humanasy Sociales，CCHS-CSIC）网络计量研究室（Cybermetrics Lab）自2004年起，每隔六个月发布一次《世界大学网络排名》（*Ranking Web or Webometrics*），目的是期望提升各大学的学术知识与资料在网络上公开出版（web publication）的程度，促进科研成果出版的开放获取，以经济、快速的知识扩散方式，提升其影响力。此项排名的指标，分为两项（Methodology，2012），共四个指标。

（1）可见度（visibility），占50%：考查各大学的影响力。数据来源为Majestic SEO和Ahrefs，统计的是网站的入链数和这些入链网站的出链数。

（2）活跃度（activity），占50%：考查下列三项指标：①表现力（presence），占1/3：被Google检索的域名内的网页数。②开放性（openness），占1/3：被Google Scholar检索的可获得性文件（如pdf，doc，docx，ppt等）的数量。③优越性（excellence），占1/3：被SCImago Research Group收录的学术论文数量。

上述各排名指标（V，P，O，E），按照特定的查询命令，针对各个大学的域名，在搜索引擎中检索，整理合并结果。按照不同比例合并指标排名，作为最终的排名WR（World Ranking）。

## 2.2 国内研究现状

目前，国内对世界一流大学进行研究比较有名的有以下几个。

### 2.2.1 武汉大学中国科学评价研究中心

在我们连续几年做中国大学评估的基础上，2006年开始做世界大学科研竞争力评价。自2006年至2012年，我们进行了5次世界大学及科研机构学科竞争力评价，从2012年开始其评价指标改由科研生产力、科研影响力、科研创新力、科研发展力和网络影响力5个部分构成。其内容和特点具体分析将在下文中展开。

2012年武汉大学中国科学评价研究中心对世界大学和机构科研竞争力进行评

价，得到了3类共32个排行榜，它们分别是：《2012年世界大学科研竞争力排行榜》、《2012年世界大学与科研机构分22个学科的科研竞争力排行榜》和《2012年世界大学科研竞争力分8个基本指标排行榜》。世界大学科研竞争力评价采用了目前最权威的高水平的数据来源工具（ESI），数据准确可靠，并且以新颖的评价理念设置了科学合理的评价体系，提供了国内目前最详尽的世界大学评价报告，不仅针对国家、机构，而且评价学科专业。

### 2.2.2 上海交通大学高等教育研究所

国内最早对世界一流大学进行系统研究的是上海交通大学高等教育研究所。1993年，上海交通大学出版社出版了国内第一本有关世界一流大学研究的专著《世界一流大学研究》，1999年出版了《攀登——我国创建世界一流大学的研究》，为我国创建世界一流大学提供了有益的、多方位的思考与借鉴（浙江大学大学评价研究课题图，2004）。

2001年，上海交通大学高等教育研究所刘念才等向教育部科学技术委员会提交了《我国名牌大学离世界一流大学有多远》（刘念才等，2002）的研究报告，指出学术声誉通过诺贝尔奖、《自然》和《科学》论文、SCI论文等可量化的国际可比性指标表达，教师质量通过诺贝尔奖、博士学位教师比例等表达。

上海交通大学高等教育研究所于2003年夏天首次在国际互联网上发布了“世界大学学术排行”（Academic Ranking of World Universities，ARWU），之后每年8月中旬进行更新。上海交通大学高等教育研究所的世界大学学术排行主要考虑大学的如下几个方面（世界大学学术排名，2011）。

1）教育质量

这一指标的权重为10%，主要考察获得诺贝尔奖和菲尔兹奖的校友折合数。这些校友包括在该大学取得学士学位、硕士学位和博士学位的人。对不同年代的获奖校友，采用每回推十年权重递减10%的方式赋予其不同的权重；对于在该大学获取两个及以上学位的校友只计算最近的一次。

2）教师质量

这项指标下设两个二级指标，即获诺贝尔奖和菲尔兹奖的教师折合数和各学科领域被引用次数最高的教师数量（HiCi）。

获得诺贝尔奖和菲尔兹奖的教师折合数指标的权重为20%，考察的是那些在校工作时获奖的人。对不同年代的获奖者赋予不同的权重，同样也是采取每回推十年权重递减10%的方式。如果一位获奖者属于不止一个机构的话，每个机构就平均分配其共同拥有的获奖人数。如果是几个人一起合作而获奖，那么就按照比例来赋予权重。

各学科领域被引用次数最高的教师数量是指一所大学在各学科领域被引用次数最高的教师总数，按汤森科技信息研究所公布的二十年来在21个领域内被引用次数最高的5000余位研究人员的情况进行统计。其权重也是20%。

3）科研成果

该项指标下面也有两个二级指标，是在《自然》和《科学》上发表论文的折合数（N & S）和被科学引文索引（SCIE）和社会科学引文索引（SSCI）收录的论文数量（PUB）。

N & S 指标的权重为 20%。考察最近 5 年内，在这两个国际杂志上发表论文的折合数量，只统计研究论文，不统计检索评论或快讯等。对不同作者单位排序赋予不同的权重，通讯作者单位的权重为 100%，第一作者单位（如果第一作者单位与通讯作者单位相同，则为第二作者单位）的权重为 50%，下一个作者单位的权重为 25%，其他作者单位的权重为 10%。

PUB 指标指的是一所大学过去一年被 SCIE 和 SSCI 收录的论文数量，只统计研究论文，不统计检索评论或快讯等。考虑到社会科学领域的学者经常以著作等形式发表其研究成果，根据实证数据，对 SSCI 收录的论文赋予 2 倍的权重。该项指标权重为 20%。

4）师均表现

该指标是一所大学的师均学术表现，由前 3 项指标得分之和除以全时（full time equivalent）教师数而得，记作 PCP，其权重为 10%。

上海交通大学的大学排名主要根据研究成绩来对研究型大学进行评价，所用的数据具有国际可比性，但仅仅考虑了诺贝尔奖和菲尔兹奖、《自然》和《科学》杂志，并赋予很高的权重，其他奖项、重要杂志没有纳入其中，使得对文科实力较强的学校的评价不够全面和公平。

### 2.2.3 浙江大学大学评论国际委员会

浙江大学的“国际大学创新力客观评价报告”课题组于 2005 年 10 月筹建了大学评价国际委员会，下设大学评价国际学术委员会、大学评价工作委员会和大学评价办公室三个机构。发布《国际大学创新力客观评价报告》，主要评价世界知名大学的创新力，对大学的选择具有较高的要求。以世界范围内具有较高知名度的协会中所列大学和现有评价结果中名次靠前的大学为研究对象（大学评价国际委员会，2007）。

浙江大学的“国际大学创新力客观评价”主要围绕创新实力、创新活力和创新影响力这三个一级指标展开。

1）创新实力

创新实力体现了一所大学创新能力的人力、物资和财力基础。此项指标在整个评价中的权重为 20%。创新实力包括 3 个二级指标：①教师中获诺贝尔奖人数，只计算获奖时在该校工作的教师，此项数据来源于诺贝尔奖网站，在创新实力中的权重为 40%。②教师人数，指在一所大学工作的教师总量，此项数据基本来源于各个国家或地区的统计网站，在创新实力中的权重为 30%。③人均科研经费，指一所大学平均每个教师获得的本国国家基金资助科研经费，各国货币根据购买力平价（PPP）进行折算。此项数据基本来源于各国或地区科研经费网站，在创新实力中的权重为30%。

2）创新活力

创新活力反映了一所大学创新能力的过程和结果。此项指标在整个评价中的权重为70%。创新活力包括7个二级指标：①在《自然》和《科学》上发表论文数，此项数据来源于ISI（Institute for Scientific Information）的SCI数据库，在创新活力中的权重为30%。②1%顶级学科数，指的是大学在ESI数据库中列出的22个学科中按被引论文计算后位列国际前1%的学科数。这项数据来源于ISI的基本科学指标数据库（Essential Science Indicators，ESI），在创新活力中的权重为15%。③高被引论文数，指的是大学在ESI数据库中进入国际前1%的被引论文数。这项数据来源于ISI的基本科学指标数据库ESI，在创新活力中的权重为15%。④ESI引文数，指的是大学在ISI的ESI数据库中的累积引文数。这项数据在创新活力中的权重为10%。⑤ESI论文数，指的是大学在ISI的ESI数据库中的累积论文数。这项数据在创新活力中的权重为10%。⑥专利数，指的是大学在ISI的DII（Derwent Innovations Index）数据库中拥有的专利数。这项数据在创新活力中的权重为10%。⑦人均ESI引文数和论文数，指的是大学在ISI的ESI数据库中的人均引文和论文数。这项数据在创新活力中的权重为10%。

3）创新影响力

创新影响力标志着一所大学的创新活动对国家或地区社会、经济、文化的发展所产生的影响。此项指标在整个评价中的权重为10%。创新影响力包括2个二级指标：①当年培养博士生数，指的是一所大学当年获得博士学位的人数。这项数据来源于各国或地区的统计网站，在创新影响力中的权重为20%。②本地指数，指的是大学对地方经济和社会发展的影响，用排名系数表示。如果一所大学在本国或地区的排名结果中名列第一位，那么就为其赋值100；如果名列第二，就赋值99；如果名列第三，就赋值98……以此类推。这项数据来源于大学所在国家或地区权威的排名结果，在创新影响力中的权重为80%。

浙江大学的“国际大学创新力客观评价”实属创新之举，在全球范围内首次对大学的创新力进行了评价，将大学评价与大学创新力建设结合起来了。它全部采用权威机构发布的客观数据，以之为基础对大学的创新实力、创新活力、创新影响力进行评价。浙江大学的“国际大学创新力客观评价”中完全没有采取主观评价如同行评议等，在一定程度上保证了数据的客观性，但对于一些难以用数据进行量化考察的方面如学术氛围等难以进行评价。

## 2.3 国内外比较研究

通过对目前国内外关于世界一流大学评价的现状进行分析，我们发现不同的评价机构运用不同的评价指标，各有特色。

《美国新闻与世界报道》、英国《泰晤士报高等教育》、《新闻周刊》都是评教分离的典范，而上海交通大学和浙江大学本身既是教育机构，又参与了评价工作，目的是为了以评促建。

《美国新闻与世界报道》、英国《泰晤士报高等教育》(2008 年以前的指标体系)注重主观数据的收集和利用，为评价大学声誉而开展同行评议，分别赋予 25%和 50%的权重。2010 年开始，英国《泰晤士报高等教育》采取新的标准进行评价，将声誉调查分为教学与研究两个方面，并削减了主观数据所占的比重，分别占 15%和 19.5%，但总体权重仍然是整个指标体系中最大的一部分。然而在“隔行如隔山”的条件下，“同行专家”的选择不易，在实际操作中要仔细甄别，避免出现外行评价内行的情况。

上海交通大学和浙江大学的评价注重客观数据的收集和分析，保证了数据的国际可比性，在评价自然科学方面比较公正，而且都侧重于对大学科研方面的评价。但他们对于人文科学方面较强的大学有失公平，因为人文科学方面的研究发文量没有自然科学大。而且，单纯的客观数据分析无法评价大学的学术氛围、校园文化等，而这些对于世界一流大学来说也是非常重要的。

总的来说，《美国新闻与世界报道》的大学评价比较全面，为学生和家长择校提供了参考性意见。英国《泰晤士报高等教育》引入了评价有关国际化程度的指标，鲜明地体现了现代世界一流大学的时代特征。《新闻周刊》在考查大学科研实力的同时也考虑了大学图书馆的藏书量，也就是触及有关大学硬件资源的评价。西班牙网络计量实验室的世界大学网络排名采用网络搜索引擎收集数据，排除了人为干扰，更为客观公正，同时又显示出了各个大学的信息化实力。上海交通大学注重对大学科研产出进行评价，浙江大学侧重于对大学创新力的评价。每个评价体系都有自身的特色和侧重点，因此，他们的评价结果也各有不同，但都为我们对世界一流大学的研究提供了很好的素材，有很多值得我们借鉴和学习的地方。

## 2.4 世界一流大学的基本特征与评价标准

对于世界一流大学的特征和评价，目前没有统一的范式，各个国家各个学派都有着不同的见解。但是世界一流大学在学术大师汇聚、科研经费充裕、科研成果卓著、优秀学生培养、办学特色鲜明等方面显示出共同的特征。建立良好的工作氛围，提供一流的教学和研究设施，为教师确定合理的工作量以使教师有充分的自主性和更多的时间从事科研工作，被认为是建设一流教师队伍的最重要的三个方面。

李岚清认为世界一流大学有以下几个共同点：第一，有杰出教育家，特别是有出色的校领导。他们有教育家的战略思想，有国际视野。第二，培养出大批优秀的人才。第三，提倡学术自由，鼓励理论创新；多用启发式、讨论式的教学方式。第四，拥有一批具有一流学术水平的优势学科。第五，非常重视研究生教育，尤其重视培养博士生。第六，是发表一流研究论文和学术著作的主力军。第七，有深厚的文化积淀。第八，以多种形式服务社会。

美国大学联合会常务副主席约翰·冯在接受访问谈到世界一流大学的建设时，提到“一所世界一流大学要有足够广泛的学科领域，基本应当涵盖所有主要的学术和人文领域”“严格一点来评价，就不仅要关注科研方面的数量和科研方面的广度，也要考虑质量，以及教师们在相关学科做了多少前沿性重要研究，在世界范围内是否处于领先地位等”（王晓阳等，2010）。

上海交通大学校长张杰院士提出“以‘世界一流大学’为建设目标的研究型大学应紧紧围绕世界一流大学的基本特征来构建核心竞争力”，他认为世界一流大学一般具有以下 10 项基本特征：追求卓越目标，服务国家战略；办学理念清晰，发展定位明确；学科门类齐全，学术声誉卓著；教师素质超群，学术大师汇聚；教学资源丰富，教学水平先进；生源质量优良，创新人才辈出；科研经费充裕，科研成果斐然；国际交流广泛，学术氛围浓厚；促进文化繁荣；引领社会进步；杰出校长掌舵；管理科学规范（张杰，2008）。

北京大学原校长、中国科学院院士许智宏认为，世界一流大学主要有 3 个标准：一是有从事一流研究工作的国际知名教授；二是有一大批影响人类文明和社会经济发展的成果；三是培养出一大批为人类文明做出很大贡献的优秀学生。只有能满足这三个条件，才能称为世界一流大学，而中国目前还没有真正的世界一流大学（许智宏，2010）。南京大学校长陈骏指出，世界一流大学主要具备三大功能：第一，作为培养与造就高素质创新人才的教学机构，帮助学生形成科学的世界观和价值观，掌握分析问题和解决问题的基本方法和技能，让他们养成终生追求真理和正义的习惯，引领社会的健康发展；第二，作为生产、传承各种知识、理论和观念的学术机构，让人类创造的优秀文化能够历久弥新，生生不息；第三，作为服务国家宏大目标和促进区域发展的社会机构，成为国家核心竞争力的重要组成部分（陈骏，2010）。华中科技大学的周光礼教授通过抽取关键要素的方式，归纳出世界一流大学共同的特质：第一，具有一流的国际声誉；第二，具有世界一流的师资队伍；第三，具有世界一流的优势学科；第四，培养出大批的精英人才；第五，具有充足而灵活的办学资源；第六，具有完善的管理构架；第七，具有较高的国际化水平。只有以先进的建设世界一流大学办学理念为指导，以大学文化建设和体制机制创新为基础，才能找到一条行之有效的建设世界一流大学的路径，最终形成“中国特色、世界水平”的一流大学发展模式和先进文化（周光礼，2010）。

将一批国内大学建设成世界一流大学是我国大学发展的远景目标，总结国内外世界一流大学的特征，我们可以得出以下结论。首先，世界一流大学绝大多数是研究型大学，无论是美国的麻省理工学院、哈佛大学，还是英国的剑桥大学、牛津大学，加拿大的多伦多大学，以及日本的东京大学、早稻田大学，国内的北京大学、清华大学，这些占据各国大学排名前列的世界一流大学实际上都是研究型大学。

其次，世界一流大学的分类标准参照美国研究型大学模式。卡内基教学促进基金会在克拉克·克尔博士的主持下，对美国高等院校的分类进行了多次修改。至第七次修改的2010年版《卡内基分类法》共有7种分类模式，即基本分类、选择性分类和5种独立（平行）分类。基本分类是按照所授学位的层次及数量，将高等院校分为副学士学位授予学院、研究型大学、博士学位授予大学、硕士学位授予学院/大学、学士学位授予学院、专业主导机构、部落学院7种基本类型。其他几种分类体系则分别是从本专科培养项目、研究生培养项目、学生类型、学制和机构规模五个层面进行分类（The Carnegie Foundation for the Advancement of Teaching）。表1是卡内基教学促进基金会2010年版关于博士学位授予机构、硕士学位授予机构和学士学位授予机构的划分标准。

**表1 卡内基2010年高等教育机构分类标准——基础分类**

| 院校大类 | 院校类别 | | 定义 |
|---|---|---|---|
| 博士型 | 研究型，非常高 | 年授予博士学位数不少于20个 | 在7项科研指标的总得分和师均得分上的综合表现很好 |
| | 研究型，高 | | 在7项科研指标的总得分和师均得分上的综合表现比较好 |
| | 博士/研究型 | | 在7项科研指标的总得分和师均得分上的综合表现一般 |
| 硕士型 | 硕士型，大规模 | 年授予硕士学位不少于50个，且年授予博士学位不到20个 | 年授予硕士学位数不少于200个 |
| | 硕士型，中等规模 | | 年授予硕士学位数100～199个 |
| | 硕士型，小规模 | | 年授予硕士学位数50～99个 |
| 学士型 | 学士型，文理 | 学士学位授予量占所有本科学位授予量的比例不低于10%，且年授予硕士学位不到50个 | 文理学科的学士学位授予量占所有学士学位授予量的比例不低于50% |
| | 学士型，多学科 | | 文理学科的学士学位授予量占所有学士学位授予量的比例不到50% |
| | 学士/副学士型 | | 学士学位授予量占所有本科学位授予量的比例不到50% |

资料来源：The Carnegie Classification of Institutions of Higher Education，2010 Edition

而一流大学则是一个动态发展中的比较性概念，它可以是高等教育机构各类型的比较，也可以是地理范围之内大学间的比较，还可能是在学科层面上的比较。欧洲古典大学、英式大学、德国模式大学都曾经是世界一流大学，美国研究型大学则是目前世界一流大学的主体。这提示我们，必须动态地看待一流大学的建设问题。

最后，实现世界一流大学目标的主要途径是建设研究型大学。国内高等教育界已经把研究型大学作为实现世界一流大学的首选大学模式（马陆亭，2007）《国家中长期科学和技术发展规划纲要（2006—2020年）》明确提出，“加快建设一批高水平大学，特别是一批世界知名的高水平研究型大学，是我国加速科技创新、建设国家创新体系的需要”。以科学研究见长的研究型大学是保持我国国际竞争力的重要战略资源。为贯彻落实《国家中长期科学和技术发展规划纲要2006—2020年》及其配套政策，加快研究型大学建设，增强高等学校自主创新能力，教育部在2007年7月10日发布了《教育部关于加快研究型大学建设增强高等学校自主创新能力的若干意见》教技〔2007〕5号。文件中指出（教育部，2007），研究型大学是国家创新体系的重要组成部分，加快建设一批研究型大学，对于加强人才培养与科学研究，提高高等教育质量，建设创新型国家具有重要意义。要努力加大投入、深化改革，优化研究型大学发展环境，同时加强领导，协同配合，促进研究型大学健康发展；加强“211工程”和“985工程”建设，到2020年，努力形成一批拥有国家重点科学研究基地、具有承担国家重大科研任务能力和广泛国际合作基础的研究型大学，使其成为培养高素质创新型人才的中心、知识创新的源头和创新文化的重要发源地。

我国创建世界一流大学计划实施十多年来取得了伟大成绩。教育部副部长郝平总结指出，这主要体现在三点：一是自主创新能力快速提升，产生了一大批具有国家标志的科研成果；二是汇聚了一大批具有国际水准的中青年学者，促进了人文社会科学的繁荣；三是学科建设有了重大突破（中国新闻网，2009）。这个成绩，是党中央、国务院的战略指导和精心部署的结果，也是社会各界特别是高等学校努力奋斗和艰辛探索的结果。2010年7月发布的《国家中长期教育改革和发展规划纲要2010—2020年》明确提出，要“加快创建世界一流大学和高水平大学的步伐，培养一批拔尖创新人才，形成一批世界一流学科，产生一批国际领先的原创性成果，为提升我国综合国力贡献力量”，这就为我国高校提出了更高的要求，高校要以科学发展观为指导，以落实《国家中长期教育改革和发展规划纲要（2010—2020年）》为动力，在认真总结基本经验的基础上，科学地评估自身世界一流大学建设的战略趋势和战略规划，进一步明确未来的战略重点。这不仅对加速推进“中国特色、世界一流”大学建设事业，而且对我们构建高等教育强国的战略实现，都具有重大的理论和现实意义。

世界一流大学的评价标准是什么，目前还没有一个统一的答案。它是动态的，而非静态的。世界一流大学没有约定俗成的固定标准，各国对大学的评价体系差异很大。但是，世界一流大学的评价体系的核心是，评价标准是多元的，没有任何一个机构可以垄断整个评价过程。同时，评价标准和评价机构本身也处于不断的被评价之中，没有任何一方可以拒绝被评价。评价机构还必须向公众说明使用评价标准的选取原则和资料来源，不能暗箱操作，不能被权力和商业利益随意操纵，否则，就会遭到公开的质疑。

世界一流大学的评价指标无疑包括定性标准和定量标准两个关键指标，定性标准是对各大学教育教学情况的主体进行描述，包括授予学位的层次、学科分布情况、学校声誉等；定量标准是对各大学授予不同层次的学位数量、一流的师资（如诺贝尔奖获得者、院士等）、高水平的论文数量、学科分布情况和专利发明做量化规定。但因为有些评价指标难以量化，或是数据来源不足、评价标准无法统一，而某些定性的指标在评价的过程中，又难免因为评价主体的复杂多变性和评价者的主观差异而导致评价的结果难如人意。尽管如此，有一些公认的评价指标可以用来衡量世界一流大学的办学水准，如高水平论文数量、一流的师资、专利发明等。

清华大学教育研究所的李越等（2002）研究认为世界一流大学的学术基准是：科研经费、SCI（含SSCI）论文数量、在《自然》和《科学》上发表的论文数量、教师中的院士人数、诺贝尔奖获得者、学术声誉。对于大学学术成就的高低大小要作精细的评价很难，不过可以从国际学术界公认的权威机构及学术刊物，如《自然》、《科学》、SCI中找到公认的指标。大学的学术水准在根本上就是教师的学术水准，而教师的学术水准同样可以通过国际学术界公认的成就标志来衡量，如诺贝尔奖、院士等。一所大学如果在以上这些方面长期都有突出的稳定表现，就会在国际学术界乃至一般民众心里留下深刻的良好印象，这就形成了一所大学的学术声誉。学术声誉看似无形，似乎是一种主观判断，实际上却是对一所大学综合实力及其影响的最深刻的反映。

清华大学校长徐遐生（2005）认为，建立世界一流大学需要杰出的教师、卓越的学术名声和工作环境、崇高的学术标准，以及优良的基础运作设施或系统。香港大学校长徐立之认为，要发展成国际大学，首先是一个“国家的大学”，并且需要具备以下四个条件：第一是老师的国际名气和教学态度；第二是学生的质量；第三是大学的设备；第四是社会的支持。

世界大学的评价应该以“提升高等教育质量”为主要导向，这是教育评价存在的意义和最终目标；同时，世界大学评价应更加注重应用价值，排名结果将会与政府、高校的决策和管理者，以及广大学生及家长的需要紧密结合，真正达到以人为本、价值导向的目的（邱均平，2010）。世界一流大学的评价应兼顾质性和量性的指标，其中量性指标主要包括科研经费、授予博士的数量、教师获奖情况、研究成果发表和引用情况，而质性指标主要用于了解一个机构的整体状况，并主要依赖大学校长们的判断力，能更好地洞察量性指标反映不到的情况，二者适当结合运用，共同达到世界一流大学建设的引导作用。

## 3 世界一流大学及学科评价的具体做法

### 3.1 评价对象和范围

本次进入《世界大学科研竞争力排行榜》的大学为美国ESI数据库中近11年来论文总被引次数排列在前1%的1738所大学。另外，ESI根据学科发展的特点等因素设置了22个学科，其中包括一个交叉学科，分学科将大学和科研机构按近11年来论文总被引次数排列，只有排在前1%的学科方可进入ESI学科排行，2013年共有2900所大学和科研机构进入ESI学科排行，其中大学1738所，科研机构1162

所。总的来说，这些大学和机构是可以满足我们评价需要的，其数量和代表性是都可以得到较好保证的。另外，由于科研机构的性质差别非常大，比如，中国科学院是综合性研究机构，而欧洲分子生物实验室基本上是专科性科研机构，这种差异普遍存在。因此，我们没有单独设立一个科研机构竞争力排行榜，而是将科研机构的评价完全放在某一学科内和大学一起进行评价，对它们的详细分析也只是分指标进行。此次评价中我们彻查了所有评价对象，把同一个机构的不同机构名称进行了合并，同时也把隶属机构名称进行了相应的合并。

## 3.2 数据来源

论文指标我们利用的是美国 ESI 数据库 2003 年 1 月 1 日至 2013 年 2 月 28 日时段的数据。专利指标我们使用的是美国 DII 数据库 2008～2012 年这近 5 年的数据，并且我们根据 DII 数据库和 ESI 学科设置的特点，把德温特创新引文索引数据库的化学、电子与电气和工程三个部门的专利分别相应划分在 ESI 的化学、物理学和工程学三个学科下面。其中有几个指标的概念解释如下：①高被引论文：是 ESI 根据论文在相应学科领域和年代中的被引频次排在前 1%以内的论文。②热门论文：某学科领域发表在最近两年间的论文在最近两个月内被引次数排在 0.1%以内的论文。③ESI 划分的 22 个学科按名称的英文字母排列依次为：农业科学、生物学与生物化学、化学、临床医学、计算机科学、经济学与商学、工程学、环境科学与生态学、地球科学、免疫学、材料科学、数学、微生物学、分子生物学和遗传学、综合交叉学科、神经科学和行为科学、药理学和毒物学、物理学、植物学与动物学、精神病学与心理学、社会科学总论、空间科学。

## 3.3 指标体系的构建

1）科研生产力

用近 11 年来发表论文数（被 ESI 收录的论文数量）这一指标来衡量，反映该大学对世界学术交流做的贡献，而且被 ESI 收录的论文都是经过同行评议的论文，各论文发表的期刊也在该学科有着显著影响，所以相对来说，这些论文都是较高质量的论文。

2）科研影响力

用近 11 年来发表论文总被引次数、高被引论文数和进入排行的学科数这三个指标来衡量。量的积累固然重要，但是也要特别注重质的方面，被引次数高低正是反映质的一个重要指标。另外，进入排行的学科数越多说明该单位的影响面越大，学术辐射范围越广泛，引起的关注就越多。

3）科研创新力

用热门论文和专利这两个指标来衡量。热门论文的产生必然说明此论文适应学科和社会发展的要求，具有很强的创新性，这是一个单位或学科富有朝气的原动力。专利本身的特点之一就是要有新颖性，是科技进步的重要体现，是转化为生产力最宝贵的知识财富之一。但是需要强调的是，我们专利的申请有学科的局限性，所以我们只在物理学、化学和工程学三个学科设置了专利指标。专利分为发明、实用新型、外观设计三种类型，这里专利作为科研创新力的一部分，本次评价中使用的专利数据为发明型专利的数量。根据我国和国际上对不同国家发明型专利的统计数据，基本上得出不同国家近 5 年来发明型专利所占比例，从而计算出其发明型专利的数量。

4）科研发展力

用高被引论文占有率这一指标来衡量，其中，高被引论文占有率 = 高被引论文数/论文发表数。这一比率越高说明该单位在以后发展中有可能生产出更多优秀的论文，有能力持久保持该学科的领先地位。

5）网络影响力

用网络排名这一指标来衡量。网络排名可以告知各大学的学术知识与资料在网络上公开出版的程度，若大学本身认为其实力排名与此网络计量排名相差甚远，则可借此促进科研成果出版的开放获取，进而提升其影响力。

对于大学和学科的评价有不同的评价体系，这些思想也在本次评价中得到体现，具体指标体系分别如表 2 和表 3 所示。

**表 2　世界大学科研竞争力评价指标体系**

| 一级指标 | 二级指标 |
|---|---|
| 科研生产力 | 收录论文数 |
| 科研影响力 | 论文被引次数<br>高被引论文数<br>进入排行学科数 |
| 科研创新力 | 发明专利数<br>热门论文数 |
| 科研发展力 | 高被引论文占有率 |
| 网络影响力 | 国内外网络排名 |

**表 3　世界科研机构（包括大学、研究院所）分 22 个学科科研竞争力评价指标体系**

| 一级指标 | 二级指标 |
|---|---|
| 科研生产力 | 收录论文数 |
| 科研影响力 | 论文被引次数<br>高被引论文数 |
| 科研创新力 | 发明专利数<br>热门论文数 |
| 科研发展力 | 高被引论文占有率 |

### 3.4 世界一流大学与学科的界定

在给出和解释评价结果之前，我们有必要界定什么是世界一流大学。此次进入评价范围的世界大学共 1738 所，我们将前 600 名（即位居全世界前千分之五的大学）定义为世界高水平大学，同时我们结合国内外一些大学对自己的定位与规划，又将世界高水平大学分为三个档次：前 100 名为世界顶尖大学，101～300 名为世界高水平著名大学，301～600 名定义为世界高水平知名大学，其中世界顶尖大学和世界高水平著名大学称为“世界一流大学”。

至于世界一流学科的界定，我们主要是根据所评 22 个学科的不同评价单位在相应学科中的排名情况而进行划分的，其标准为某学科排名前 10%以内的科研单位为该学科世界一流学科。世界一流学科也划分为三个档次：某学科前 1%（含 1%）的科研单位的学科为世界顶尖学科；1%～5%（含 5%）的为世界高水平著名学科；5%～10%（含 10%）的为世界高水平知名学科。

## 4 结果分析

本次评价最终得到 37 个排行榜，对排行榜中的数据进行分析，得到 7 个分析表格，其中表 4 是 2013 年科研竞争力位居前 30 强的国家（地区）；表 5 是 2013 年世界一流大学的国别分布情况；表 6 是 2013 年世界一流大学前 10 强和部分中国大学分指标排名情况；表 7 是 2013 年与 2012 年中国科研实力具体指标对比分析；表 8 是 2013 年世界大学的学科分布表（前 10 强和部分中国大学）；表 9 是 2013 年中国大学或科研机构进入世界一流学科排名情况。通过对以上评价结果的分析与思考，我们得到以下一些结论和启示。

### 4.1 中国的整体科研实力有显著提升

本次的国家（地区）科研竞争力评价以所有大学作为统计样本，得出国家（地区）科研竞争力排行榜。科研竞争力方面，美国牢牢占据了榜首位置，几乎各个指标的得分都位居首位，显示了绝对领先的科研水平，紧随其后的是英国、日本、德国、加拿大、中国内地、意大利、法国、澳大利亚（表 4，图 1）。中国内地位于第 5 位，中国台湾列第 13 位，中国香港列第 26 位。与 2012 年排名结果相比，中国内地大学的科研竞争力总排名再次上升，中国台湾从去年排名的第 15 位上升到第 13 位，但中国香港由去年的第 21 位下降到今年的第 26 位。从各个具体指标来看，中国高校的发表论文得分和专利得分继续保持较高水平，论文被引得分、高被引论文得分、高被引占有率得分和热门论文得分依然相对偏低。说明中国大学的科研生产力和科研影响力发展不均衡，科研生产力迅速提高，但科研影响力的进步不是特别显著。因此，中国大学未来努力的重点和方向是在保持科研产出数量稳定的前提下适当偏重于质量，以求得科研生产力和科研影响力的同步发展。

表 4 2013 年国家（地区）科研竞争力前 30 强

| 排名 | 国家/地区 | 发表论文得分 | 论文被引得分 | 专利得分 | 高被引论文得分 | 高被引占有率得分 | 热门论文得分 | 总分 |
|---|---|---|---|---|---|---|---|---|
| 1 | 美国 | 100.00 | 100.00 | 100.00 | 100.00 | 100.00 | 100.00 | 100.00 |
| 2 | 英国 | 95.48 | 95.03 | 92.72 | 94.82 | 95.64 | 95.07 | 95.24 |
| 3 | 日本 | 95.77 | 94.41 | 99.21 | 92.79 | 93.30 | 92.92 | 94.81 |
| 4 | 德国 | 95.22 | 94.57 | 93.45 | 94.05 | 95.12 | 93.68 | 94.75 |
| 5 | 中国 | 96.02 | 93.06 | 99.23 | 92.62 | 92.89 | 93.03 | 94.53 |
| 6 | 加拿大 | 93.81 | 92.98 | 90.78 | 92.80 | 95.26 | 92.84 | 93.41 |
| 7 | 意大利 | 93.88 | 92.80 | 88.33 | 92.35 | 94.73 | 93.20 | 93.20 |
| 8 | 法国 | 92.68 | 91.86 | 95.16 | 92.00 | 95.59 | 92.47 | 92.77 |
| 9 | 澳大利亚 | 92.77 | 91.54 | 90.75 | 91.23 | 94.70 | 91.33 | 92.19 |
| 10 | 韩国 | 93.00 | 90.52 | 100.00 | 89.68 | 92.86 | 90.59 | 92.08 |
| 11 | 西班牙 | 92.75 | 91.17 | 91.22 | 90.53 | 94.00 | 90.96 | 91.90 |
| 12 | 荷兰 | 91.98 | 91.45 | 88.11 | 90.74 | 95.01 | 90.25 | 91.57 |
| 13 | 中国台湾 | 91.78 | 89.19 | 97.27 | 87.96 | 92.30 | 88.91 | 90.63 |
| 14 | 瑞士 | 90.38 | 90.26 | 89.38 | 89.60 | 95.47 | 89.86 | 90.53 |
| 15 | 瑞典 | 91.23 | 90.59 | 81.94 | 88.85 | 93.79 | 88.94 | 90.27 |
| 16 | 巴西 | 91.78 | 88.88 | 89.20 | 87.25 | 91.55 | 88.10 | 89.86 |
| 17 | 印度 | 90.98 | 88.49 | 89.27 | 87.17 | 92.27 | 87.66 | 89.45 |
| 18 | 俄罗斯 | 91.02 | 87.75 | 83.86 | 87.43 | 92.51 | 88.58 | 89.19 |
| 19 | 土耳其 | 90.88 | 87.68 | 85.48 | 87.23 | 92.44 | 88.65 | 89.15 |
| 20 | 丹麦 | 89.21 | 88.67 | 85.73 | 87.61 | 94.58 | 87.89 | 88.92 |
| 21 | 以色列 | 89.29 | 88.19 | 87.08 | 87.67 | 94.56 | 87.80 | 88.90 |
| 22 | 比利时 | 89.39 | 88.62 | 87.72 | 86.78 | 93.49 | 88.14 | 88.87 |
| 23 | 芬兰 | 89.27 | 88.42 | 81.66 | 87.75 | 94.67 | 88.33 | 88.79 |
| 24 | 波兰 | 89.92 | 87.38 | 81.45 | 86.55 | 92.69 | 88.02 | 88.38 |
| 25 | 奥地利 | 88.83 | 87.90 | 85.38 | 86.91 | 94.22 | 86.29 | 88.27 |
| 26 | 中国香港 | 88.81 | 87.27 | 88.75 | 87.02 | 94.36 | 85.87 | 88.26 |
| 27 | 新加坡 | 88.41 | 86.93 | 87.24 | 86.60 | 94.33 | 87.20 | 88.01 |
| 28 | 挪威 | 88.22 | 87.13 | 82.32 | 86.87 | 94.83 | 87.42 | 87.87 |
| 29 | 希腊 | 88.57 | 86.74 | 80.14 | 85.92 | 93.42 | 86.85 | 87.48 |
| 30 | 葡萄牙 | 88.17 | 86.22 | 86.97 | 85.88 | 93.79 | 85.78 | 87.42 |

## 4.2 中国大学离世界一流大学仍然有较大差距

中国大学取得的进步令人欣喜，但是世界大学都在前进，不进则退，进步速度慢了也等于是在退步。从表 5 可以看出，美国、英国、日本、澳大利亚、加拿大这 5 个国家囊括了近 3/4 的排名前 100 的世界顶尖大学，61%的排名前 200 和 55%的排名前 300 的世界高水平著名大学。由此可见，这 5 个国家拥有全世界绝大多数的优秀的科研机构，也有着雄厚的科研实力和强大的科研影响力。与 2012 年的评价结果相比，中国进入前 100 名的顶尖大学由原来的一所变为了现在的 4 所，除了原来的北京大学、清华大学和浙江大学外，上海交通大学也在这次评价中进入了前 100 名，这一结果无疑是令人欣喜的。前 200 名有 7 所，分别是浙江大学（59）、北京大学

(69)、清华大学（70)、上海交通大学（88)、复旦大学（127)、中国科学技术大学(169)、南京大学（176)；前 300 名中国有 14 所，除去前 7 所外，还有中山大学(204)、山东大学（245)、四川大学（268)、哈尔滨工业大学（275)、华中科技大学(282)、吉林大学（296)、武汉大学（297)；前 600 名的大学有 34 所。与 2012 年的评价结果相比，我们惊喜地发现，中国内地进入各个分类段的高校数都有不同程度的增加，而且进入前 300 名的高校的排名也有了不同程度的提升。但是，从表 4 中我们可以发现，位居前 300 名的中国高校还是少之又少，多数高校的排名都在 600 名之后，甚至 800 名之后，也就是说我国的整体科研水平在世界范围内仍然处于中等偏下水平。优秀科研团队和创新体系的建设对于提升我国整体科研水平，特别是建设一批世界一流水平的大学和科研机构仍然具有强大的推进作用（图 1)。

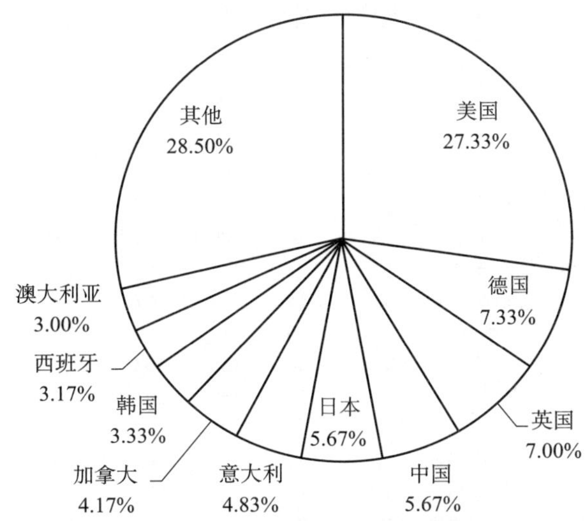


图 1　2013 年科研竞争力前 600 名大学国家和地区分布图

**表 5　2013 年科研竞争力前 600 名大学国别分布与比例**

| 国家/地区 | 前 100 名 | | 前 200 名 | | 前 300 名 | | 前 400 名 | | 前 500 名 | | 前 600 名 | |
|---|---|---|---|---|---|---|---|---|---|---|---|---|
| | 数量/所 | 比例/% | 数量/所 | 比例/% | 数量/所 | 比例/% | 数量/所 | 比例/% | 数量/所 | 比例/% | 数量/所 | 比例/% |
| 美国 | 48 | 48.00 | 81 | 40.50 | 103 | 34.33 | 125 | 31.25 | 146 | 29.20 | 164 | 27.33 |
| 英国 | 8 | 8.00 | 16 | 8.00 | 26 | 8.67 | 34 | 8.50 | 37 | 7.40 | 42 | 7.00 |
| 日本 | 6 | 6.00 | 9 | 4.50 | 12 | 4.00 | 16 | 4.00 | 24 | 4.80 | 34 | 5.67 |
| 澳大利亚 | 5 | 5.00 | 7 | 3.50 | 8 | 2.67 | 9 | 2.25 | 13 | 2.60 | 18 | 3.00 |
| 加拿大 | 5 | 5.00 | 9 | 4.50 | 16 | 5.33 | 21 | 5.25 | 23 | 4.60 | 25 | 4.17 |
| 中国 | 4 | 4.00 | 7 | 3.50 | 14 | 4.67 | 21 | 5.25 | 26 | 5.20 | 34 | 5.67 |
| 意大利 | 4 | 4.00 | 8 | 4.00 | 12 | 4.00 | 19 | 4.75 | 23 | 4.60 | 29 | 4.83 |
| 荷兰 | 4 | 4.00 | 6 | 3.00 | 9 | 3.00 | 11 | 2.75 | 12 | 2.40 | 12 | 2.00 |
| 德国 | 3 | 3.00 | 13 | 6.50 | 26 | 8.67 | 34 | 8.50 | 41 | 8.20 | 44 | 7.33 |
| 韩国 | 2 | 2.00 | 4 | 2.00 | 6 | 2.00 | 11 | 2.75 | 17 | 3.40 | 21 | 3.50 |
| 中国台湾 | 1 | 1.00 | 2 | 1.00 | 2 | 0.67 | 4 | 1.00 | 9 | 1.80 | 9 | 1.50 |
| 中国香港 | 0 | 0.00 | 2 | 1.00 | 2 | 0.67 | 5 | 1.25 | 5 | 1.00 | 6 | 1.00 |

另外，我们又对世界一流大学行列中的前10强和所有中国大学的分指标排名情况做了进一步的统计分析（表6），中国的一流大学和世界顶尖大学之间在整体上还存在着巨大的差距，尤其是在总被引、高被引和热引这些质量指标上还有非常大的提升空间。不过，与同档次的世界大学相比，浙江大学、清华大学、台湾大学和上海交通大学的专利指标还是名列前茅的；浙江大学、清华大学、台湾大学、北京大学及上海交通大学的发文数量也是值得肯定的；清华大学的发文量和热门论文数量同样也是值得褒奖的。但是，总体来看，这些优势还很微弱，而且不够稳定，在能够真正充分体现科研竞争力的主体指标实力方面仍有很大欠缺，有待进一步提高。

**表6　2013年世界一流大学前10强和部分中国大学分指标排名情况**

| 总排名 | 中文名称 | 国家/地区 | 发文排名 | 总被引排名 | 高被引排名 | 热引排名 | 专利排名 | 网络排名 | 高被引率排名 | 学科数排名 |
|---|---|---|---|---|---|---|---|---|---|---|
| 1 | 哈佛大学 | 美国 | 1 | 1 | 1 | 1 | 43 | 1 | 38 | 1 |
| 2 | 斯坦福大学 | 美国 | 11 | 5 | 2 | 2 | 23 | 2 | 43 | 1 |
| 3 | 华盛顿大学（西雅图） | 美国 | 7 | 3 | 3 | 6 | 35 | 25 | 62 | 1 |
| 4 | 密歇根大学 | 美国 | 4 | 6 | 7 | 10 | 24 | 4 | 114 | 1 |
| 5 | 加州大学洛杉矶分校 | 美国 | 6 | 4 | 6 | 3 | 981 | 7 | 85 | 1 |
| 6 | 约翰·霍普金斯大学 | 美国 | 5 | 2 | 8 | 8 | 42 | 32 | 135 | 1 |
| 7 | 东京大学 | 日本 | 2 | 16 | 26 | 34 | 1 | 49 | 398 | 25 |
| 8 | 加州大学伯克利分校 | 美国 | 13 | 7 | 4 | 5 | 982 | 8 | 46 | 1 |
| 9 | 麻省理工学院 | 美国 | 39 | 14 | 5 | 7 | 12 | 3 | 32 | 25 |
| 10 | 多伦多大学 | 加拿大 | 3 | 10 | 17 | 20 | 334 | 36 | 183 | 1 |
| 59 | 浙江大学 | 中国 | 38 | 140 | 150 | 152 | 5 | 99 | 948 | 213 |
| 68 | 台湾大学 | 中国台湾 | 44 | 110 | 166 | 202 | 18 | 39 | 939 | 82 |
| 69 | 北京大学 | 中国 | 49 | 111 | 85 | 97 | 94 | 98 | 533 | 114 |
| 70 | 清华大学 | 中国 | 41 | 141 | 165 | 68 | 10 | 77 | 955 | 265 |
| 88 | 上海交通大学 | 中国 | 51 | 181 | 182 | 153 | 20 | 86 | 945 | 213 |
| 119 | 香港大学 | 中国香港 | 118 | 129 | 132 | 124 | 356 | 82 | 415 | 46 |
| 127 | 复旦大学 | 中国 | 99 | 188 | 152 | 141 | 52 | 161 | 633 | 213 |
| 169 | 中国科学技术大学 | 中国 | 1024 | 1193 | 1046 | 1172 | 215 | 1559 | 1120 | 1224 |
| 176 | 南京大学 | 中国 | 109 | 213 | 215 | 185 | 77 | 203 | 824 | 299 |
| 183 | 香港中文大学 | 中国香港 | 157 | 191 | 189 | 294 | 234 | 198 | 553 | 147 |
| 187 | 台湾成功大学 | 中国台湾 | 127 | 276 | 350 | 163 | 32 | 195 | 1204 | 265 |
| 204 | 中山大学 | 中国 | 159 | 272 | 248 | 160 | 315 | 233 | 755 | 213 |
| 245 | 山东大学 | 中国 | 153 | 320 | 303 | 376 | 64 | 92 | 1012 | 369 |
| 268 | 四川大学 | 中国 | 150 | 342 | 408 | 443 | 70 | 659 | 1291 | 406 |
| 275 | 哈尔滨工业大学 | 中国 | 184 | 405 | 332 | 225 | 40 | 303 | 1024 | 515 |
| 282 | 华中科技大学 | 中国 | 163 | 369 | 425 | 315 | 62 | 428 | 1305 | 459 |
| 296 | 吉林大学 | 中国 | 173 | 311 | 325 | 579 | 76 | 221 | 1035 | 515 |
| 297 | 武汉大学 | 中国 | 226 | 337 | 368 | 495 | 111 | 156 | 1006 | 334 |

## 4.3　我国高质量的论文数量与世界科研强国相比仍然差距较大

中国高被引和热门论文的数量不断增加，但是与世界科研强国仍有很大差距。从表 7 可以看出，中国内地高被引论文指标位居第 6 位，热门论文位居第 5 位，与 2012 年相比，高被引论文以及热门论文的排名都有所上升，说明我国的科研影响力正在持续增大，尤其是热门论文排名与去年相较上升了 4 位，这在很大程度上说明了随着我国科技的快速发展，国家领导及科研人员对于创新力日益重视，创新意识不断增强，我国论文的创新性也在不断提高。在看到我国科研取得较大成绩和进步的同时，也可以看到在 ESI 总发文、高被引论文、热门论文这几个高质量论文数量的指标里，中国内地与世界科研强国——美国的差距仍然比较大，尤其是高被引论文的相对差距最为明显。例如，排在世界前 20 名的大学高被引论文数量都在 1000 篇以上，其中哈佛大学的高被引数量高达 5331 篇，但是在中国内地排名第一的北京大学的高被引论文仅为 504 篇。排在世界前 20 名的大学热门论文数量也是北京大学的 2～10 倍。尽管中国的高被引论文和热门论文的数量和位次得到较大的提高，但是高被引论文占有率较低，如 2012 年中国内地的高被引论文占有率位居世界第 64 位。这些都表明中国论文的质量仍需要大幅度提升。高质量论文数少，说明我们国家在国际上影响力较大的科学家少，生产大量的创新知识的人才少。这些对于我们国家的长期发展是非常不利的，中国要出一流的科学家甚至诺贝尔奖获得者，如果没有高质量论文和一流成果的保障是不可能实现的。所以现在就要注重优秀人才的培养和储备，在政策、机制、资金、环境等方面给以保障，以改变中国现在科研的被动局面。

**表 7　中国各地区科研实力具体指标对比分析**

| 地区 | 发文排名 | | | 总被引排名 | | | 高被引论文排名 | | |
|---|---|---|---|---|---|---|---|---|---|
| | 2012 年 | 2013 年 | 变化 | 2012 年 | 2013 年 | 变化 | 2012 年 | 2013 年 | 变化 |
| 内地 | 5 | 2 | ↑3 | 8 | 5 | ↑3 | 7 | 6 | ↑1 |
| 香港 | 24 | 27 | ↓3 | 22 | 26 | ↓4 | 19 | 23 | ↓4 |
| 台湾 | 15 | 14 | ↑1 | 16 | 15 | ↑1 | 21 | 15 | ↑6 |

| 地区 | 专利排名 | | | 热门论文排名 | | | 高被引占有率排名 | | |
|---|---|---|---|---|---|---|---|---|---|
| | 2012 年 | 2013 年 | 变化 | 2012 年 | 2013 年 | 变化 | 2012 年 | 2013 年 | 变化 |
| 内地 | 3 | 3 | → | 9 | 5 | ↑4 | 53 | 64 | ↓11 |
| 香港 | 11 | 15 | ↓4 | 31 | 32 | ↓1 | 22 | 31 | ↓9 |
| 台湾 | 5 | 5 | → | 22 | 15 | ↑7 | 63 | 78 | ↓15 |

注：表 6 中的“↑”表示上升，“→”表示无变化，“↓”表示下降。

## 4.4　我国创新型研究成果离世界科研强国还有很大距离

一个国家的专利水平和热门论文数量都反映了该国在世界上的科研创新能力。在 2013 年评价过程中，我们同样采用了发明型专利数量作为专利指标的数据，目的是为了真实地反映出一个国家、大学或机构的科研创新能力。

从表 7 可以看出，中国内地在专利总量上的排名与去年相同，排名第三，但总数却上升了很多，进一步缩小了与专利产出大国——美国的差距。然而，中国内地在热门论文指标上的排名同样相较于去年上升了四位，排在第 9 位，从数量上来看，也几乎增加了一倍，同样也在一定程度上缩小了与美、英、德等国家的差距。在取得可喜成绩的同时，我们也同样看到我国与世界科研强国之间仍存在巨大的差距，尤其是热引论文，在绝对数量上连美国的十分之一都不到，而专利也与之相差上千。因此，从我国的科研产出总量和创新型科研成果可以看出，创新型科研成果所占比例相对较小，这与我国建设创新型国家以及世界一流大学和科研机构还相距甚远，是一个需要长期努力和加强建设的方向。

### 4.5 世界一流学科的建设仍需大力加强

在此次评价中，中国内地进入 ESI 排名的学科数 22 个，22 个学科均有中国内地的大学或科研机构进入 ESI 排名。这是值得肯定和可喜可贺的！2012 年评价中中国内地进入 ESI 排名的学科也为 22 个，22 个学科也都有学校和科研机构进入排名，可见我国在学科建设当中所付出的巨大努力。而在进入 ESI 排名中的中国内地大学中，只有清华大学的工程学和材料科学两个学科都排在了世界前十位。同时，我国内地在农业科学、化学、工程学、地球科学、材料科学等 17 个学科都进入了世界一流学科的行列，尤其在生物学与生物化学、化学、工程学、环境科学与生态学、地球科学、材料科学、物理学，以及植物学与动物学、计算机科学、药理学与毒物学 10 个学科还有顶尖机构存在，不过除了清华的工程学及材料科学、哈尔滨工业大学及上海交通大学的工程学，以及浙江大学的化学，其他所有的顶尖学科机构均被中国科学院独揽；同时在科研机构的统计中，中国科学院在发文量、被引总量、热门论文数、高被引论文数和进入 ESI 学科数 5 个指标上都是名列前茅的，这些都深刻印证了它作为国内第一研究巨头的科研竞争实力。

但是，从表 8、表 9 可以看出，中国内地的高校在学科建设上仍然表现较弱。虽然 22 个学科均有中国内地的大学或科研机构进入 ESI 排名，但是对于每所大学或科研机构而言，进入 ESI 学科排名的学科数量很少，绝大多数的中国内地大学都只有 1～9 个学科进入 ESI 排名。而且除去中国科学院、清华大学和北京大学有排在世界前 10 的学科外，其他大学或科研机构未有进入世界前十的学科。同时，总排名前 10 位的世界一流大学其学科都很齐全，并且每个学科影响力都很大，如哈佛大学有 14 个学科位于世界前 10 强，斯坦福大学有 14 个学科位于世界前 10 强等。这都在一定程度上说明中国内地的高校在学科建设上仍然表现较弱。表 9 虽然显示出我国许多大学或科研机构在材料科学、工程学等学科都有不俗的表现，但是却也反映出我国内地，以及香港、台湾地区学科建设普遍集中的问题，当然也可能是马太效应所致。但是在发展这些优势学科的同时，高校管理人员、教育发展监管人员也要主要在分子生物学与遗传学、经济学与商学、精神病学与心理学、空间科学、免疫学、社会科学、神经科学与行为科学，以及临床医学这 8 个学科中，我国内地还没有高校进入世界一流学科的行列。要真正实现与世界一流大学与学科的接轨，中国内地的科研创新和学科建设仍任重而道远。

**表 8　2013 年世界大学学科分布表（前 10 强与部分中国大学）**

| 排名 | 大学名称 | 国家／地区 | 进入 ESI 的学科 | | 前 10 的学科 | |
|---|---|---|---|---|---|---|
| | | | 数量/所 | 比例/% | 数量/所 | 比例/% |
| 1 | 哈佛大学 | 美国 | 22 | 100 | 14 | 63.64 |
| 2 | 斯坦福大学 | 美国 | 22 | 100 | 14 | 63.64 |
| 3 | 华盛顿大学（西雅图） | 美国 | 22 | 100 | 8 | 36.36 |
| 4 | 密歇根大学 | 美国 | 22 | 100 | 5 | 22.73 |
| 5 | 加州大学洛杉矶分校 | 美国 | 22 | 100 | 6 | 27.27 |
| 6 | 约翰·霍普金斯大学 | 美国 | 22 | 100 | 5 | 22.73 |
| 7 | 东京大学 | 日本 | 21 | 95 | 6 | 28.57 |
| 8 | 加州大学伯克利分校 | 美国 | 22 | 100 | 11 | 50.00 |
| 9 | 麻省理工学院 | 美国 | 21 | 95 | 10 | 47.62 |
| 10 | 多伦多大学 | 加拿大 | 22 | 100 | 3 | 13.64 |
| 59 | 浙江大学 | 中国 | 15 | 68 | 0 | 0.00 |
| 68 | 台湾大学 | 中国台湾 | 19 | 86 | 0 | 0.00 |
| 69 | 北京大学 | 中国 | 18 | 82 | 0 | 0.00 |
| 70 | 清华大学 | 中国 | 13 | 59 | 2 | 15.38 |
| 88 | 上海交通大学 | 中国 | 15 | 68 | 0 | 0.00 |
| 119 | 香港大学 | 中国香港 | 20 | 91 | 0 | 0.00 |
| 127 | 复旦大学 | 中国 | 15 | 68 | 0 | 0.00 |
| 169 | 中国科学技术大学 | 中国 | 1 | 5 | 0 | 0.00 |
| 176 | 南京大学 | 中国 | 12 | 55 | 0 | 0.00 |
| 183 | 香港中文大学 | 中国香港 | 17 | 77 | 0 | 0.00 |
| 187 | 台湾成功大学 | 中国台湾 | 13 | 59 | 0 | 0.00 |
| 204 | 中山大学 | 中国 | 15 | 68 | 0 | 0.00 |
| 245 | 山东大学 | 中国 | 10 | 45 | 0 | 0.00 |
| 268 | 四川大学 | 中国 | 9 | 41 | 0 | 0.00 |
| 275 | 哈尔滨工业大学 | 中国 | 7 | 32 | 0 | 0.00 |
| 282 | 华中科技大学 | 中国 | 8 | 36 | 0 | 0.00 |
| 296 | 吉林大学 | 中国 | 7 | 32 | 0 | 0.00 |
| 297 | 武汉大学 | 中国 | 11 | 50 | 0 | 0.00 |
| 309 | 香港科技大学 | 中国香港 | 13 | 59 | 0 | 0.00 |
| 316 | 香港城市大学 | 中国香港 | 10 | 45 | 0 | 0.00 |
| 317 | 香港理工大学 | 中国香港 | 10 | 45 | 0 | 0.00 |
| 324 | 台湾“清华大学” | 中国台湾 | 8 | 36 | 0 | 0.00 |
| 330 | 南开大学 | 中国 | 9 | 41 | 0 | 0.00 |
| 346 | 大连理工大学 | 中国 | 7 | 32 | 0 | 0.00 |
| 359 | 西安交通大学 | 中国 | 8 | 36 | 0 | 0.00 |
| 360 | 东南大学 | 中国 | 7 | 32 | 0 | 0.00 |
| 370 | 兰州大学 | 中国 | 11 | 50 | 0 | 0.00 |
| 387 | 台湾“中央大学” | 中国台湾 | 7 | 32 | 0 | 0.00 |
| 389 | 北京师范大学 | 中国 | 11 | 50 | 0 | 0.00 |
| 398 | 中南大学 | 中国 | 6 | 27 | 0 | 0.00 |
| 404 | 天津大学 | 中国 | 5 | 23 | 0 | 0.00 |

续表

| 排名 | 大学名称 | 国家/地区 | 进入ESI的学科 | | 前10的学科 | |
|---|---|---|---|---|---|---|
| | | | 数量/所 | 比例/% | 数量/所 | 比例/% |
| 405 | 台湾交通大学 | 中国台湾 | 6 | 27 | 0 | 0.00 |
| 409 | 厦门大学 | 中国 | 8 | 36 | 0 | 0.00 |
| 410 | 同济大学 | 中国 | 7 | 32 | 0 | 0.00 |
| 424 | 台湾“中山大学” | 中国台湾 | 9 | 41 | 0 | 0.00 |
| 427 | 台湾中兴大学 | 中国台湾 | 11 | 50 | 0 | 0.00 |
| 436 | 台湾阳明大学 | 中国台湾 | 8 | 36 | 0 | 0.00 |
| 444 | 长庚大学 | 中国台湾 | 7 | 32 | 0 | 0.00 |
| 447 | 中国农业大学 | 中国 | 7 | 32 | 0 | 0.00 |
| 489 | 华东理工大学 | 中国 | 4 | 18 | 0 | 0.00 |
| 501 | 上海大学 | 中国 | 5 | 23 | 0 | 0.00 |
| 504 | 华南理工大学 | 中国 | 5 | 23 | 0 | 0.00 |
| 521 | 华东师范大学 | 中国 | 7 | 32 | 0 | 0.00 |
| 525 | 香港浸会大学 | 中国香港 | 10 | 45 | 0 | 0.00 |
| 527 | 重庆大学 | 中国 | 4 | 18 | 0 | 0.00 |
| 536 | 湖南大学 | 中国 | 6 | 27 | 0 | 0.00 |
| 538 | 中国医学科学院&北京协和医学院 | 中国 | 6 | 27 | 0 | 0.00 |
| 564 | 苏州大学 | 中国 | 5 | 23 | 0 | 0.00 |
| 592 | 中国海洋大学 | 中国 | 7 | 32 | 0 | 0.00 |

**表9 2013年中国大学或科研机构进入世界一流学科排名情况**

| 排名 | 机构名称 | 档次 | 地区 | 学科 | 排名 | 机构名称 | 档次 | 地区 | 学科 |
|---|---|---|---|---|---|---|---|---|---|
| 1 | 中国科学院 | 顶尖 | 内地 | 材料科学 | 102 | 武汉大学 | 知名 | 内地 | 化学 |
| 5 | 清华大学 | 顶尖 | 内地 | 材料科学 | 1 | 中国科学院 | 顶尖 | 内地 | 环境科学与生态学 |
| 21 | 复旦大学 | 著名 | 内地 | 材料科学 | 5 | 中国科学院 | 顶尖 | 内地 | 计算机科学 |
| 28 | 上海交通大学 | 著名 | 内地 | 材料科学 | 11 | 清华大学 | 著名 | 内地 | 计算机科学 |
| 37 | 浙江大学 | 知名 | 内地 | 材料科学 | 17 | 上海交通大学 | 著名 | 内地 | 计算机科学 |
| 40 | 哈尔滨工业大学 | 知名 | 内地 | 材料科学 | 168 | 上海交通大学 | 知名 | 内地 | 临床医学 |
| 50 | 大连理工大学 | 知名 | 内地 | 材料科学 | 8 | 中国科学院 | 著名 | 内地 | 农业科学 |
| 55 | 北京大学 | 知名 | 内地 | 材料科学 | 17 | 中国农业大学 | 著名 | 内地 | 农业科学 |
| 62 | 吉林大学 | 知名 | 内地 | 材料科学 | 23 | 浙江大学 | 著名 | 内地 | 农业科学 |
| 66 | 北京科技大学 | 知名 | 内地 | 材料科学 | 4 | 中国科学院 | 顶尖 | 内地 | 生物学与生物化学 |
| 1 | 中国科学院 | 顶尖 | 内地 | 地球科学 | 12 | 中国科学院 | 著名 | 内地 | 数学 |
| 17 | 中国地质科学院 | 著名 | 内地 | 地球科学 | 14 | 北京大学 | 知名 | 内地 | 数学 |
| 30 | 中国地质大学 | 知名 | 内地 | 地球科学 | 12 | 中国科学院 | 著名 | 内地 | 微生物学 |
| 40 | 西北大学 | 知名 | 内地 | 地球科学 | 1 | 中国科学院 | 顶尖 | 内地 | 物理学 |
| 36 | 中国科学院 | 知名 | 内地 | 分子生物学与遗传学 | 38 | 中国科学技术大学 | 著名 | 内地 | 物理学 |
| 1 | 中国科学院 | 顶尖 | 内地 | 工程学 | 61 | 北京大学 | 知名 | 内地 | 物理学 |

续表

| 排名 | 机构名称 | 档次 | 地区 | 学科 | 排名 | 机构名称 | 档次 | 地区 | 学科 |
|---|---|---|---|---|---|---|---|---|---|
| 6 | 清华大学 | 顶尖 | 内地 | 工程学 | 3 | 中国科学院 | 顶尖 | 内地 | 药理学与毒物学 |
| 11 | 哈尔滨工业大学 | 顶尖 | 内地 | 工程学 | 1 | 中国科学院 | 顶尖 | 内地 | 植物学与动物学 |
| 13 | 上海交通大学 | 顶尖 | 内地 | 工程学 | 57 | 中国农业大学 | 知名 | 内地 | 植物学与动物学 |
| 21 | 浙江大学 | 著名 | 内地 | 工程学 | 73 | 浙江大学 | 知名 | 内地 | 植物学与动物学 |
| 47 | 东南大学 | 著名 | 内地 | 工程学 | 80 | 中国农业科学院 | 知名 | 内地 | 植物学与动物学 |
| 56 | 西安交通大学 | 著名 | 内地 | 工程学 | 84 | 华中农业大学 | 知名 | 内地 | 植物学与动物学 |
| 64 | 华中科技大学 | 知名 | 内地 | 工程学 | 2 | 中国科学院 | 著名 | 内地 | 综合交叉学学科 |
| 70 | 中国科学技术大学 | 知名 | 内地 | 工程学 | 58 | 台湾大学 | 知名 | 台湾 | 材料科学 |
| 81 | 大连理工大学 | 知名 | 内地 | 工程学 | 67 | 台湾“清华大学” | 知名 | 台湾 | 材料科学 |
| 103 | 电子科技大学 | 知名 | 内地 | 工程学 | 14 | 台湾成功大学 | 顶尖 | 台湾 | 工程学 |
| 114 | 北京大学 | 知名 | 内地 | 工程学 | 28 | 台湾大学 | 著名 | 台湾 | 工程学 |
| 1 | 中国科学院 | 顶尖 | 内地 | 化学 | 46 | 台湾交通大学 | 著名 | 台湾 | 工程学 |
| 13 | 浙江大学 | 顶尖 | 内地 | 化学 | 89 | 台湾“清华大学” | 知名 | 台湾 | 工程学 |
| 18 | 北京大学 | 著名 | 内地 | 化学 | 91 | 高雄海洋科技大学 | 知名 | 台湾 | 工程学 |
| 19 | 清华大学 | 著名 | 内地 | 化学 | 78 | 台湾大学 | 知名 | 台湾 | 化学 |
| 24 | 复旦大学 | 著名 | 内地 | 化学 | 127 | 台湾大学 | 知名 | 台湾 | 临床医学 |
| 26 | 南京大学 | 著名 | 内地 | 化学 | 44 | 香港浸会大学 | 知名 | 香港 | 材料科学 |
| 44 | 中国科学技术大学 | 著名 | 内地 | 化学 | 14 | 香港大学 | 著名 | 香港 | 地球科学 |
| 47 | 吉林大学 | 著名 | 内地 | 化学 | 16 | 香港城市大学 | 顶尖 | 香港 | 工程学 |
| 49 | 南开大学 | 著名 | 内地 | 化学 | 22 | 香港理工大学 | 著名 | 香港 | 工程学 |
| 58 | 华东理工大学 | 知名 | 内地 | 化学 | 54 | 香港大学 | 著名 | 香港 | 工程学 |
| 65 | 中山大学 | 知名 | 内地 | 化学 | 55 | 香港科技大学 | 著名 | 香港 | 工程学 |
| 80 | 大连理工大学 | 知名 | 内地 | 化学 | 119 | 香港中文大学 | 知名 | 香港 | 工程学 |
| 85 | 厦门大学 | 知名 | 内地 | 化学 | 115 | 香港大学 | 知名 | 香港 | 临床医学 |

## 4.6 世界一流大学与学科的特征和评价标准值得我们重新审视

从表 8 和表 9 来看，排名前 10 位的大学的学科都很齐全，并且每个学科影响力都很大。这和我们平时所见所闻的一些情况不太相符，比如，麻省理工学院，大家一般认为它学科比较单一，以理工科为主，但是我们从原始数据和评价结果来看，它有着齐全的学科体系，并且每个学科都排名较靠前。由此可见，学科互补也是很重要的，中国大学合理的合并是有道理的，有利于创建世界一流大学。同时，在当今繁荣的网络环境下，世界一流大学的网站建设也是非常重要的，这次新增加的网络影响力这一指标可用来进一步考察各学校的声望情况、科研成果的开放获取程度等。麻省理工学院、哈佛大学、斯坦福大学、康奈尔大学，以及加州大学伯克利分校的网络排名是最为靠前的，中国内地排名最好的是清华大学，但仍在 70 名以外，与其总排名还有一定的差距，距排名最前的台湾大学（38 名）差距甚远。可见，中国内地在网络建设、成果公开、社会声誉等方面还需要进一步加大力度，努力向世界一流大学全方位靠近。总之，世界一流大学应该具有明显的综合性、前沿性、创新性和开放性等特征，必须是高水平的、高影响力的研究型大学。

## 5 结 语

武汉大学科学评价研究中心从美国汤森路透科技信息集团出版的《基本科学指标》（ESI）数据库获取原始数据，对世界一流大学与科研机构的学科竞争力进行了科学合理、客观公正的评价研究和综合分析。

从评价结果上来看，在过去的一年里，中国内地的大学和科研机构取得了巨大的进步和成果。较之于 2012 年，中国内地的科研竞争实力排名发生了变化，由 2012 年的第六位上升为 2013 年的第五位，而且中国香港和中国台湾地区的科研竞争力排名也有了不同程度的提升。在短短一年时间里，中国的科研竞争力能取得如此大的进步，的确是令人欣喜的。2013 年，中国进入世界一流大学排行榜的学校无论从数量上还是从排名上，都发生了显著的变化。2012 年，中国内地只有一所学校世界排名进入前 100，而 2013 年，中国内地共有四所大学进入世界前 100 名。而且总体上大部分学校在世界一流大学排行榜中的名次都有所上升，少数学校上升幅度非常明显。在学科建设方面，中国内地进入 ESI 排名的学科数较 2012 年增多，22 个学科均有中国内地的大学或科研机构进入 ESI 排名，尤其是存在在材料学、工程学等 16 个学科领域进入世界一流学科层次的高校或科研机构，并在 10 个学科名列世界顶尖。由此可见，无论是在进入世界一流大学及学科排行的学校名次上还是比例上，中国在 2013 年都比 2012 年有所进步，而且相对于世界科研强国的差距在逐渐缩小。

不过，世界一流大学与科研机构竞争力评价也存在着一些不足。一个最大的不足就是：评价更关注于科技方面，对于人文社会科学方面考虑得较少。目前的指标体系对于那些理工类大学更有利，尤其是那些在化学、电气，以及机械工程方面有着很强实力的学校在专利这个指标上可能表现得就很突出，而那些人文类学校或科研机构在这个指标上的表现可能就有些差强人意。我们也将在后期的研究中对这些问题和不足进行逐步改进和完善。

## 参考文献

陈骏．2010. 推进开放办学战略建设世界一流大学．中国高等教育，(15/16)：20-23.

大学评价国际委员会．2007. 2007 年国际大学创新力客观评价报告．高等教育研究，28（6）：29-32.

教育部 . 2007. 教育部关于加快研究型大学建设、增强高等学校自主创新能力的若干意见 . http：//www. jyb. cn/xwzx/gdjy/xxgl/t20070718 _ 99574. htm ［2013-07-05］.

李越，叶赋桂，蓝劲松 . 2002. 跻身世界一流大学的学术基准 . 教育发展研究，（12）：50-53.

刘念才，程莹，刘莉，等. 2002. 我国名牌大学离世界一流大学有多远. 高等教育研究，23（2）：19-24.

马陆亭 . 2007-01-08. 当今“世界一流大学”建设基本模式与经验 . 中国教育报，第 5 版 .

邱均平 . 2010. 大学评价进入历史新阶段 . http：//news. xinmin. cn/rollnews/2010/12/04/8103993. html ［2013-07-05］.

邱均平，赵蓉英，王伟军，等 . 2013. 世界一流大学与科研机构学科竞争力评价研究报告（2012 年度）. 北京：机械工业出版社 .

世界大学学术排名 . 2011. 2011 世界大学排名 . http：//www. shanghairanking. cn/ARWU-Methodology-2011. html ［2013-07-05］.

王晓阳，刘宝存，李婧 . 2010. 世界一流大学的定义、评价与研究——美国大学联合会常务副主席约翰·冯（John Vaugh）访谈录 . 比较教育研究，（1）：13-19.

徐遐生 . 2005. 如何建立世界一流研究型大学 . http：//hotnews. cc. nthu. edu. tw/view. asp? ID=783 ［2013-07-05］.

许智宏 . 2010. 北大原校长：中国目前没有世界一流大学　建设急功近利 . http：//edu. ifeng. com/news/detail _ 2010 _ 04/15/526043 _ 0. shtml ［2013-07-05］.

张杰 . 2008. 世界一流大学一般具有 10 项基本特征 . http：//edu. people. com. cn/GB/145827/145949/145955/145962/9173610. html ［2013-07-05］.

浙江大学大学评价研究课题组 . 2004. 世界一流大学研究引论 . 评价与管理，（3）：24-30.

中国新闻网 . 2009. 中国“创建世界一流大学”计划实施十年成绩显著 . http：//www. chinanews. com. cn/edu/edu-zcdt/news/2009/09-28/1890813. shtml ［2013-07-05］.

周光礼 . 2010. 世界一流大学的特质 . 中国高等教育，（12）：44-47.

Baty P. 2013. Global rankings system methodology reflects universities’ core missions. http：//www. timeshighereducation. co. uk/story. asp? sectioncode=26&storycode=413382&c=1 ［2013-07-05］.

Baty P. 2013. 世界大学排名的历史、方法和影响 . http：//www. nseac. com/html/135/214384. html ［2013-07-05］.

Morse R. 2013. Methodology：undergraduate ranking criteria and weights. http：//www. usnews. com/articles/education/best-colleges/2010/08/17/methodology-undergraduate-ranking-criteria-and-weights-2011. html? PageNr=1 ［2013-07-05］.

The Carnegie Foundation for the Advancement of Teaching. 2010. Standard listings. http：//classifications. carnegiefoundation. org/lookup-listings/standard. php ［2013-07-05］.

# 4-6 基于目标管理的地方本科高校二级学院绩效评价体系研究与实践——以Z大学为例

朱苏永[①] 诸葛洋[①] 谢京华[①]

**摘 要**：地方本科高校二级学院绩效评价体系主要由指标项、指标权重和指标评价标准三部分构成。高校要根据评价体系的设计原则和办学定位、特色及目标，筛选确定二级学院绩效评价指标项，测算指标权重，制定评价标准，最终计算二级学院综合绩效得分。Z大学通过实施基于目标管理的二级学院绩效评价体系，有效引导二级学院树立目标意识和绩效观念，落实了校院两级管理目标责任制，提升了办学活力和核心竞争力。

**关键词**：目标管理；地方本科高校；绩效评价

地方本科高校是指隶属于各省（自治区、直辖市），以地方财政供养为主，承担着为地方（行业）培养人才、提供服务的本科层次的高等学校（应用技术大学（学院）联盟和地方高校转型发展研究中心，2013）。目前，我国有1000余所地方本科高校，是我国高等教育体系中数量最多的群体之一。相对而言，地方本科高校在人、财、物等教育资源获取方面处于劣势地位（蔡袁强等，2010），面临着更加激烈的办学竞争，因此，必须树立更强的目标意识和绩效观念，构建基于目标管理的二级学院绩效评价体系，充分调动基层办学主体的积极性、主动性和创造力，从而提升地方本科高校办学活力和核心竞争力。

## 1 绩效评价体系的设计原则

（1）目标管理、绩效导向。地方本科高校二级学院绩效评价体系要以学校整体办学目标为导向，以既定的发展规划和任务指标为依据，重点评价二级学院实际工作绩效及其目标完成情况，弱化对其具体办学过程的考核，以定量评价为主，以定性评价为辅。这样既可以提高评价工作效率，确保评价结果公平，同时又保障了二级学院的办学自主性。

（2）全面系统、重点突出。二级学院绩效评价是一项综合性、系统性的工作，要全面系统地反映作为本科高校的二级学院在人才培养、科学研究、社会服务、师

① 浙江大学宁波理工学院，宁波，315100。

资队伍、办学管理等各方面的履职情况，同时又要体现地方性高校服务地方定位以及自身的学科专业特色、建设发展目标等实际情况，通过绩效评价政策引导二级学院集中有限的办学资源，实现中心工作、重点目标的突破。

(3) 数据客观、易于操作。二级学院绩效评价体系是一个量化评价系统，既要确保评价指标原始数据的来源权威、客观、准确，又要使得各项指标的数据具有可测性，便于获取、统计、分析和评价。当指标的全面性与数据的权威性、可测性出现矛盾时，评价体系设计时则要舍弃部分指标。

(4) 统筹兼顾、体现公平。大部分地方本科高校内部二级学院之间具有学科性质差异、体量大小差异、基础强弱差异等问题，二级学院绩效评价体系要统筹兼顾各方面的差异性，通过评价体系的合理设计和科学处理，尽量消除由此产生的不公平性，实现不同类型二级学院的评价结果具有横向可比性。

## 2　绩效评价指标项的筛选

根据地方本科高校的实际特点，以及其人才培养、科学研究、社会服务等方面的基本职能，提出二级学院绩效评价的备选指标，并根据指标体系设计的基本原则，通过专家咨询访谈、问卷调查等方式，筛选确定 Z 大学二级学院绩效评价指标项为一级指标 8 项，二级指标 42 项。

(1) 师资建设。师资队伍是地方本科高校的核心资源，是提升学校核心竞争力的关键因素。师资建设指标主要评价二级学院师资队伍建设的规模、结构、质量等要素，包括专职专任教师占编率、专职专任教师教授比例、专职专任教师副教授比例、专职专任教师博士比例、博士后研究人员占编率、外籍专家教师人数、出国进修教师比例、人才工程等 8 个二级指标。

(2) 学科建设。学科发展水平是评价一所本科高校办学水平和办学实力最重要的指标之一，更是作为学科组织的二级学院最为重要的办学职能。学科建设指标主要评价二级学院学科建设与研究生教育的成效，包括创新团队、学科创新平台、研究生建设项目、重点学科建设等 4 个二级指标。

(3) 人才培养。人才培养是大学的根本使命，地方本科高校更要确立人才培养的中心地位，着力培养适应地方需求的优秀人才。人才培养指标主要评价二级学院在招生、学生培养过程，以及毕业等各个环节的成效，包括毕业生正常毕业率、毕业生国内外深造率、毕业生双证率、毕业生一次就业率、学生竞赛获奖、本科招生生源、国际交流学生人数等 7 个二级指标。

(4) 专业建设。专业是大学开展人才培养的基本单元，抓好专业建设是提高人才培养质量的基础性工作。专业建设指标主要评价二级学院在专业建设、教学基本建设、教学改革等方面的成效，包括重点专业建设、教研教改项目及论文、重点教材建设、课程建设、教学成果获奖、教学荣誉称号、国际交流与合作办学平台等 7 个二级指标。

(5) 科学研究。科学研究是本科高校的基本职能之一。科学研究指标主要评价二级学院教师的科研创新能力，包括纵向项目立项数、重大横向项目立项数、科研成果获奖、学术专著、发表论文、发明专利等 6 个二级指标。

（6）社会服务。地方本科高校肩负着为地方经济社会发展提供服务的历史使命。社会服务指标主要评价二级学院服务社会发展的能力和成效，包括外源到款科研经费、成果转化、培训经费等 3 项二级指标。

（7）管理约束。管理约束是指高校对二级学院加强日常管理，维护良好的办学秩序提出的约束性指标，主要从二级学院发生受约束性行为的情况进行反向评价。包括安全稳定责任事故、违法违纪行为、教学事故、违反学术道德规范行为等 4 项二级指标。

（8）特色创新。特色创新是指高校对二级学院推进改革创新方面的鼓励性指标，主要评价二级学院在人才培养、教育教学改革、科研与社会服务、党建与思政、文化建设、对外交流、服务师生等方面的创新做法及取得的成效，包括教育创新、协同创新、管理服务创新等 3 个二级指标。

## 3　绩效评价指标权重的测算

根据筛选确定的 8 项一级指标和 42 项二级指标，选择 Z 大学校领导、中层干部、教师、学生等不同层面代表和绩效考评专家共 59 人，就每个指标采用 T. L. Saaty 1～9 标度评分，运用 SPSS 12.0 建立数据库，通过 AHP 法建立指标体系的层次结构模型，采用 MATLAB 6.5 计算出各指标权重（表 1）。

## 4　绩效评价标准的制定

各指标项评价标准因指标性质而异，并根据绩效评价体系的设计原则，采取不同的评价办法。Z 大学二级学院绩效评价体系各指标项评价标准可分为五类，实际操作中部分指标项则采用多种评价标准相结合的办法进行综合评价。

（1）目标管理评价法。地方本科高校二级学院绩效评价体系是基于目标管理导向的，是落实目标责任制的重要体现。因此，依据既定的任务指标（包括五年规划指标和年度计划指标）考核各二级学院的目标完成率，是 Z 大学二级学院绩效评价体系中各指标最常用的评价标准。评价方法为指标项的实际完成值与目标值的比值，如 B2 专职专任教师教授数等指标。

（2）年度增长评价法。考虑到各二级学院之间基础强弱差异等因素，同时为对上年基础较差而当年进步较快的学院给予激励，Z 大学二级学院绩效评价体系中部分指标采用年度增长率进行评价，或是将年度增长率与目标完成率两类评价办法进行综合，如 B1 专职专任教师占编率等指标。

（3）等级赋值评价法。部分指标项中不同等级项目体现了不同绩效水平，如 B12 重点学科建设等指标。因此，需要对不同等级项目进行赋值，即对各二级学院取得的项目分为国家级、省部级、地厅级等不同等级进行赋值，所有赋分累加后为该项指标的评价得分。

**表 1　Z 大学二级学院绩效评价指标层及各指标权重**

| 一级指标 | | 二级指标 | | 分层权重 | 综合权重 |
|---|---|---|---|---|---|
| A1 | 师资建设（0.200） | B1 | 专职专任教师占编率 | 0.250 | 0.050 |
| | | B2 | 专职专任教师教授比例 | 0.150 | 0.030 |
| | | B3 | 专职专任教师副教授比例 | 0.150 | 0.030 |
| | | B4 | 专职专任教师博士比例 | 0.150 | 0.030 |
| | | B5 | 博士后研究人员占编率 | 0.040 | 0.008 |
| | | B6 | 外籍专家教师人数 | 0.030 | 0.006 |
| | | B7 | 出国进修教师比例 | 0.080 | 0.016 |
| | | B8 | 人才工程 | 0.150 | 0.030 |
| A2 | 学科建设（0.120） | B9 | 创新团队 | 0.290 | 0.035 |
| | | B10 | 学科创新平台 | 0.290 | 0.035 |
| | | B11 | 研究生建设项目 | 0.130 | 0.016 |
| | | B12 | 重点学科建设 | 0.290 | 0.035 |
| A3 | 人才培养（0.250） | B13 | 毕业生正常毕业率 | 0.150 | 0.038 |
| | | B14 | 毕业生国内外深造率 | 0.150 | 0.038 |
| | | B15 | 毕业生双证率 | 0.150 | 0.038 |
| | | B16 | 毕业生一次就业率 | 0.150 | 0.038 |
| | | B17 | 学生竞赛获奖 | 0.100 | 0.025 |
| | | B18 | 本科招生生源 | 0.200 | 0.050 |
| | | B19 | 国际交流学生人数 | 0.100 | 0.025 |
| A4 | 专业建设（0.150） | B20 | 重点专业建设 | 0.190 | 0.029 |
| | | B21 | 教研教改项目及论文 | 0.150 | 0.023 |
| | | B22 | 重点教材建设 | 0.120 | 0.018 |
| | | B23 | 课程建设 | 0.150 | 0.023 |
| | | B24 | 教学成果获奖 | 0.150 | 0.023 |
| | | B25 | 教学荣誉称号 | 0.120 | 0.018 |
| | | B26 | 国际交流与合作办学平台 | 0.120 | 0.018 |
| A5 | 科学研究（0.120） | B27 | 纵向项目立项数 | 0.190 | 0.023 |
| | | B28 | 重大横向项目立项数 | 0.180 | 0.022 |
| | | B29 | 科研成果获奖 | 0.180 | 0.022 |
| | | B30 | 学术专著 | 0.180 | 0.022 |
| | | B31 | 发表论文 | 0.180 | 0.022 |
| | | B32 | 发明专利 | 0.090 | 0.011 |
| A6 | 社会服务（0.080） | B33 | 外源到款科研经费 | 0.400 | 0.032 |
| | | B34 | 成果转化 | 0.300 | 0.024 |
| | | B35 | 培训经费 | 0.300 | 0.024 |
| A7 | 管理约束（0.040） | B36 | 安全稳定责任事故 | 0.330 | 0.013 |
| | | B37 | 违法违纪行为 | 0.170 | 0.007 |
| | | B38 | 教学事故 | 0.170 | 0.007 |
| | | B39 | 违反学术道德规范行为 | 0.330 | 0.013 |
| A8 | 特色创新（0.040） | B40 | 教育创新 | 0.540 | 0.022 |
| | | B41 | 协同创新 | 0.300 | 0.012 |
| | | B42 | 管理服务创新 | 0.160 | 0.006 |

（4）反向扣分评价法。Z大学二级学院绩效评价体系各指标项中的管理约束指标，如B36安全稳定责任事故等4个指标项，均采用反向扣分评价法，即未发生受约束行为的情况得100分，而每发生一起受约束行为则需要扣除相应分值。

（5）定性评价法。Z大学二级学院绩效评价体系各指标项中的特色创新指标，如B40教育创新等3个指标项，均采用定性评价法，即由评价专家对各二级学院的特色创新工作进行等级评价，并依据相应等级给予量化赋值后计算该指标的评价得分。

## 5 绩效评价得分的计算

二级学院绩效评价得分的计算分为两个步骤，一是根据各指标评价标准计算并处理其绩效得分，二是根据各指标项得分计算绩效评价综合得分。

（1）各指标项绩效评价得分处理。各指标项绩效评价得分是对二级学院单个指标工作绩效情况的反映。各指标项之间属性不同、评价标准不同，因此各指标评价的实际得分之间不具有直接可比性，需要通过一定的数学方法处理，使得各指标的绩效评价得分均调整到0～100（即所有被评价二级学院中，得分最高的为100分，得分最低的为0分）。数学处理方法为：

设：$k$学院$i$指标实际得分从低到高为：$S_{i1}$，…，$S_{ik}$，…，$S_{im}$；

　　$k$学院$i$指标处理后得分从低到高为：0，…，$P_{ik}$，…，100。

则：$k$学院$i$指标绩效评价得分$P_{ik}=100\times(S_{ik}-S_{i1})\div(S_{im}-S_{i1})$。

（2）绩效评价综合得分计算。绩效评价综合得分是对二级学院整体工作绩效及其目标完成情况的综合反映。由于各二级学院之间具有差异性，根据评价体系设计原则，部分二级学院的部分指标项可不列入评价范围。因此，二级学院综合绩效评价仅对列入评价范围的有效指标项计算加权平均分。计算方法为：

设：$k$学院$i$指标绩效评价得分为$P_{ik}$，$i$指标权重为$W_i$，有效指标项为$n$

则：$k$学院绩效评价综合得分$Q_k=\frac{\sum_1^n P_{ik}\times W_i}{\sum_1^n W_i}$。

## 6 小　结

（1）地方本科高校二级学院绩效评价体系的构建。地方本科高校二级学院绩效评价体系是反映二级学院各办学指标之间关系及其重要程度而建立的量化系统，主要由指标项、指标权重和指标评价标准三部分构成。地方本科高校可根据评价体系的设计原则和办学定位、特色及发展目标，筛选确定二级学院绩效评价指标项，测算各指标权重，制定各指标评价标准，评价得出二级学院综合绩效得分。

（2）Z大学二级学院绩效评价体系的实践。Z大学基于目标管理的二级学院绩效评价体系从2012年开始实施，每年开展一次年度绩效评价，主要依据是“十二五”事业发展规划及其办学指标分解方案，旨在通过二级学院年度绩效评价考核，激发二级学院办学主动性和积极性，确保Z大学“十二五”事业发展规划确定的任务目标如期完成。评价体系实施2年来，其评价结果比较客观地反映了各二级学院年度办学的实际情况。同时，各二级学院围绕绩效评价体系确立的政策导向，分析诊断办学中的差距与不足，有针对性地改进工作并提高办学绩效。Z大学的评价实践表明，基于目标管理的二级学院绩效评价体系在引导二级学院树立目标意识和绩效观念，落实校院两级管理目标责任制，提升办学活力和核心竞争力等方面发挥了重要的导向、诊断、改进和激励作用。

## 参考文献

蔡袁强，戴海东，翁之秋．2010．地方本科院校办学面临的困惑与对策——以温州大学为研究对象．高等工程教育研究，(1)：96-97.

应用技术大学（学院）联盟，地方高校转型发展研究中心．2013．地方本科院校转型发展实践与政策研究报告．http：//www. moe. edu. cn/ewebeditor/uploadfile/2014/01/07/20140107102329855. doc［2013-11-01］．

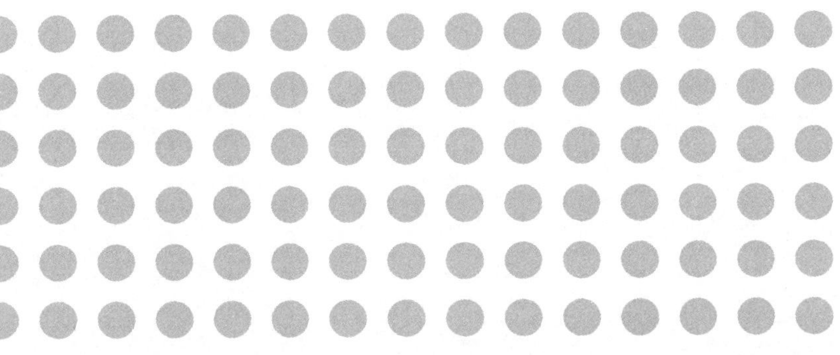

# 五、科学计量学在中国

# 5-1 Visualization Analysis on the International Influence of the Father of Scientometrics: Commemorating Price's 30th Anniversary of His Death

Liu Zeyuan[①], Liang Yongxia[①②], Chen Yue[①]

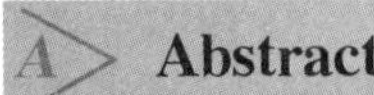

It is the 30th death anniversary of the father of scientometrics Derek Price on September 3, 2013. D. Price has remarkable achievements in the history of science, science of science, scientometrics and science citation network fields. We take Price's publishing and cited papers which indexed by Web of Science as source data and analyze Price's academic achievements and their international impact by CiteSpace visualization techniques. The results exhibit a wide range of research interests and tireless spirit of exploration of Price. The cited times remain high 30 years after his death which shows the broad scope and long time of his influence. It is highlighting the development and a high degree of practicality of his theory as well as academic permanent charming.

**Keywords**: Derek Price; international influence; scientometrics; the science of science; knowledge mapping

## 1 Introduction: Main Achievements Masterpiece

Derek John de Solla Price (1922. 1. 22-1983. 9. 3) is the world-famous physicist

① WISE Lab, Dalian University of Technology, P. R. China, Dalian.

② National Science Library, Chinese Academy of Sciences, P. R. China, Beijing.

and historian of science, and information scientist , and he is generally acknowledged the father of scientometrics.

D. Price is knowledgeable, has a wide range of interests, has published extensively, and has outstanding contributions. It is until 1981, he had published 9 monographs and 232 papers which about science of science and scientometrics, and two monographs and 14 articles to be processed. He is to accomplish something in science policy, science structure, scientific ethics, sociology of science, scientific histology (scientific cooperation), etc. , but the most prominent achievement is undoubtedly the foundational contribution for scientometrics. His groundbreaking achievements in scientometrics should thanks to his superb mathematical skills, a solid base of history of science, and deep theory heritage of science of science. E. Garfield, the inventor of SCI believes that, in the historical process from science of science to scientometrics, Price is pioneering figures of scientomertrics formation who elevates science of science to quantitative analysis phase after J. D. Bernal (1901-1971), who is the founder of science of science (Garfield, 2009). Price's academic achievements and masterpiece as follows.

(1) Contributions for history of science. He specializes in ancient science, the ancient scientific instruments and the history of modern physics. In 1959, he inspected the magic of ancient Greek Antikythera Mechanism, for the first time, he speculated that it is an astronomical mechanical computer which can predict the specific location in the zodiac of sun and moon any day (Price, 1959). It is confirmed until later by the detailed features of CT processing analysis of the X-ray chromatographic techniques. Its representatives include *Science Since Babylon* (Price, 1961).

(2) Contributions for science of science. Theoretically, he profoundly reveals the inevitability of the science of science which is a new discipline. He proposes singular epitaxial multidisciplinary definition for science of science and clarifies that its nature is a second-order subject of first-order importance. He also elucidates the three chief types of analysis of science of science: quantitative studies, theoretical models, and studies of policy and administration. Its representatives include *The Science of Science* (Price, 1964).

(3) Contributions for scientometrics. He advances the exponential law and the law of logistic for the growth of science, as well as the step-development law to which the former will eventually shift . He puts forward the Price Index giving an expression to the scientific literature half-life and the Price's law reflecting the distribution of scientific productivity, etc. Its representatives include *Little Science, Big Science* (Price, 1963). The book takes "prelude of science of science" as title, indicating clearly that large-scale modern science allows us from the past small science

to enter the era of big science, but it still follows the same law of development according to quantitative analysis of the science.

(4) Contributions for science network analysis. It is change for scientometrics from the statistical analysis of historical documents to the citation analysis as mainstream. The change includes the citation network based on SCI database that has power-law distributions and being the first example of a scale-free network. The change also proposes a mathematical theory of the growth of citation networks which is now called a preferential attachment process as based measurement. The masterpiece is *Networks of Scientific Papers* (Price, 1965). In this paper, Price's network analysis insights go far ahead of the later thinking of the social network analysis and complex network analysis, and have laid the foundation for scientomerics towards mapping knowledge and knowledge visualization.

Price has been dead for 30 years, but his writings and ideas still have a profound impact on the international academic peers. To memorize the 30th anniversary of the death of the great father of scientometrcis, and carry forward hts scientomertics thinking and scientific spirit, this paper tries to take Price's publishing and cited papers as source data and analyze Price's academic achievements and their international impact by CiteSpace visualization techniques.

## 2 Data Source and Analysis

The data is from the Web of Science (including SCI-E, SSCI, A&HCI citation databases). As to the retrieval time we select 1900-2012. The study takes two kinds of data. The first is to analyze Price's published papers. The search strategy is entering the name of the author, including Price DJD, Price DD, Price DJ, and Price D, etc. After identification and selection, we have 44 articles in the Web of Science. The second is to use the function of cited references in Web of Science, we got 2681 records citing Price's writings.

The Chinese data is from CNKI (the statistical analysis and visualization analysis about Price's articles are temporarily omitted). We have data cleaning and specification including Price's English name, different wording book title as well as other non-standard data.

Price has published more than 200 papers in his life, but indexed by Web of Science only 44 of which 38 Price as the first author and 6 as the co-author. Figure 1 shows the annual distribution of Price's published articles, and the highest publishing year is 1965, 1969 and 1971. Up to the time of his death, Price always publishes papers.

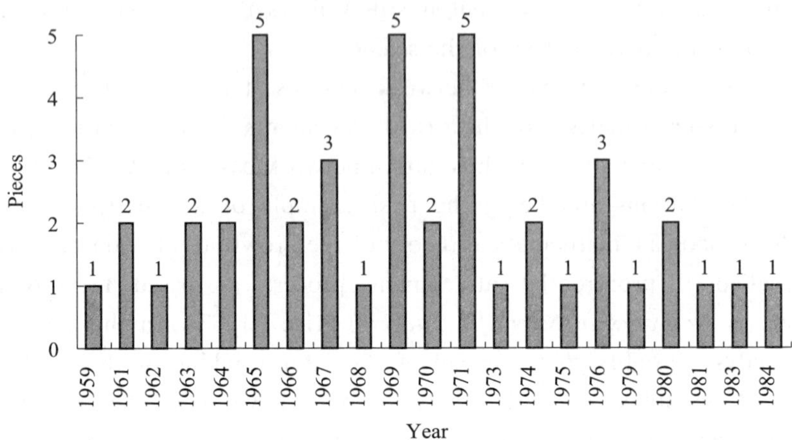


Figure 1 Annual distribution of Price's published articles (in Web of Science, 1959-1984)

Price papers are distributed in 30 journals, most of which are scientific top journals, such as *Nature*, *Science*, or top journals of information science, science of science and renowned society journal. The themes of papers are also very broad and occupy 23 subjects in Web of Science that the most important fields of the history of science, history of philosophy of science, computer science, and the library and information science.

The times of citing Price's writings show Price's academic influence, in a certain sense, mainly reflecting the following extension and development of academic peers on Price's academic thinking and views. Figure 2 is the annual distribution of citing Price's articles and books. It is can be divided into three stages from 1956 to 2012 when Price's papers have been cited in Web of Science. The first stage is from 1956 to 1969 when the citation has been in a gradual upward trend. The second stage is from 1970 to middle 2000, after the bibliometrics and scientometrics two disciplines by its proper name in 1969, the citations each year between 40-60 shocks and down to 30 times in 2000, then began to continue to rise to 70 times. In 2007, the citation begins to rise rapidly and breaks the 100 mark after the founding of the new journal of *Journal of Informetrics*. It is worth noting that the citing of Price's writings remains unabated after Price passed away and display its far-reaching academic influence.

The 2681 articles of citing Price's writings are from 73 countries and regions, showing the broad influence of academic thinking of Price. Figure 3 is the 21 countries and regions' distribution of citing Price's articles and books. The first one is the United States (1, 248 articles), while China ranks first seven (63 articles).

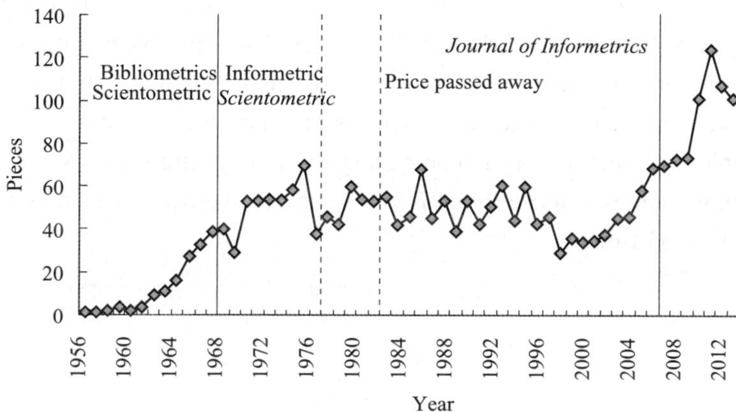


Figure 2　Annual distribution of citing Price's articles and books (in Web of Science, 1956-2012)

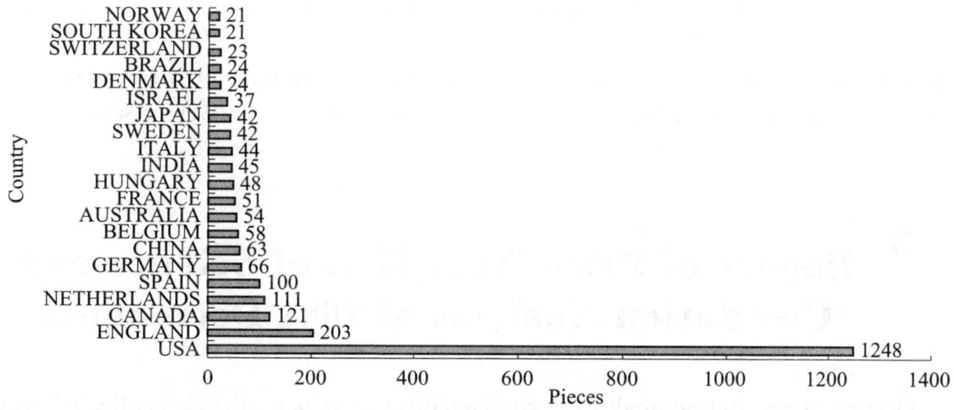


Figure 3　The national and regional distribution of literatures cited more than 20 papers of Price (in Web of Science)

The authors who have ever cited Price's papers are found wildly, and the authors who cited Price's papers more than 5 times have reached 80 persons. Among them, E. Garfield holds the first place who has cited Price's papers more than 80 times, and followed by L. Leydesdorff who has cited Price's papers more than 40 times. The outstanding leaders in the field of scientometrics, such as W. Glanzel, A. Schubert, L. Egghe, H. Small, R. Rousseau, Chen Chaomei, T. Braun, and B. C. Griffith, all have ever cited Price's papers more than 15 times.

By the view of cited times of Price' papers, *Little Science*, *Big Science* (Price, 1963) has been cited more than 940 times by Web of Science, cited no fewer than 2974 times by Google Scholar; *Networks of Scientific Papers* (Price, 1965) has been cited more than 770 times by Web of Science and 1438 times by Google Scholar. Figure 4 indicates that the cited times of Price' papers maintain a relatively higher level every year, and especially it is so thought-provoking that the cited times have been keeping rising since 2003. The reason lies in the lucky coincidence for Price's complex network theory encountering the new century. Figure 5 indicates the total cited times achieved more than 27 633 times and average cited times of each item also reached 35. 56 times.

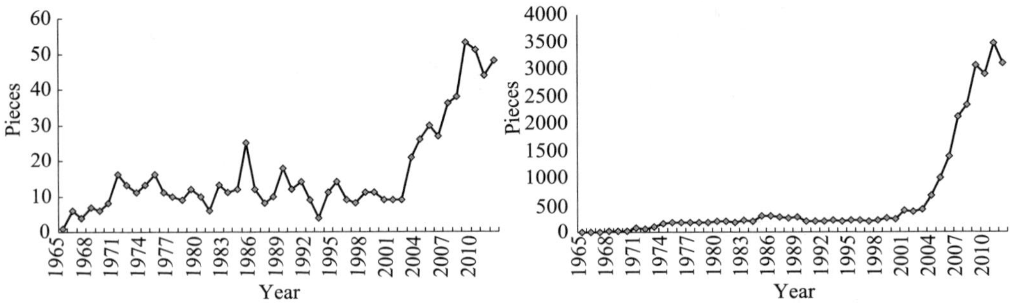


Figure 4 Cited history of *Networks of Scientific Papers* (in Web of Science)

Figure 5 Cited times of *Networks of Scientific Papers* (in Web of Science)

## 3 Papers of Price from Web of Science and Co-citation Analysis of 2681 Documents

Document co-citation analysis is conducted on 44 papers of Price collected from Web of Science by slicing the time 1959-1984 into 9 periods with CiteSpace, and each period covers top 30 documents. Figure 6 shows there are 20 clusters and their knowledge groups in the result network which indicates the main research areas and their evolution based the achievements from Price himself and citations from others. Around *Little Science*, *Big Science* (Price, 1963), there are 7 clusters which are also the core research areas and leading contributions of Price. There are 13 clusters scattering in the periphery which present several research areas Price focuses on.

Further more, 2681 references citing Price' works were analyzed by co-citation analysis and we sliced time into 6 years as one period from 1951-2012. Each period covered top 30 papers. The result is in Figure 7 which shows 187 nodes and 270 edges in the network.

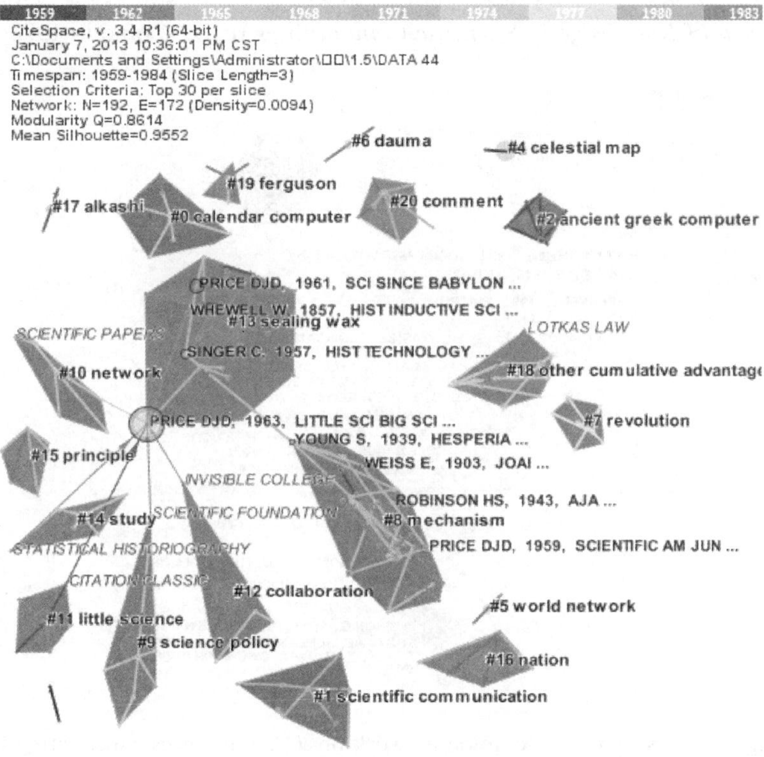


Figure 6 Document co-citation network of Price

Figure 7 indicates that the top right cluster is about cultural influences in the middle of 1950s. 4 knowledge groups can be distinguished from the largest component according to the color change of edges from the deep blue to red which reflects the evolution progress of academic exploration of Price's works：

(1) The bottom knowledge group colored by deep blue is about the discussion of scientific revolution from the view of history of science，which views *Science Since Babylon* of Price in 1961 and *Structure of Scientific Revolutions* of Kuhn in 1962 as the knowledge base.

(2) The light blue knowledge group is the exploration of rules ofthe growth of science and scientists，and scientific collaboration which is based on *Little Science*，*Big Science* of Price in 1963，*Collaboration in an Invisible College* of Price in 1966，and *The Matthew Effect in Science* of R. K. Merton in 1968.

(3) The green knowledge group in the middle is the study on the document co-citation network and collaboration network which is centered around works of *Networks of Scientific Papers* of Price in 1964，*The Structure of Scientific Literatures*

*I*: *Identifying and Graphing Specialties* of H. Small in 1974, and *Invisible Colleges*: *Diffusion of Knowledge in Scientific Communities* of D. Crane in 1972.

Figure 7 Document co-citation network of 2681 references which citing Price's work

(4) The top left knowledge group colored by orange connected with the third group which focuses on the research about the power-law distribution of citation network asthe knowledge base of papers of Price in 1979 and referred to Lotka's law. Simultaneously, the application of social network analysis and complex network in citation network is also demonstrated here referred to *Collective Dynamics of "Small-World" Networks* of D. J. Watts in 1998 (Watts and Strogatz, 1998).

## 4 Conclusions

(1) As the father or scientometrics, Price has made outstanding contributions in the areas of history of science, science of science, sociology of science, especially in scientometrics and scientific citation network in which *Science Since Babylon*, *Little Science*, *Big Science*, *The Science of Science*, and *Networks of Scientific Papers* are the most classic interdisciplinary works.

(2) Through the statistics and citation analysis of papers of Price and the works citing them, the extensive research interests and the indefatigable exploration are

displayed. The citations of his works distribute widely in 73 countries and areas, and last 55 years. Even during the 30 years after he passed away, the citations are still high which manifests the evolvability and practicality of his theories, and the permanent academic charm.

(3) The co-citation network of the documents citing Price's works indicates that the research areas of Price have been changed continuously which are from history of science to the statistics of literatures, from scientific citation analysis to citation network analysis by tracing his academic research footprint. Especially, in 1964 after *Networks of Scientific Papers* published 30 years , its cited frequency had a new incensement, which can be found in the prosperity of complex network research in the orange knowledge group. Meanwhile, this is closely related to the extensive usage of internet which has changed the ways of research so that we can enter the period of big science and big data, which has formed a new paradigm that data intensive research driven.

## References

Chen C. 2006. CiteSpace II: detecting and visualizing emerging trends and transient patterns in scientific literature. *Journal of the American Society for Information Science and Technology*, 57 (3): 359-377.

Garfield E. 2009. From the science of science to scientometrics: visualizing the history of science with HistCite software. *Journal of Informetrics*, 3 (3): 173-179.

Price D J D. 1959. An ancient Greek computer. *Scientific American*, 200 (6): 60-67.

Price D J D. 1961. *Science Since Babylun*. New Haven: Yale University Press.

Price D J D. 1963. *Little Science, Big Science*. New York: Columbia University Press.

Price D J D. 1964. The science of science. *In*: Goldsmith M, Mackay A L. *The Science of Science*. Souvenir Press. (Chinese, 1985: 227-245. ) See also: *Science and Public Policy*, 1989, 16 (3): 152-158.

Price D J D. 1965. Networks of scientific papers. *Science*, 149 (3683): 510-515.

Price D J D. 1976. A general theory of bibliometric and other cumulative advantage processes. *Journal of the American Society for Information Science*, 27: 292-306.

Price D J D, Beaver D. 1966. Collaboration in an invisible college. *American Psychologist*, 21 (11): 1011-1018.

Watts D J, Strogatz S H. 1998. Collective dynamics of "small-world" networks. *Nature*, 393 (6684): 440-442.

# 5-2 The Rhythm of Dr. R. Rousseau's Academic Research

**Liang Liming** ①, **Zhong Zhen** ②

## Abstract

Dr. Rousseau is one of the most productive and influential scientists in scientometrics and informetrics. His rhythm of the academic research is worth studying. In this study we create a new indicator by combining the publication indicator and rhythm indicator. Based on Rousseau's publication-citation matrix, Rousseau's rhythm of academic research is created and compared with Professor Dr. Tibor Braun's.

**Keywords**: R. Rousseau; rhythm; academic research

## 1 Rhythm Indicators for Measuring the Evolution of Science

### 1. 1 Study on the Rhythm Indicators

In 2002 we launched a study on the academic career of a scientist with double identities, namely Tibor Braun, Hungarian chemist and scientometrician. Our aim was to find a quantitative relationship between Braun's chemicalcareer and

① Institute for Science, Technology and Society, Henan Normal University, Xinxiang, 453007, China. Email: liangliming 1949@sina. com.

② School of Management, Henan University of Technolgy, Zhenzhou, 450001.

scientometric career. The hypothesis was that the rhythm of his chemical career and the rhythm of his scientometric career might be complementary.

At that time we did not wantto restrict ourselves to the traditional publication or citation indicators using only observation-based data for describing the ups and downs in Braun's productivity and influence. We were looking for an intelligent way of combining the publication and citation indicators, as well as a way of comparing observation-based and expectation-based citations.

The results ofthe study on Braun's academic career were not published before 2010 (Liang and Hou, 2010). However, the new indicators and methods created in that study were published (Liang, 2005; Liang et al., 2006; Liang, 2007) and developed (Liang and Rousseau, 2007; Egghe et al., 2008; Liang and Rousseau, 2010).

Dr. Ronald Rousseau is the supervisor of my doctoral thesis at Antwerp University. After four and a half years hard work, we created 6 clusters of rhythm indicators, including 12 groups and 24 rhythm indicators, forming a system of rhythm indicators. The mathematical properties, the methodological significance and the applications of the rhythm indicators were explored and discussed. This work won the Emerald/EFMD Outstanding Doctoral Research Awards (2007).

Among the six clusters of rhythm indicators the $R$-cluster is the first one we proposed based on a general publication-citation matrix, including four rhythm indicators: $R$, $R'$, $r$ and $r'$. The $R$ and $R'$ indicators are constructed from the viewpoint of "cited year", and $r$ and $r'$ the viewpoint of "citing year". $R$ and $r$ are constructed based on a triangle citation window, and $R'$ and $r'$ the parallelogram citation window. $R$ is the basic rhythm indicator and the most effective rhythm indicator.

## 1.2 How to Create the R Indicator Based on a General Publication-Citation Matrix?

A general publication-citation matrix (p-c matrix for short) consists of two parts, the publication data and the citation data. Table 1 is an example of the general p-c matrix, covering 9 publication years and 9 citation years. $P_i$ denotes the number of publications in publication year $i$. $C_{ij}$ denotes the number of citations received in the year $j$ by items published in the year $i$.

**Table 1　A general publication-citation matrix**

| | Publication year $i$ | Number of publication $P_i$ | | Citing year $j$ and number of citations $C_{ij}$ | | | | | | | | |
|---|---|---|---|---|---|---|---|---|---|---|---|---|
| | | | | 2003 | 2004 | 2005 | 2006 | 2007 | 2008 | 2009 | 2010 | 2011 |
| | | $i/j$ | $P_i$ | $P_1$ | $P_2$ | $P_3$ | $P_4$ | $P_5$ | $P_6$ | $P_7$ | $P_8$ | $P_9$ |
| Publication year $i$ and number of publications $P_i$ | 2003 | 1 | $P_1$ | $C_{11}$ | $C_{12}$ | $C_{13}$ | $C_{14}$ | $C_{15}$ | $C_{16}$ | $C_{17}$ | $C_{18}$ | $C_{19}$ |
| | 2004 | 2 | $P_2$ | | $C_{22}$ | $C_{23}$ | $C_{24}$ | $C_{25}$ | $C_{26}$ | $C_{27}$ | $C_{28}$ | $C_{29}$ |
| | 2005 | 3 | $P_3$ | | | $C_{33}$ | $C_{34}$ | $C_{35}$ | $C_{36}$ | $C_{37}$ | $C_{38}$ | $C_{39}$ |
| | 2006 | 4 | $P_4$ | | | | $C_{44}$ | $C_{45}$ | $C_{46}$ | $C_{47}$ | $C_{48}$ | $C_{49}$ |
| | 2007 | 5 | $P_5$ | | | | | $C_{55}$ | $C_{56}$ | $C_{57}$ | $C_{58}$ | $C_{59}$ |
| | 2008 | 6 | $P_6$ | | | | | | $C_{66}$ | $C_{67}$ | $C_{68}$ | $C_{69}$ |
| | 2009 | 7 | $P_7$ | | | | | | | $C_{77}$ | $C_{78}$ | $C_{79}$ |
| | 2010 | 8 | $P_8$ | | | | | | | | $C_{88}$ | $C_{89}$ |
| | 2011 | 9 | $P_9$ | | | | | | | | | $C_{99}$ |

Based on the p-c matrix we give the definition of the $R$ indicator from the perspective of "cited year" using the triangle method.

In general we will consider a time period of length $n$. The new indicator $R$ is defined as a time series of ratios $R_i = O_i / E_i$, $i = 1, \cdots, n$. The denominator $E_i$ is an expected citation value, while the numerator $O_i$ is an actual (observed) citation value. The sequence $R_1$, $R_2$, $\cdots$, $R_n$ is called the $R$-sequence.

The $R$-sequence is a time series of the actor's (e. g., journal's, institution's, nation's, field's, scientist's) performance, showing how the observed values ($O_i$) approach to or deviate from the corresponding expected values ($E_i$). It will be interpreted as an aspect of the internal rhythm of science.

The symbols $O_i$ and $E_i$ are defined as follows:

$$O_i = \sum_{j=i}^{n} C_{ij}$$

$$E_i = P_i \sum_{k=1}^{n-i+1} C_k$$

Here $C_k$ denotes the average number of citations per paper in the $k$-$th$ year after its publication ($k = 1, \cdots, n$, where $k = 1$ refers to a paper's publication year). $C_k$ is the key measure in the construction of the $R$-indicator.

$$C_k = \frac{\sum_{j=1}^{n-k+1} C_{j,j+k-1}}{\sum_{j=1}^{n-k+1} P_j}$$

We refer to this method of calculating $R_i$ as the triangle method, because all the $C_{ij}$ used in the formulae cover a triangular area in the p-c matrix.

The rearrangement equality can be shown as follows:

$$\sum_{i=1}^{n} O_i = \sum_{i=1}^{n} E_i$$

## 1.3 Dr. R. Rousseau's Rhythm of Academic Research

On the memorable occasion when we celebrate the 15th anniversary of Dr. R. Rousseau's visiting and working in China, I would like to study Rousseau's rhythm of academic research to show my sincere respect to my professor. My former doctoral student Zhong Zhen helped me to retrieve and identify Rousseau's publication and citation data in Web of Science.

# 2 Rousseau's Publication and Citation Data in Web of Science

Professor Rousseau is a productive and influential scientist. Since 1976 he has published 225 publications covered by Web of Science and received more than 2 400 citations. When we study the rhythm indicator only articles, proceeding papers, notes and letters are selected as samples, and the period is 1976-2012, leading to a total of 204 publications and 2410 citations. Table 2 is his p-c matrix.

# 3 Methodology: Combining the Publication and Rhythm Indicators

In this exploration two indicators are used. The first is the publication indicator $P_i$, the second is the rhythm indicator $R_i$. The two indicators are combined as $P_i + \alpha R_i$. Here, the publication indicator and rhythm indicator are given the same weight, that means:

$$\alpha \sum_{i=1}^{n} R_i = \sum_{i=1}^{n} P_i$$

# 4 Rousseau's Rhythm of Academic Research Measured by Indicator $P_i + \alpha R_i$

Figure 1 shows the curves of Rousseau's indicators $P_i$ and $\alpha R_i$, and makes a comparison between them. From Figure 1 we observed that, usually, the two

**Table 2　Rousseau's publication-citation matrix（1976-2012，WoS）**

| | 项目 | $P_i$ | $C_{ij}$ | | | | | | | | | | | | | | | | | | | | | | | | | | | | | | | | | | | |
|---|---|---|---|---|---|---|---|---|---|---|---|---|---|---|---|---|---|---|---|---|---|---|---|---|---|---|---|---|---|---|---|---|---|---|---|---|---|---|
| | | | 1976 | 1977 | 1978 | 1979 | 1980 | 1981 | 1982 | 1983 | 1984 | 1985 | 1986 | 1987 | 1988 | 1989 | 1990 | 1991 | 1992 | 1993 | 1994 | 1995 | 1996 | 1997 | 1998 | 1999 | 2000 | 2001 | 2002 | 2003 | 2004 | 2005 | 2006 | 2007 | 2008 | 2009 | 2010 | 2011 | 2012 |
| | 1976 | 1 | 0 | 1 | 1 | 0 | 0 | 0 | 0 | 0 | 0 | 0 | 0 | 0 | 0 | 0 | 0 | 0 | 0 | 0 | 0 | 0 | 0 | 0 | 0 | 0 | 0 | 0 | 0 | 0 | 0 | 0 | 0 | 0 | 0 | 0 | 0 | 0 | 0 |
| | 1977 | 2 | | 0 | 1 | 2 | 0 | 2 | 0 | 1 | 0 | 0 | 0 | 0 | 0 | 0 | 0 | 0 | 0 | 0 | 0 | 1 | 0 | 0 | 0 | 0 | 0 | 0 | 0 | 0 | 0 | 0 | 0 | 0 | 0 | 0 | 0 | 0 | 0 |
| | 1978 | 0 | | | 0 | 0 | 0 | 0 | 0 | 0 | 0 | 0 | 0 | 0 | 0 | 0 | 0 | 0 | 0 | 0 | 0 | 0 | 0 | 0 | 0 | 0 | 0 | 0 | 0 | 0 | 0 | 0 | 0 | 0 | 0 | 0 | 0 | 0 | 0 |
| | 1979 | 2 | | | | 0 | 0 | 3 | 0 | 1 | 0 | 0 | 0 | 0 | 0 | 0 | 0 | 0 | 0 | 0 | 0 | 0 | 0 | 0 | 0 | 0 | 0 | 0 | 0 | 0 | 0 | 0 | 0 | 0 | 0 | 0 | 0 | 0 | 0 |
| | 1980 | 0 | | | | | 0 | 0 | 0 | 0 | 0 | 0 | 0 | 0 | 0 | 0 | 0 | 0 | 0 | 0 | 0 | 0 | 0 | 0 | 0 | 0 | 0 | 0 | 0 | 0 | 0 | 0 | 0 | 0 | 0 | 0 | 0 | 0 | 0 |
| | 1981 | 3 | | | | | | 0 | 0 | 1 | 0 | 0 | 0 | 0 | 0 | 0 | 0 | 0 | 0 | 0 | 0 | 0 | 0 | 0 | 0 | 0 | 0 | 0 | 0 | 0 | 0 | 0 | 0 | 0 | 0 | 0 | 0 | 0 | 0 |
| | 1982 | 0 | | | | | | | 0 | 0 | 0 | 0 | 0 | 0 | 0 | 0 | 0 | 0 | 0 | 0 | 0 | 0 | 0 | 0 | 0 | 0 | 0 | 0 | 0 | 0 | 0 | 0 | 0 | 0 | 0 | 0 | 0 | 0 | 0 |
| | 1983 | 2 | | | | | | | | 0 | 0 | 0 | 0 | 0 | 0 | 0 | 0 | 0 | 0 | 0 | 0 | 0 | 0 | 0 | 0 | 0 | 0 | 0 | 0 | 0 | 0 | 0 | 0 | 0 | 0 | 0 | 0 | 0 | 0 |
| | 1984 | 1 | | | | | | | | | 0 | 0 | 0 | 0 | 0 | 0 | 0 | 0 | 0 | 0 | 0 | 0 | 0 | 0 | 0 | 0 | 0 | 1 | 1 | 0 | 0 | 0 | 0 | 0 | 0 | 0 | 0 | 0 | 0 |
| | 1985 | 1 | | | | | | | | | | 0 | 0 | 0 | 0 | 0 | 0 | 0 | 0 | 0 | 0 | 1 | 0 | 0 | 1 | 1 | 0 | 0 | 1 | 0 | 0 | 0 | 0 | 0 | 0 | 0 | 0 | 0 | 0 |
| | 1986 | 1 | | | | | | | | | | | 0 | 1 | 2 | 0 | 0 | 0 | 0 | 0 | 0 | 0 | 0 | 0 | 0 | 0 | 0 | 0 | 1 | 1 | 0 | 2 | 0 | 0 | 0 | 0 | 1 | 0 | 0 |
| | 1987 | 2 | | | | | | | | | | | | 0 | 0 | 2 | 3 | 0 | 2 | 1 | 0 | 0 | 0 | 0 | 1 | 1 | 2 | 2 | 2 | 1 | 1 | 5 | 0 | 0 | 0 | 3 | 0 | 2 | 0 |
| | 1988 | 3 | | | | | | | | | | | | | 0 | 3 | 5 | 1 | 3 | 1 | 1 | 1 | 2 | 2 | 2 | 1 | 0 | 0 | 1 | 0 | 0 | 1 | 0 | 0 | 1 | 0 | 0 | 0 | 1 |
| | 1989 | 3 | | | | | | | | | | | | | | 0 | 0 | 0 | 1 | 0 | 0 | 0 | 1 | 0 | 1 | 0 | 3 | 1 | 1 | 1 | 2 | 2 | 3 | 2 | 4 | 4 | 8 | 5 | 6 |
| | 1990 | 3 | | | | | | | | | | | | | | | 0 | 2 | 3 | 2 | 1 | 0 | 1 | 1 | 0 | 3 | 1 | 0 | 2 | 2 | 2 | 6 | 1 | 1 | 0 | 1 | 0 | 1 | 1 |
| | 1991 | 2 | | | | | | | | | | | | | | | | 0 | 9 | 1 | 0 | 1 | 1 | 0 | 2 | 3 | 4 | 2 | 0 | 2 | 1 | 2 | 1 | 3 | 0 | 2 | 1 | 0 | 2 |
| | 1992 | 7 | | | | | | | | | | | | | | | | | 0 | 6 | 5 | 3 | 3 | 3 | 4 | 4 | 1 | 6 | 7 | 1 | 6 | 11 | 6 | 4 | 2 | 7 | 4 | 2 | 6 |
| $C_{ij}$ | 1993 | 4 | | | | | | | | | | | | | | | | | | 0 | 1 | 2 | 2 | 1 | 3 | 4 | 2 | 3 | 6 | 0 | 3 | 7 | 2 | 1 | 1 | 0 | 3 | 0 | 1 |
| | 1994 | 4 | | | | | | | | | | | | | | | | | | | 0 | 2 | 3 | 2 | 2 | 5 | 4 | 6 | 5 | 2 | 2 | 5 | 3 | 3 | 2 | 3 | 3 | 2 | 2 |
| | 1995 | 4 | | | | | | | | | | | | | | | | | | | | 0 | 2 | 3 | 8 | 5 | 6 | 3 | 4 | 0 | 1 | 5 | 1 | 1 | 1 | 3 | 4 | 5 | 1 |
| | 1996 | 6 | | | | | | | | | | | | | | | | | | | | | 1 | 6 | 13 | 13 | 7 | 5 | 8 | 2 | 7 | 15 | 3 | 3 | 7 | 8 | 6 | 6 | 6 |
| | 1997 | 4 | | | | | | | | | | | | | | | | | | | | | | 0 | 7 | 1 | 4 | 3 | 1 | 3 | 3 | 2 | 0 | 6 | 3 | 6 | 9 | 3 | 4 |
| | 1998 | 8 | | | | | | | | | | | | | | | | | | | | | | | 3 | 5 | 6 | 4 | 2 | 4 | 3 | 5 | 2 | 6 | 5 | 5 | 8 | 5 | 3 |
| | 1999 | 6 | | | | | | | | | | | | | | | | | | | | | | | | 3 | 7 | 3 | 11 | 6 | 11 | 5 | 7 | 5 | 4 | 13 | 3 | 6 | 9 |
| | 2000 | 9 | | | | | | | | | | | | | | | | | | | | | | | | | 1 | 10 | 18 | 3 | 9 | 16 | 7 | 7 | 19 | 12 | 19 | 19 | 15 |
| | 2001 | 10 | | | | | | | | | | | | | | | | | | | | | | | | | | 2 | 8 | 4 | 11 | 8 | 4 | 1 | 7 | 5 | 3 | 3 | 1 |
| | 2002 | 6 | | | | | | | | | | | | | | | | | | | | | | | | | | | 3 | 4 | 17 | 24 | 12 | 28 | 26 | 39 | 35 | 28 | 27 |
| | 2003 | 5 | | | | | | | | | | | | | | | | | | | | | | | | | | | | 4 | 12 | 14 | 17 | 17 | 20 | 19 | 21 | 17 | 15 |
| | 2004 | 11 | | | | | | | | | | | | | | | | | | | | | | | | | | | | | 0 | 11 | 16 | 12 | 14 | 12 | 8 | 8 | 6 |
| | 2005 | 10 | | | | | | | | | | | | | | | | | | | | | | | | | | | | | | 8 | 9 | 10 | 13 | 20 | 15 | 15 | 12 |
| | 2006 | 11 | | | | | | | | | | | | | | | | | | | | | | | | | | | | | | | 4 | 25 | 32 | 30 | 28 | 19 | 24 |
| | 2007 | 9 | | | | | | | | | | | | | | | | | | | | | | | | | | | | | | | | 5 | 41 | 59 | 43 | 33 | 25 |
| | 2008 | 11 | | | | | | | | | | | | | | | | | | | | | | | | | | | | | | | | | 3 | 24 | 35 | 18 | 8 |
| | 2009 | 17 | | | | | | | | | | | | | | | | | | | | | | | | | | | | | | | | | | 13 | 30 | 20 | 24 |
| | 2010 | 8 | | | | | | | | | | | | | | | | | | | | | | | | | | | | | | | | | | | 1 | 19 | 19 |
| | 2011 | 12 | | | | | | | | | | | | | | | | | | | | | | | | | | | | | | | | | | | | 2 | 33 |
| | 2012 | 16 | | | | | | | | | | | | | | | | | | | | | | | | | | | | | | | | | | | | | 17 |

indicators $P_i$ and $\alpha R_i$ are mutually complementary in the peak years of one indicator, especially in the year with the maximum value.

The years when $\alpha R_i$ is much higher than $P_i$: 1991, 2002, 2003, 2007.

The years when $\alpha R_i$ is much lower than $P_i$: 1981, 1998, 2001, 2004, 2009.

(2012 is an exception, both $\alpha R_i$ and $P_i$ are very high.)

This observation is similar to the relationship between Professor Braun's $P_i$ and his $\alpha R_i$ (Liang and Hou, 2010). In that paper $\alpha R_i$ is denoted as $N\text{-}R_i$ (Figure 2, Figure 3). For Braun, it is true that the two indicators $P_i$ and $\alpha R_i$ are mutually complementary in the peak years of one indicator, especially in the year with the maximum value, no matter what the field is.

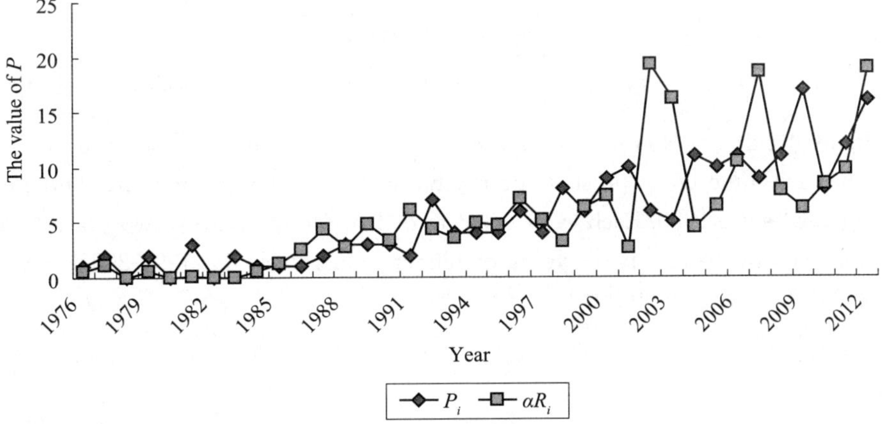


Figure 1 Rousseau: comparison of $P_i$ and $\alpha R_i$

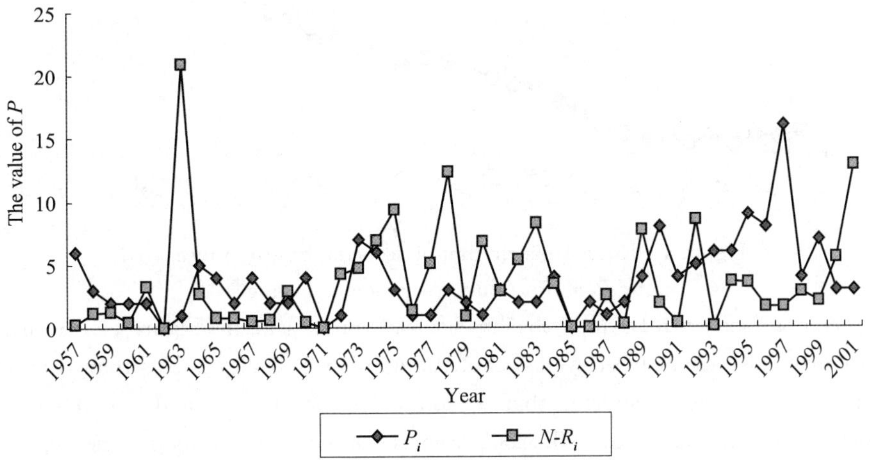


Figure 2 Braun: comparison of $P_i$ and $\alpha R_i$ in chemistry

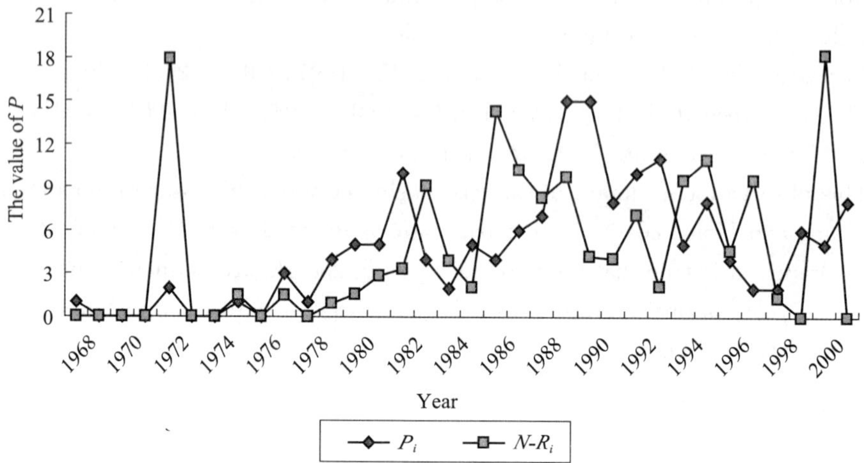


Figure 3 Braun: comparison of $P_i$ and $\alpha R_i$ in scientometrics

Figure 4 exhibits Rousseau's rhythm of academic research measured by $P_i + \alpha R_i$. From the early time up to present the rhythm curve had been increasing. Three cycles have emerged since 1999. Each cycle consists of five (four) years: two years of lower value followed by three (two) years of higher value. Cycle 1: 1999-2003; Cycle 2: 2004-2008; Cycle 3: 2009-2012.

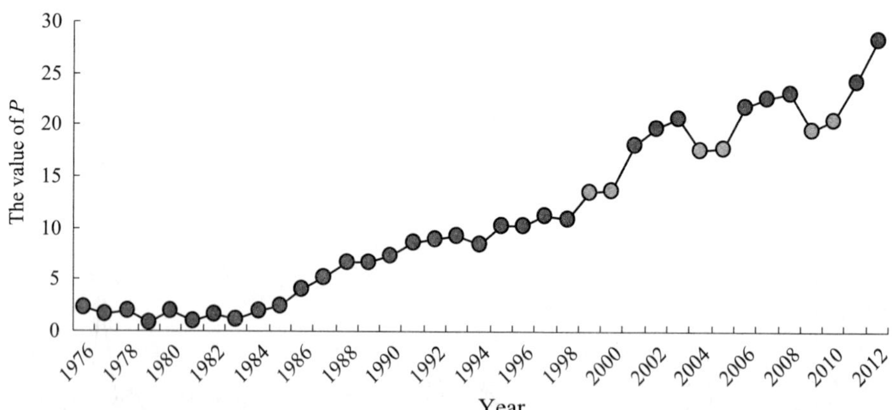


Figure 4 Rousseau's rhythm of academic research measured by $P_i + \alpha R_i$ (three-year moving average)

In comparison with Braun's rhythm of academic research (Figure 5), Rousseau's rhythm curve also consists of cycles, but the number of cycles is less than Braun's and the length of a cycle is shorter than Braun's. For Braun four and a half cycles had emerged during the period of 1957-2001. Each cycle consists of about nine years: four years of lower value followed by five years of higher value. Cycle 1: 1957-1965; Cycle 2: 1966-1974; Cycle 3: 1975-1982; Cycle 4: 1983-1991; the last half cycle: 1992-

1995.

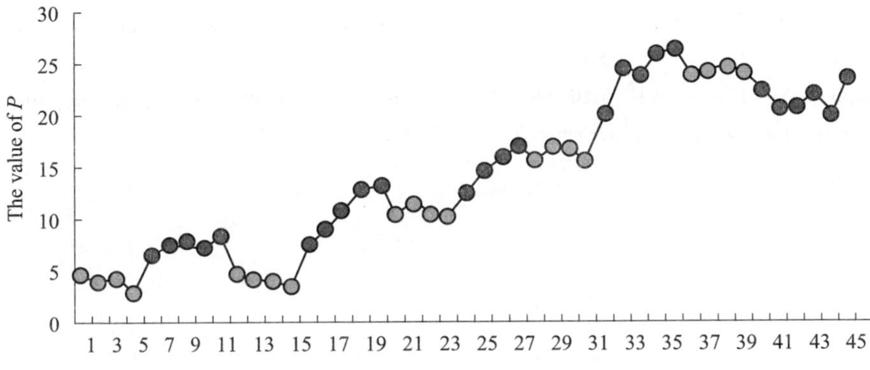


Figure 5 Braun's rhythm of academic research measured by $P_i + \alpha R_i$ (five-year moving average)

In this study we just created Dr. Rousseau's rhythm of academic research and compared it with Dr. Braun's rhythm. Rousseau is one of the most productive and influential scientists in scientometrics and informetrics. His rhythm of the academic research is worth further study. In such an analysis of Rousseau's rhythm of academic research the following aspects could be emphasized: the topics of the publications, the highly cited papers and the research collaboration.

## Acknowledgement

The work presented in this paper is sponsored by the National Natural Science Foundation of China (grant number 71373252).

## References

Egghe L, Liang L M, Rousseau R. 2008. Fundamental properties of rhythm sequences. *Journal of the American Society for Information Science and Technology*, 59: 1469-1478.

Liang L M. 2005. The *R*-sequence: a relative indicator for the rhythm of science. *Journal of the American Society for Information Science and Technology*, 56: 1045-1049.

Liang L M. 2007. Revealed similarities between the journals *Nature* and *Science*: using a new cluster of rhythm indicators. *In*: Torres-Salinas D, Moed H F (eds.), *Proceedings of ISSI* 2007 (pp. 504-513). Madrid: CSIC.

Liang L M, Hou C H. 2010. Scientist's rhythm of academic research presented by combining the publication and citation indicators—a case study on Tibor Braun. *Journal of the*

*China Society for Scientific and Technical Information*, 29 (6): 1116-1124.

Liang L M, Rousseau, R. 2007. Transformations of basic publication-citation matrices. *Journal of Informetrics*, 1: 249-255.

Liang L M, Rousseau R, Fei S. 2006. A rhythm indicator for science and the rhythm of science. *Scientometrics*, 68: 535-544.

Liang L M, Rousseau R. 2010. Measuring a journal's input rhythm based on its publication-reference matrix. *Journal of Informetrics*, 4: 201-209.

# 5-3 The Core-tail Profile of Ronald Rousseau on Chinese Collaboration 1998-2013

Fred Y. Ye [1], Helen F. Xue[2], Star X. Zhao[3]

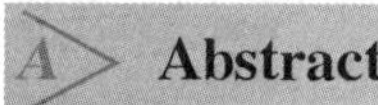

## Abstract

In commemoration of Ronald Rousseau's cooperation with Chinese scholars for 15 years, the article contributes an interesting profile of Ronald Rousseau from 1998 to 2013, under the framework of $h$-core and $h$-tail, which is just a representative work of the authors in collaboration with him. The data and indicators show that Chinese collaborations contribute 3 articles in Rousseau's total $h$-core and 11 articles in Chinese $h$-core.

**Keywords**: $h$-index; $h$-core; $h$-tail; core-tail; Ronald Rousseau

## 1 Introduction

The year 2013 marks the 15th anniversary since Ronald Rousseau initiated his cooperation with Chinese scholars. In this special year, an international seminar on collaboration commemoration between Professor Ronald Rousseau and China, proposed by Chinese communities, is hosted by the Professional Board of the Scientometrics and Informetrics of the Chinese Association for Science of Science and S&T Policy. During the past 15 years ever since his first visit to China in 1998, Professor Rousseau has become a leader working closely with Chinese scholars and built good friendships with Chinese colleagues.

We are honored to be the second/third generation scholars linking with Ronald

① School of Information Management, Nanjing University, Nanjing, China. yye@nju. edu. cn.
② Library of Zhejiang University, Zhejiang University, Hangzhou, China.
③ Department of Information Resource Management, Zhejiang University, Hangzhou, China.

Rousseau in China. Among our collaborations, the core-tail framework is our representative work. Applying the framework, we present a profile of Ronald Rousseau, in special commemoration of his Chinese collaboration for the period of 1998-2013.

# 2 Data and Method

## 2.1 Data

We retrieved data in Thomson Reuters' Web of Science according to the following search strategy: author = Rousseau R and address = Belgium, with time span = 1998-2013 and database = SCI-expanded, SSCI, CPCI-S, CPCI-SSH.

The citation report shows that there are totally 161 publications with 1864 citations (excluding self-citations as 1625) and CpP (Citations per Publication) 11.58 with $h$-index 23. If we choose database = SCI-expanded, SSCI, the report shows that there are totally 151 publications with 1859 citations (excluding self-citations as 1630) and CpP 12.31 with the same $h$-index 23. So, we keep 161 publications with the 10 conference papers included.

Meanwhile, if we set the following search strategy: Author = Rousseau R and address = China, with the same time span = 1998-2013 and database = SCI-expanded, SSCI, CPCI-S, we have results of $h$-index as 11, which indicates that Rousseau's total $h$-index is 23 while his Chinese collaboration's $h$-index is 11 over the past 15 years (1998-2013).

From the searched data, we can see that Chinese collaborations contribute 3 articles in Rousseau's total $h$-core and 11 articles in Chinese $h$-core, as shown in Table 1.

**Table 1 Representatives of Chinese collaborations in rousseau's publications** (1998-2013)

| Location | Publications with Chinese | Source | Citations |
|---|---|---|---|
| Total $h$-core | Jin B H, Liang L M, Rousseau R & Egghe L (2007). The $R$-and AR-indices: complementing the $h$-index | *Chinese Science Bulletin*, 52 (6), 855-863 | 158 |
| | Ren S L & Rousseau R (2002). International visibility of Chinese scientific journals | *Scientometrics*, 53 (3), 389-405 | 44 |
| | Wang Y, Wu YS, Pan YT, Rousseau R (2005). Scientific collaboration in China as reflected in co-authorship | *Scientometrics*, 62 (2), 183-198 | 25 |

Continued

| Location | Publications with Chinese | Source | Citations |
|---|---|---|---|
| | Rousseau R & Ye F Y (2008). A proposal for a dynamic $h$-type index | *Journal of the American Society for Information Science and Technology*, 59 (11), 1853-1855 | 18 |
| | Ren S L & Rousseau R. (2004). The role of China's English-language scientific journals in scientific communication | *Learned Publishing*, 17 (2), 99-104 | 15 |
| | Liu Y & Rousseau R (2009). Properties of Hirsch-type indices: the case of library classification categories | *Scientometrics*, 79 (2), 235-248 | 14 |
| | Hu X & Rousseau R (2009). A comparative study of the difference in research performance in biomedical fields among selected Western and Asian countries | *Scientometrics*, 81 (2), 475-491 | 14 |
| Total $h$-tail and Chinese $h$-core | Rousseau R Jin B H, Yang N H et al. (2001). Observations concerning the two-and three-year synchronous impact factor, based on the Chinese Science Citation Database | *Journal of Documentation*, 57 (3), 349-357 | 13 |
| | Fang Y & Rousseau R (2001). Lattices in citation networks: an investigation into the structure of citation graphs | *Scientometrics*, 50 (2), 273-287 | 13 |
| | cHu X, Rousseau, R & Chen J (2010). In those fields where multiple authorship is the rule, the $h$-index should be supplemented by role-based $h$-indices | *Journal of Information Science*, 36 (1), 73-85 | 12 |
| | Ye F Y & Rousseau R (2010). Probing the $h$-core: an investigation of the tail-core ratio for rank distributions | *Scientometrics*, 84 (2), 431-439 | 11 |
| | Liu Y & Rousseau R (2008). Definitions of time series in citation analysis with special attention to the $h$-index | *Journal of Informetrics*, 2 (3), 202-210 | 11 |

## 2.2 Method

Separated by $h$-index (Hirsch, 2005; Alonso et al., 2009), the total number of publications (P) can be divided into two sub-sets: the first $h$-publications, denoted by $P_h$, referring to the $h$-core (Rousseau, 2006), and the other publications, as $P_t$, referring to the $h$-tail (Ye and Rousseau, 2010; Chen et al., 2013), and $P_z$, referring to the uncited ones (Liu et al., 2013).

Let $C$ indicate the total number of citations and $C_h$ denote the number of citations received by the publications in the $h$-core. Using the concept of excess citations (Zhang, 2009) and symbols $e^2$, $h^2$, $t^2$ as citations of excess area ($C_e$), $h$-area and tail-area (Zhang, 2013), we have:

$$C = e^2 + h^2 + t^2 \geqslant 0 \tag{1}$$

$$C_h = e^2 + h^2 \geqslant 0 \tag{2}$$

Since $e \geqslant 0$ and $h = P_h$, we see that

$$C_h \geqslant P_h^2 \tag{3}$$

and

$$\sqrt{C_h} = R \geqslant h = P_h \tag{4}$$

where $R$ denotes the $R$-index (Jin et al., 2007).

Figure 1 shows Rousseau's core-tail distribution with the above data of 1998-2013.

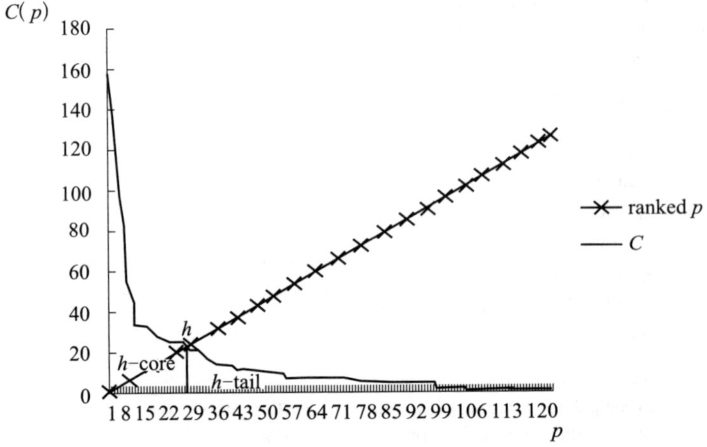


Figure 1 Core-tail distribution of Ronald Rousseau (1998-2013)

# 3 Results and Comments

## 3.1 Description

Combining Table 1 and Figure 1, with referring to formula (1) and (2), we see that Chinese collaborations give following contributions in Rousseau's output and impact during 1998 to 2013.

There are 23 publications in his total $h$-core, and only 3 articles are written in Chinese, where the top 2 belong to the excess area.

Most Chinese collaborative papers fall in his total $h$-tail, where 8 articles belong to his Chinese $h$-core. Some are distributed in the uncited area, without any citation.

If we only consider Rousseau's Chinese collaboration contributions, the $h$-index is 11. The number difference of $h$-index $23 - 11 = 12$ indicates the research gap between the international and Chinese level.

From the perspective of informetrics, we would like to compute the related informetric indicators of Ronald Rousseau (1998 - 2013), which are summarized in Table 2.

**Table 2 The related informetric indicators of Ronald Rousseau** (1998-2013)

| Data | | | | Indicators | | | |
|---|---|---|---|---|---|---|---|
| $P$ | $P_h$ | $P_t$ | $P_z$ | $h^2$ | $e^2$ | $t^2$ | $R^2$ |
| 161 | 23 | 103 | 35 | 529 | 655 | 680 | 1184 |
| $C$ | $C_h$ | $C_t$ | $C_e$ | $h$ | $e$ | $t$ | $R$ |
| 1, 864 | 1, 184 | 680 | 655 | 23 | 25. 59 | 26. 08 | 34. 41 |
| | | | | | 9 | 8 | 1 |

His academic matrix $V$ and trace $T$ (Ye and Leydesdorff, 2013), are computed with the results shown as follows:

$$V_{\text{Rousseau}} = \begin{pmatrix} 3.29 & 65.89 & 7.61 \\ 150.13 & 248.07 & 230.16 \\ 146.84 & 182.17 & 222.55 \end{pmatrix} \quad (5)$$

$$T_{\text{Rousseau}} = 3.29 + 248.07 + 222.55 = 473.91 \quad (6)$$

## 3.2 Curve Fitting

The curve in Figure 1 can be fitted with a power-law function, which is just a popular distribution in informetrics, and the fitting result is shown in Figure 2.

The result shows that the relation between $P$ and $C$ satisfies Heaps' law or Herdan's law (Egghe, 2007):

$$C = \alpha P^{\beta} \quad (7)$$

If we consider uncited publications together, the shifted Lotka/Zipf function (Egghe and Rousseau, 2012) could also be applied.

Above results provide us interesting references.

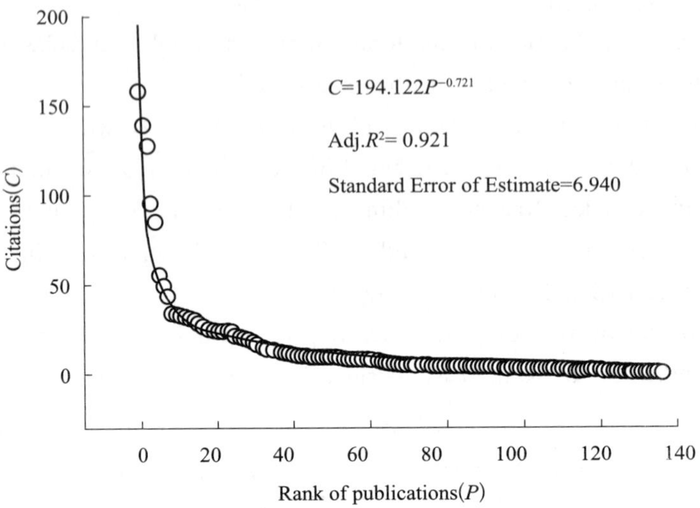


Figure 2 The curve fitting of the core-tail function of Ronald Rousseau (1998-2013)

# 4 Conclusions

With informetric methods and indicators, this short article just provides an interesting profile of Ronald Rousseau in commemoration of his cooperation with Chinese scholars for 15 years, where Rousseau's P-C distribution follows Heaps' law or Herdan's law. The data and indicators suggest that Chinese collaborations contribute 3 articles in Rousseau's total $h$-core and 11 articles in Chinese $h$-core, while he contributes a lot to Chinese scholars. We appreciate Ronald Rousseau's assistances.

## Acknowledgement

We acknowledge Ronald Rousseau's collaborations since 2008.

## References

Alonso S, Cabrerizo F J, Herrera-Viedma E, et al. 2009. $h$-Index: a review focused in its variants, computation and standardization for different scientific fields. *Journal of Informetrics*, 3: 273-289.

Chen D-Z，Huang M-H，Ye F Y. 2013. A probe into dynamic measures for $h$-core and $h$-tail. *Journal of Informetrics*，7（1）：129-137.

Egghe L. 2007. Untangling Herdan's law and Heaps' law：mathematical and informetric arguments. *Journal of American Society for Information Science and Technology*，58（5）：702-709.

Egghe L，Rousseau R. 2012. Theory and practice of the shifted Lotka function. *Scientometrics*，91（1）：295-301.

Hirsch J E. 2005. An index to quantify an individual's scientific research output. *Proceedings of the National Academy of Sciences of the USA*，102（46）：16569-16572.

Jin B，Liang L，Rousseau R，et al. 2007. The $R$- and AR-indices：complementing the $h$-index. *Chinese Science Bulletin*，52（6）：855-863.

Liu J Q，Rousseau R，Wang M S，et al. 2013. Ratios of $h$-cores，$h$-tails and uncited sources in sets of scientific papers and technical patents. *Journal of Informetrics*，7（1）：190-197.

Rousseau R. 2006. New developments related to the Hirsch index. *Science Focus*，1：23-2（in Chinese）. An English translation is available online at http：//eprints. rclis. org/6376/.

Ye F Y，Leydesdorff L. 2013. The "academic trace" of the performance matrix：a mathematical synthesis of the $h$-index and the integrated impact indicator（I3）. *Journal of the American Society for Information Science and Technology*（in press），arXiv1307. 3616.

Ye F Y，Rousseau R. 2008. The power law model and total career $h$-index sequences. *Journal of Informetrics*，2（4）：288-297.

Ye F Y，Rousseau R. 2010. Probing the $h$-core：an investigation of the tail-core ratio for rank distributions. *Scientometrics*，84：431-439.

Zhang C T. 2009. The $e$-index，complementing the $h$-index for excess citations. *PLoS ONE*，4：e5429.

Zhang C T. 2013. A novel triangle mapping technique to study the $h$-index based citation distribution. *Scientific Reports*，3：1023.